THEORY AND DESIGN FOR MECHANICAL MEASUREMENTS

RICHARD S. FIGLIOLA
Clemson University

DONALD E. BEASLEY
Clemson University

WILEY

JOHN WILEY & SONS

NEW YORK / CHICHESTER / BRISBANE / TORONTO / SINGAPORE

Library of Congress Cataloging-in-Publication Data:

Figliola, Richard S.
    Theory and design for mechanical measurements / Richard S.
  Figliola, Donald E. Beasley.
    Includes bibliographical references and index.
    ISBN 0-471-61994-9 (cloth)
    1. Mensuration.   I. Beasley, Donald E.   II. Title.
  T50.F54   1991
  681′.2—dc20   90-45035
                                                                      CIP

Printed in the United States of America

10 9 8 7 6 5 4 3 2 1

CONVERSION FACTORS

MASS

$1.0\ lb_m = 453.59237\ g$
$1.0\ slug = 32.174\ lb_m$
$1.0\ kg = 2.2046\ lb_m$

LENGTH

$1.0\ inch = 2.54\ cm$
$1.0\ m = 3.208\ ft = 39.37\ inches$
$1.0\ cm = 0.01\ m = 0.3937\ in = 0.0323\ ft$
$1.0\ mm = 0.001\ m = 1 \times 10^{-3}\ m$
$1.0\ \mu m = 0.000001\ m = 1 \times 10^{-6}\ m$
$1.0\ nm = 0.000000001\ m = 1 \times 10^{-9}\ m$
$1.0\ km = 1000\ km = 0.612\ miles$
$1.0\ miles = 5280\ ft$

TIME

$1.0\ min = 60\ s$
$1.0\ h = 60\ min$
$1.0\ day = 8.64 \times 10^4\ s$

FORCE

$1.0\ N = 1\ kg\text{-}m/s^2 = 1 \times 10^5\ dynes$
$1.0\ lb = 4.44822\ N$

PRESSURE

$1.0\ Pa = 1\ N/m^2$
$\quad\quad = 1.4504 \times 10^{-4}\ lb/in^2$
$1.0\ lb/in^2 = 6894.76\ N/m^2$
$1.0\ atm = 14.696\ lb/in^2 = 760\ Torr$
$1.0\ bar = 14.505\ lb/in^2$
$\quad\quad = 1 \times 10^5\ N/m^2$
$\quad\quad = 1 \times 10^8\ dynes/cm^2$
$1.0\ inch\ Hg = 3376.8\ N/m^2$
$1.0\ inch\ H_2O = 248.8\ N/m^2$
$\quad\quad = 0.0362\ lb/in^2$

AREA

$1.0\ m^2 = 10.76\ ft^2$
$1.0\ cm^2 = 1 \times 10^{-4}\ m^2 = 0.155\ in^2$

VOLUME

$1.0\ cm^3 = 1\ l = 1 \times 10^{-3}\ m^3$
$\quad\quad = 0.2642\ gallons$

$1.0\ gallon = 231.0\ in^3$
$1.0\ ft^3 = 0.0283\ m^3$

POWER

$1.0\ W = 1.0\ J/s = 860.42\ cal/hr$
$1.0\ hp = 745.7\ W = 550.0\ ft\text{-}lb/s$
$1.0\ kW = 1 \times 10^3\ W = 3412\ BTU/hr$
$1.0\ BTU/hr = 778.16\ ft\text{-}lb/hr$

ENERGY

$1.0\ J = 1.0\ N\text{-}m = 1 \times 10^7\ ergs$
$1.0\ erg = 1\ dyne\text{-}cm$
$1.0\ cal = 4.1855\ J$
$1.0\ BTU = 778.16\ ft\text{-}lb$
$\quad\quad = 252.16\ cal = 1055.06\ J$

VISCOSITY

$1.0\ N\text{-}s/m^2 = 0.672\ lb_m/ft\text{-}s$

SPECIFIC HEAT

$1.0\ kJ/kg\text{-}°C = 0.23884\ BTU/lb_m\text{-}°F$

THERMAL CONDUCTIVITY

$1.0\ W/m\text{-}°C = 0.5778\ BTU/hr\text{-}ft\text{-}°F$

HEAT TRANSFER COEFFICIENT

$1.0\ W/m^2\text{-}°C = 0.1761\ BTU/hr\text{-}ft^2\text{-}°F$

PHYSICAL CONSTANTS

Standard Acceleration of Gravity
$\quad g = 9.80665\ m/s^2 = 32.1742\ ft/s^2$
Speed of Light
$\quad c = 2.998 \times 10^8\ m/s$
Plank's Constant
$\quad h_p = 6.626 \times 10^{-34}\ J\text{-}s$
Stefan-Boltzmann Constant
$\quad \sigma = 5.673 \times 10^{-8}\ W/m^2\text{-}K^4$
$\quad\quad = 0.1712 \times 10^{-8}\ BTU/lb_m\text{-}mole\text{-}°R$
Universal Gas Constant
$\quad \overline{R} = 8.3143\ J/gmole\text{-}K$
$\quad\quad = 1.9859\ BTU/lbmole\text{-}°R$

PREFACE

The overall goals of an instrumentation and measurements course for engineers are shaped by a variety of applications including controls, quality assurance, performance testing, design, and research. In this text, we have adopted three key objectives:

1. To provide a fundamental background in the theory of engineering measurements and measurement system performance.
2. To convey the principles and practice for the design of measurement systems, including the role of statistics and uncertainty analysis in design.
3. To establish the physical principles and practical techniques used to measure those quantities most important for engineering applications.

Implicit in these objectives is the assumption that certain aspects of measurement systems, such as dynamic response and signal reconstruction, can be generalized, whereas other aspects are best treated in the context of the measurement of physical quantities such as acceleration or temperature.

Chapters 1 to 6 provide a fundamental background in measurement systems. The first chapter provides an introduction to the important concepts of standards and calibration, and introduces some aspects of design of engineering experiments. Error sources in engineering instruments are introduced and classified. An awareness of the design process for measurement systems and calibration procedures motivates the introduction of the statistical nature of physical variables and for uncertainty analysis in Chapters 4 and 5.

Chapters 2 and 3 are an introduction to the generalized description of signals, including their characterization through frequency and amplitude content. The development of Fourier coefficients emphasizes the role of amplitude and frequency content of signals in defining instrument performance. The primary conclusion of Chapter 2, that signals may be characterized as complex periodic waveforms, provides the necessary background both for the generalized description of instrument response in Chapter 3 and for the generalized discrete sampling concepts in Chapter 6. Many students will have previously examined first- and second-order system response in the context of vibrations or dynamic systems. However, it has been our experience that the application of these concepts to measurement systems reinforces the material while allowing the development of practical descriptive parameters of measurement systems such as bandwidth and rise time.

Chapters 4 and 5 are directed toward developing a design methodology for measurement systems based on statistics and uncertainty analysis. Chapter 4 establishes the statistical background necessary for engineers to understand random variables and to quantify the range of expected values for a measured variable. The importance of statistics for mechanical engineering is increasing; it is our philosophy that the measurement of physical variables is an excellent context for the introduction of statistics.

A measurement is not complete unless an uncertainty interval and confidence level is established. Recently established standards reflect the growing awareness of the role of uncertainty analysis in measurements, from the design stage to final reporting. Statistical process control can also be understood in the light of precision contributions to uncertainty. Chapter 5 establishes a methodology for performing uncertainty analyses based on a variety of information, ranging from manufacturers' specifications to interlaboratory comparisons. That chapter provides a working knowledge of uncertainty analysis through carefully selected examples. The statistical nature of measured variables and uncertainty analysis are integrated throughout the remaining chapters in the text, providing contextual examples for a number of common measurement systems.

The expanding role of digital data acquisition in process monitoring and quality assurance requires engineers to have a working knowledge of digital concepts and discrete sampling procedures. We have adopted a basic approach to this subject that avoids introducing any particular hardware or architecture. Rather, our objective is to convey those aspects of the processing of both analog, discrete time, and digital signals that allow the implementation of data acquisition. In this context, Chapter 6 describes basic electrical measurements that form the basis for all types of data acquisition and allow interfacing with the majority of modern transducers. Criteria for discrete sampling techniques are developed, and the role of the discrete Fourier transform in signal reconstruction is emphasized.

The remaining chapters (Chapters 7–11) describe the principles and practice for measuring important mechanical engineering variables, such as temperature, pressure, displacement and strain. The goals of each of these chapters are to provide an understanding of the physical principles used to measure these variables, and to provide sufficient practical information for an engineer properly to select components of a measurement system. These chapters attempt to identify the most common and fundamental engineering variables, and to integrate the concepts of uncertainty analysis and design of measurement systems throughout the treatment. As such, the material is not intended as a survey of the most current technology, nor is it intended to be comprehensive in its coverage of specific transducers. References are provided that will lead the interested reader to appropriate sources for more detailed or advanced treatments. These chapters should serve as a useful reference for the measurement and instrumentation laboratory.

We have organized the text discussion so that flexibility is allowed in the organization of a course in measurements. The first six chapters form the basis for measurement system design and provide an introduction to the key elements of measurement systems that are effectively generalized. As such, some coverage of this material, in a measurements course or a previous course, would enhance the effectiveness of the remaining chapters. Chapters 7 to 11 are written so that they can be used independently, and may be ordered in such a way as to accommodate laboratory schedules or instructor preferences.

We express our appreciation to John H. Lienhard, Massachusetts Institute of Technology, Theodore L. Bergman, The University of Texas at Austin, Charles J. Hurst, Virginia Polytechnic and State University, Richard B. Peterson, Oregon State University, and Galen B. King, Purdue University, for their constructive criticism of the manuscript. We are grateful to our wives, Suzanne and Leigh, for their support and their critical editorial input on the various preliminary manuscripts. We also thank Jamil A. Khan and Yvette Walters for their assistance with this project. Our colleagues and students at Clemson have influenced our development of this material in many ways; their input is most gratefully acknowledged.

Richard S. Figliola
Donald E. Beasley
Clemson, South Carolina

CONTENTS

CHAPTER 1
BASIC CONCEPTS OF MEASUREMENT METHODS   1

1.1 Introduction   1
1.2 General Measurement System   2
1.3 Experimental Test Plan   4
    Random Tests   6
    Replication   10
    Concomitant Methods   11
1.4 Calibration   11
    Static Sensitivity   12
    Range   13
    Accuracy   13
    Sequence Calibration   17
    Random Calibration   17
1.5 Standards   20
    Mass   20
    Time   21
    Length   21
    Temperature   22
    Electrical Dimensions   22
    Derived Units   22
    Hierarchy of Standards   23
1.6 Measurement Overview   25
1.7 Summary   26
References   27
Nomenclature   27
Problems   28

CHAPTER 2
STATIC AND DYNAMIC CHARACTERISTICS
OF SIGNALS   31

2.1 Introduction   31
2.2 Input/Output Signal Concepts   31
    Classification of Waveforms   32

2.3   Waveforms and Fourier Analysis   37
        Fourier Series and Coefficients   40
        Fourier Coefficients for Functions Having Arbitrary Periods   42
2.4   Fourier Transform and the Frequency Spectrum   50
2.5   Summary   57
References   58
Nomenclature   58
Problems   59

CHAPTER 3
MEASUREMENT SYSTEM BEHAVIOR   63
3.1   Introduction   63
3.2   General Model for a Measurement System   63
3.3   Special Cases of the General System Model   67
        Zero-Order Systems   67
        First-Order Systems   69
        Second-Order Systems   81
3.4   Transfer Functions   90
3.5   Phase Linearity   93
3.6   Multiple-Function Inputs   95
3.7   Coupled Systems   96
3.8   Summary   99
References   99
Nomenclature   99
Problems   100

CHAPTER 4
PROBABILITY AND STATISTICS   105
4.1   Introduction   105
4.2   Statistical Measurement Theory   106
        Probability Density Functions   106
4.3   Infinite Statistics   110
4.4   Finite Statistics   117
        Standard Deviation of the Means   119
        Pooled Statistics   122
4.5   Regression Analysis   123
        Least-Squares Regression Analysis   124
4.6   Data Outlier Detection   131
4.7   Number of Measurements Required   133
4.8   Summary   135
References   136
Nomenclature   136
Problems   136

CHAPTER 5
UNCERTAINTY ANALYSIS   141

5.1   Introduction   141
5.2   Measurement Errors   141
5.3   Error Sources   143
       Calibration Errors   143
       Data Acquisition Errors   143
       Data Reduction Errors   144
5.4   Bias and Precision Errors   144
       Bias Error   144
       Precision Error   146
5.5   Uncertainty Analysis: Error Propagation   147
       Propagation of Error   147
5.6   Design-Stage Uncertainty Analysis   150
5.7   Multiple-Measurement Uncertainty Analysis   155
       Propagation of Elemental Errors   155
       Propagation of Uncertainty to a Result   162
5.8   Single-Measurement Uncertainty Analysis   170
       Zero-Order Uncertainty   171
       Higher-Order Uncertainty   171
       Nth-Order Uncertainty   172
5.9   Summary   177
References   178
Nomenclature   178
Problems   179

CHAPTER 6
ELECTRICAL DEVICES, SIGNAL PROCESSING, AND
DATA ACQUISITION   185

6.1   Introduction   185
6.2   Signal Characteristics   186
       Effects of Signal-Averaging Period   190
6.3   Analog Devices: Current Measurement   190
       Direct Current   191
       Alternating Current   194
6.4   Analog Devices: Voltage Measurements   196
       Analog Meters   196
       Oscilloscope   197
       Potentiometer   198
6.5   Analog Devices: Resistance Measurements   201
       Ohmmeter Circuits   202
       Bridge Circuits   202
6.6   Loading Errors and Impedance Matching   210
       Loading Errors for Voltage Dividing Circuit   210
       Interstage Loading Errors   212
6.7   Analog Signal Conditioning: Amplifiers   215
6.8   Analog Filters   219
       Butterworth Filter Design   221

Bessel Filter Design   224
6.9   Sampling Concepts   225
      Sample Rate   225
      Alias Frequencies   227
      Amplitude Ambiguity   231
6.10   Digital Devices: Bits and Words   233
6.11   Digital Devices: Voltage Measurements   235
      Digital-to-Analog Converter   235
      Analog-to-Digital Converter   236
      Digital Voltmeters   243
6.12   Digital Devices: Data Acquisition Systems   244
      System Components   245
6.13   Summary   250
References   250
Nomenclature   251
Problems   252

CHAPTER 7
TEMPERATURE MEASUREMENTS   259
7.1   Introduction   259
      Historical Background   259
7.2   Temperature Standards and Definition   260
7.3   Thermometry Based on Thermal Expansion   264
      Liquid-in-Glass Thermometers   264
      Bimetallic Thermometers   265
7.4   Electrical Resistance Thermometry   266
      Resistance Temperature Detectors   267
      Thermistors   275
7.5   Thermoelectric Temperature Measurement   282
      Fundamental Thermocouple Laws   285
      Basic Temperature Measurement with Thermocouples   286
      Thermocouple Standards   287
      Thermocouple Voltage Measurement   289
      Multiple-Junction Thermocouple Circuits   296
7.6   Radiative Temperature Measurements   298
      Radiation Fundamentals   299
      Radiation Detectors   301
      Radiative Temperature Measurements   302
      Optical Fiber Thermometers   303
7.7   Physical Errors in Temperature Measurement   304
      Conduction Errors   307
      Radiation Errors   309
      Recovery Errors in Temperature Measurement   313
7.8   Summary   316
References   316
Nomenclature   317
Problems   318

CHAPTER 8
PRESSURE AND VELOCITY MEASUREMENTS   325

8.1   Introduction   325

8.2   Pressure Concepts   325

8.3   Pressure Reference Instruments   329
    McLeod Gauge   329
    Barometer   330
    Manometers   331
    Deadweight Testers   336

8.4   Pressure Transducers   338
    Bourdon Tube   339
    Bellows and Capsule   340
    Diaphragms   343

8.5   Pressure Measurements in Moving Fluids   348

8.6   Design and Installation: Transmission Effects   350
    Gases   352
    Liquids   354
    Heavily Damped Systems   354

8.7   Fluid Velocity Measuring Systems   354
    Pitot–Static Pressure Probe   355
    Thermal Anemometry   358
    Doppler Anemometry   360
    Selection of Velocity Measuring Methods   364

8.8   Summary   365
References   *365*
Nomenclature   *366*
Problems   *367*

CHAPTER 9
FLOW MEASUREMENTS   371

9.1   Introduction   371

9.2   Historical Comments   371

9.3   Flow Rate Concepts   372

9.4   Volume Flow Rate through Velocity Determination   374

9.5   Pressure Differential Meters   376
    Obstruction Meters   377
    Orifice Meter   380
    Venturi Meter   382
    Flow Nozzles   383
    Sonic Nozzles   389
    Obstruction Meter Selection   391
    Laminar Flow Elements   395

9.6   Insertion Volume Flow Meters   397
    Electromagnetic Flow Meters   397
    Vortex Shedding Meters   399
    Rotameters   402
    Turbine Meters   403
    Positive Displacement Meters   404

9.7   Mass Flow Meters   405
        Thermal Flow Meter   405
        Coriolis Flow Meter   406
9.8   Flow Meter Calibration and Standards   410
9.9   Summary   410
References   411
Nomenclature   411
Problems   412

**CHAPTER 10
METROLOGY, DISPLACEMENT, AND
MOTION MEASUREMENTS   417**
10.1   Introduction   417
10.2   Dimensional Measurements: Metrology   417
        Principles of Linear Measurement   418
        Optical Methods   423
10.3   Displacement Measurements   426
        Potentiometers   426
        Linear Variable Differential Transformers   427
10.4   Measurement of Mass   429
10.5   Measurement of Acceleration and Vibration   435
        Seismic Transducer   436
        Transducers for Shock and Vibration Measurement   441
10.6   Summary   443
References   443
Nomenclature   443
Problems   444

**CHAPTER 11
STRAIN MEASUREMENT   447**
11.1   Introduction   447
11.2   Stress and Strain   447
11.3   Resistance Strain Gauges   450
        Metallic Gauges   451
        Semiconductor Strain Gauges   457
11.4   Strain Gauge Electrical Circuits   458
11.5   Practical Considerations for Strain Measurement   462
        Apparent Strain   462
        Bridge Constant   465
        Construction and Installation   469
        Analysis of Strain Gauge Data   472
11.6   Optical Strain Measuring Techniques   475
        Basic Characteristics of Light   476
        Photoelastic Measurement   477
        Moiré Methods   480

11.7 Summary   481
References   482
Nomenclature   483
Problems   483

APPENDIX A
DATA ANALYSIS SOFTWARE   487
A.1   Least-Squares Regression Analysis Program   487
A.2   Discrete Fourier Transform   489

APPENDIX B
A GUIDE FOR TECHNICAL WRITING   493

APPENDIX C
PROPERTY DATA   503

INDEX   510

CHAPTER 1
BASIC CONCEPTS OF MEASUREMENT METHODS

1.1 INTRODUCTION

We make simple measurements everyday. We measure body weight on a scale or read the temperature of an outdoor thermometer. Most people put little thought into the selection of instruments for these very simple measurements. The type of instruments and techniques for everyday measurements has long been established by custom, and the outcome of these measurements is not important enough to merit much attention to features such as accuracy. But when the stakes become greater, the selection of measurement equipment and techniques and the interpretation of the measured data can demand considerable attention.

Measurements are common in engineering; they provide a basis for judgments about

- Process information
- Quality assurance
- Process control

Measurement systems are important tools for engineers. This textbook provides an introduction to the theory and design of measurement systems.

The primary objective in any measurement is to establish the value or the tendency of some variable or parameter. This is, of course, based on the value or the tendency suggested by the measurement device. But just how does one establish the relationship between the real value of a variable and the value actually measured? In other words, how can a measurement plan be devised so that the measurement provides the unambiguous information sought? How can a measurement system be used so that the engineer can easily interpret the measured data and be confident of the data's values? There are many considerations that need to be addressed to answer these basic but most important questions.

1.2   GENERAL MEASUREMENT SYSTEM

A *measurement* is an act of assigning a specific value to a physical variable. That physical variable becomes the *measured variable*. A measurement system is a tool used for this quantification of the physical variable. As such, it is used to extend the abilities of the human senses that, while they can detect and recognize different degrees of roughness, length, sound, color, and smell, are limited and are not very adept at assigning specific values to sensed variables. A general template for a measurement system is illustrated in Figure 1.1. Basically such a system consists of part or all of four general stages:

1. Sensor-transducer stage
2. Signal-conditioning stage
3. Output stage
4. Feedback-control stage

These stages form the bridge between the input information, that is, information acquired from the physical variable being measured, and the measuring system output which is used to infer the value of the physical variable measured.

The *sensor* is a physical element that employs some natural phenomenon by which it senses the variable being measured. The *transducer* converts this sensed information into a detectable signal form, which might be electrical, mechanical, optical, or otherwise. The goal is to convert the sensed information into a form that can be easily quantified. For example, the liquid contained within the bulb on the common bulb thermometer of Figure 1.2 exchanges energy with its sur-

FIGURE 1.1   Components of a general measurement system.

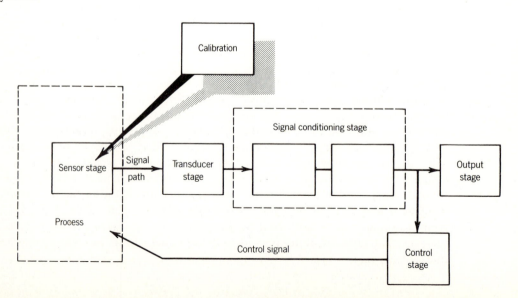

FIGURE 1.2   Components of bulb thermometer
equivalent to sensor, transducer, and output stages.

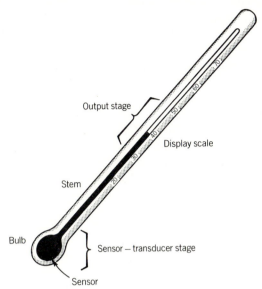

roundings until the two are in thermal equilibrium. At that point they are at
the same temperature. The phenomenon of thermal expansion of the liquid
results in its movement up and down the stem, converting the thermal information into a mechanical translation. The liquid in the bulb acts as the sensor.
By forcing the expanding liquid into a narrow capillary, this measurement system
transforms the sensed thermal information into a mechanical displacement.
Hence, the bulb internal design acts as a transducer. It is not always possible
to distinguish between the sensor and transducer stages in an instrument, and
in such instances the terms can be used interchangeably.

Sensor selection, placement, and installation are particularly important, because the input to the measurement system *is* the information sensed by the
sensor. Accordingly, the interpretation of all information passed through and
indicated by the system depends on that which is actually sensed by the sensor.

Signal conditioning equipment takes the transducer signal and modifies it to
a desired magnitude. This is an optional intermediate stage that might be used
to perform tasks such as increasing the magnitude of the signal through amplification, removing portions of the signal through some filtering technique,
and/or providing mechanical or optical linkage between the transducer and the
output stage, for example, converting a translational displacement of a sensor
into a rotational displacement of a pointer. This stage can consist of one or more
devices.

The *output* stage takes the signal and provides an indication of the value of
the measurement. The output equipment might be a simple readout display or
might contain devices that can record the signal for later analysis. Examples of
these devices are tape recorders, chart recorders, and computer disk drives. The
scale of our bulb thermometer in Figure 1.2 serves as the output stage of that
measurement system.

In those measurement systems involved in process control, a fourth stage, the *feedback-control stage*, contains a controller that interprets the measured signal and makes a decision regarding the control of the process. This decision results in a change in a process parameter that affects the magnitude of the sensed variable. In simple controllers, this decision is based on the magnitude of the signal of the sensed variable, usually whether it exceeds some high or low set point, a value set by the system operator. At the extreme end of the sophistication scale, a signal from a measurement system can be used as an input to an "expert system" controller that, through an artificial intelligence scheme, determines the optimum set conditions for the process. A simple measurement system with control stage is a household furnace thermostat. The operator fixes the set point for temperature on the thermostat display, and the furnace is activated as the local temperature at the thermostat, as determined by the sensor within the device, rises or falls about the set point.

1.3   EXPERIMENTAL TEST PLAN

Engineering measurements are not as simple as turning on the equipment and reading the numbers. Relevant information can only be extracted from test data that are obtained from a well-thought-out measurement plan. This plan will include:

1. An identification of pertinent process variables and parameters.
2. A well-thought-out measurement pattern for performing the test on the process.
3. The selection of a measurement technique and required equipment.
4. A data analysis plan.

Experimental design includes the development of a measurement plan. Such a plan will assist the engineer in achieving an optimization between time and cost expenditure regarding acquiring and analyzing test data versus information obtained. In this section, concepts related to (1) and (2) are introduced.

The identification of relevant process parameters and variables is the first step of experimental design measurement strategy. In addition to the targeted measured variable, other variables may be pertinent to the measured process. All known process variables should be listed and evaluated for any possible cause and effect relationships. If a change in one variable will not affect the value of another variable, the two are considered independent of each other. But a suspected dependency between variables mandates certain controls that must be designed into the test plan. A variable that can be changed independently of other variables is known as an *independent variable*. A variable that is affected by changes in one or more other variables is known as a *dependent variable*.

The control of variables is important. A variable is said to be *controlled* if it can be held at a constant value during a measurement of some other variable. Complete control of a variable would imply that it can be held perfectly constant and to an exact prescribed value, but such complete control of a variable is very rare. Instead we will use the adjective "controlled" to refer to a variable that can be held fixed in, at least, a nominal sense. The cause and effect relationship between an independent variable and a dependent variable is found by applying a controlled value of the independent variable while measuring the dependent variable.

A parameter is defined by a functional relationship between two or more variables.[1] A parameter that has an effect on the behavior of the measured variable is called a *control* parameter. Available methods for establishing control parameters based on known process variables include similarity and dimensional analysis techniques (e.g., [1–3]), and physical laws. A control parameter is completely controlled if it can be held at a constant value during a set of measurements.

Variables that are not or cannot be controlled during a measurement but that affect the value of the variable measured are called *extraneous variables*. Even in a measurement plan in which all process variables appear controlled, external uncontrolled variables may affect the measured variable, the controlled variables, and the measurement system in an unpredictable manner. Extraneous variables will introduce differences between measured values of the same variable obtained under seemingly controlled conditions and may imply a false trend in the behavior of that variable.

As an example, the flow rate, Q, developed by a fan depends on rotational speed, n, and the diameter, d, of the fan. A control parameter for this group of variables, found by similarity methods, is the fan flow coefficient, $C_1 = Q/nd^3$. For a given fan (d is controlled and fixed), if speed is controlled, the fan flow rate associated with that speed can be measured and the flow coefficient can be determined. However, extraneous variables that are not included in the control parameter will also affect fan operation. Under the seemingly fixed operating condition of constant fan speed (the setting is not touched), uncontrolled variations in the surrounding temperature and voltage line frequency fluctuations will affect fan motor speed and, therefore, cause variations in the fan flow rate measured. Hence, fan speed would be controlled only in a nominal sense because of extraneous influences. The output signal from the flow rate sensor would show variations consistent with the behavior of the extraneous variables.

The effects that extraneous variables produce on the measured signal can be delineated into noise and interference. *Noise* is a random variation of the value of the measured signal as a consequence of the variation of the extraneous variables. Examples include normal random variations in environmental conditions that affect the measured variable and/or the measurement system, thermal noise (Johnson noise) due to the random temperature-induced excitation of electrons within current carrying conductors, and electronic shot noise common in transistors because of random fluctuations in the rate at which carriers diffuse across transistor junctions. A completely controlled variable contains no noise.

Interference produces undesirable deterministic effects on the measured value because of extraneous variables. One form is that of a sinusoidal wave superimposed onto a measured signal path. Examples include well-defined deterministic variations in environmental conditions, local ac power line noise (60 or 50

---

[1]It can also be a single "variable," which is not functionally related to other variables. An example is time in many tests.

Hz) or fluorescent lighting arc noise (120 or 100 Hz), and electromagnetic (emi) and radio-frequency (rf) interference due to nearby motors, transmitters, welding equipment, or transformers (several megahertz). Hum and acoustic feedback in public address and audio systems are ready examples of interference effects that are superimposed onto a desirable signal.

An undesirable situation arises if the period of the interference is longer than the period over which the measurement is made, because the interference may superimpose a false trend in the behavior of the measured variable. An important goal of a measurement plan should be to perform the measurement in a manner that will break up such trends so that they appear as random variations in the data set. Although this will increase the scatter in the measured values of a data set, it is far more important to eliminate false trends, which will otherwise lead to a misinterpretation of the measured data. Random data scatter can be handled statistically. Randomization methods are available that can be easily incorporated into the measurement plan and will minimize or eliminate such trends. Several are discussed in the paragraphs that follow.

Random Tests

Consider the situation in which the dependent variable, y, is a function of several independent variables, x_a, x_b, ... However, the measurement of y will also be influenced by several extraneous variables, z_j, where $j = 1, 2, \ldots$, such that $y = f(x_a, x_b, \ldots ; z_j)$. To find the dependence of y on x_a, all other independent variables should be held fixed while x_a is varied in a controlled manner. The dependence of y on x_b can be found in a similar manner by the variation of x_b, and so forth. Although the influence of the z_j variables on these tests can not be eliminated, the possibility of their introducing a false trend on y can be minimized by a proper test strategy.

A *random test* is defined by a measurement plan that sets a random order as to the value of the independent variable applied. Trends normally introduced by the coupling of a relatively slow and uncontrolled variation in the extraneous variables with a sequential application in values of the independent variable applied will be broken up. Furthermore, each change in the controlled variable will become truly independent of the previous one. In a random test, any measured value is equally likely to contain any possible effect of the extraneous variables. This type of plan is effective for the local control of extraneous variables that change in a continuous manner.

EXAMPLE 1.1

Consider the pressure calibration system shown in Figure 1.3 in which a sensor–transducer is exposed to a known pressure, p. The transducer, powered by an external supply, converts the sensed signal into a voltage that is measured by a voltmeter. The measurement plan is to control the applied pressure by the measured displacement of a piston that is used to compress a gas contained within the piston–cylinder chamber. The gas chosen closely obeys the ideal gas law. Hence, piston displacement, x, which sets the chamber volume, $V = (x \cdot \text{area})$, is easily related to chamber pressure. Identify the independent and dependent variables in the calibration and possible extraneous variables.

FIGURE 1.3 Pressure calibration rig.

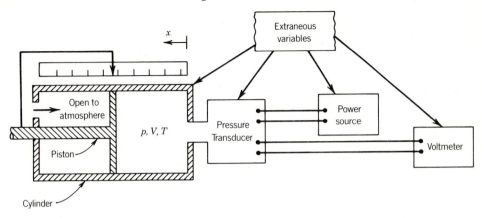

KNOWN

Pressure calibration system of Figure 1.3

FIND

Independent, dependent, and extraneous variables

SOLUTION

The control parameter for this problem can be formed from the ideal gas law: pV/T = constant, where T is the gas temperature. An independent variable in the calibration is the piston displacement that sets the volume. This variable can be controlled by locking the piston into position. From the ideal gas law, gas pressure will also be dependent on temperature, and therefore temperature is also an independent variable. However, T and V are not in themselves independent according to the control parameter. Since volume is to be varied through variation of piston displacement, T and V can be controlled provided a mechanism is incorporated into the scheme to maintain a constant gas temperature within the chamber. This will also maintain chamber area constant, a relevant factor in controlling the volume. In that way, the applied variations in V will be the only effect on pressure, as desired. The dependent variable is the chamber gas pressure. The pressure sensor is exposed to the chamber gas pressure and, hence, this is the pressure sensed by it. Other extraneous variables would include noise effects due to the room temperature, z_1, and line voltage variations, z_2, which would affect the excitation voltage from the power supply and the performance of the voltmeter. Connecting wires between devices will act as an antenna and possibly will introduce interference, z_3, superimposed onto the electrical signal. Their magnitudes can be reduced by proper electrical shielding.

Hence, $p = f(V, T; z_1, z_2, z_3)$, where $V = f_1(x, T)$.

COMMENT

Even though we might try to keep the gas temperature constant, any slight variations in the gas temperature would affect the volume and pressure and, hence, will act as an additional extraneous variable!

EXAMPLE 1.2

Develop a test plan that will minimize the effects of the extraneous variables in Example 1.1.

KNOWN

$p = f(V, T; z_1, z_2, z_3)$, where $V = f_1(x; T)$. Control variable V is changed. Dependent variable p is measured.

FIND

Randomize the extraneous variables

SOLUTION

We intend to vary volume, control gas temperature, and measure pressure. An important feature of all test plans is a strategy that minimizes the superposition of false trends onto the data set by the extraneous variables. Since z_1, z_2, and z_3 and any inability to hold the gas temperature constant are continuous extraneous variables, their influence on p can be randomized by a random test. This entails shuffling the order by which V is applied. Say that we set six values of volume: V_1, V_2, V_3, V_4, V_5, and V_6, where the subscripts correspond to an increasing sequential order of the respective values of volume. Any random order will do fine. One possibility found by using the random function features of a hand-held calculator is

$$V_2 \quad V_5 \quad V_1 \quad V_4 \quad V_6 \quad V_3$$

Discrete extraneous variables present a different sort of problem. The use of different instruments, different test operators, and different climatic conditions are examples of discrete extraneous variables that can affect the outcome of a measurement. Randomizing a test plan to minimize discrete influences can be done efficiently through the use of experimental design using random blocks. A *block* consists of M data sets of the measured variable obtained under the same value of controlled variable, but between each set the extraneous variable has been allowed to change. This enables some amount of local control over the discrete extraneous variable. Many strategies for randomized blocks exist, as do advanced statistical methods for data analysis (e.g., [4–6]). Consider the following examples.

EXAMPLE 1.3

The manufacture of a particular composite material requires mixing a percentage by weight of binder with resin to produce a gel. The gel is used to impregnate a fiber to produce the composite material in a manual process called the lay-up. The strength, σ, of the finished material depends on the percent binder in the gel. However, the strength may also be lay-up operator dependent.

Formulate a test pattern by which the strength to percent binder–gel ratio relationship under production conditions can be established.

KNOWN

$\sigma = f$ (binder; operator)

ASSUMPTION

Strength is affected only by binder and operator

FIND

Test plan to randomize effects of operator

SOLUTION

The dependent variable, σ, is to be tested against the independent variable, percent binder–gel ratio. The operator is an extraneous variable in actual production. As a simple test, we could test the relationship between three binder–gel ratios, A, B, and C, and measure strength. We could also choose three typical operators (z_1, z_2, and z_3) to produce N separate composite test samples for each of the three binder–gel ratios. This gives the 3-block test pattern:

Block				
1	A:	z_1	z_2	z_3
2	B:	z_1	z_2	z_3
3	C:	z_1	z_2	z_3

In the analysis of the test, the results of each block will include each operator's influence as a variation. We can assume that the order of the operators within each block is unimportant. But if only the samples from one operator are considered, the results may show a trend consistent with the lay-up technique of that operator. The test plan above will randomize the influence of any one operator on the strength test results.

EXAMPLE 1.4

Suppose following lay-up, the composite material of Example 1.3 is allowed to cure at a controlled but elevated temperature. We wish to develop a relationship between the binder–gel ratio and the cure temperature and strength. Develop a suitable test plan.

KNOWN

$\sigma = f$(binder, temperature; operator)

ASSUMPTION

Strength is affected only by binder, temperature, and operator

FIND

Test plan to randomize effect of operator

SOLUTION

We develop a simple plan to test for the dependence of composite strength on the independent variables of binder–gel ratio and cure temperature. Choosing three temperatures, T_1, T_2, and T_3, along with three binder–gel ratios, A, B, and C and three operators, z_1, z_2, and z_3, we can set up a 3×3 test matrix representing a single randomized block. It is important to randomize the order of the block such that no operator runs the same test combination more than once. This eliminates the influence of any one operator on a particular binder–gel ratio, temperature test.

	z_1	z_2	z_3
A	T_1	T_2	T_3
B	T_2	T_3	T_1
C	T_3	T_1	T_2

COMMENT

We could have proceeded as in Example 1.3 and set up three randomized blocks for ratio and three for temperature for a total of 18 separate tests. But the suggested test plan matrix not only randomizes the extraneous variable but has reduced the number of tests by one-half. Such a matrix is referred to as a Latin square.

If we wanted to include our ability to control the independent variables in the test data variations, we could duplicate the Latin-square test several times to build up a significant data base. Such a duplication is referred to as a replication.

Replication

In general, the estimation of the value of a variable improves as the number of measurements of that variable made under controlled conditions increases. Repeated measurements made under the same operating conditions are called *repetition*. Repetition enables the engineer to quantify the temporal behavior of the measured variable and the extent of the variation in the measured data set using statistical methods that we introduce in Chapter 4. But if repeated measurements are made without ever changing the independent variable, the repeated measurements will not be truly independent measurements because the extraneous variables will not necessarily be completely randomized. However, the extraneous variables can be completely randomized by replication. *Replication* refers to an independent duplication of a set of measurements under similar controlled conditions.

Repetition provides an estimation of the variation to be expected in the measured variable under a fixed operating condition. On the other hand, replication provides an estimation of the variation in the measured variable with the variation in the test procedure (controllability of the independent variables). For continuous extraneous variables, the time period between replications should be longer than the period of the variation in the extraneous variables. For discrete extraneous variables, replication is performed by duplicate runs of random blocks (see Comment, Example 1.4).

Concomitant Methods

All measurement plans should incorporate the use of concomitant methods. The goal is to obtain two or more estimates for the result, each based on a different method, which can be compared as a check for agreement. This may affect the experimental design in that additional variables may need to be measured. Or the different method could be analysis that estimates an expected value of the measurement (e.g., using a modal analysis to check a set of vibration measurements on a structural element). Suppose we needed to establish the volume of a cylindrical rod of known material. We could simply measure the diameter and length of the rod to compute this. Alternatively, we could measure the weight of the rod and compute volume based on the specific weight of the material. The second method complements the first and provides an important check on the first estimate.

1.4   CALIBRATION

After consideration has been given to the experimental plan, attention can be turned toward calibration. A *calibration* is the act of applying a known value of input to the selected measurement system for the purpose of observing the system output. The known value used for the calibration is called the *standard*. The relationship between the input to the measurement system and the system output is established during a calibration of the measurement system. Since the calibration is itself a measurement, a proper experimental plan should be formulated to identify the appropriate independent and dependent variables to be measured, as well as minimizing the effects of extraneous variables.

By successive application of a range of known values for the input and observation of the system output, a direct calibration curve can be developed for the measurement system. On such a curve the input, x, is plotted on the abscissa against the measured output, y, on the ordinate, as indicated in Figure 1.4. In a calibration the input value should be a controlled independent variable, while the measured output value becomes the dependent variable of the calibration.

A calibration curve forms the logic by which a measurement system's indicated output can be interpreted during an actual measurement, when the value input to the system is not known a priori. For example, the calibration curve is the basis for fixing the output display scale on a measurement system, such as that of Figure 1.2. Alternatively, a calibration curve can be used as part of developing a functional relationship, an equation known as a *correlation*, between input and output. A correlation will have the form $y = f(x)$ and is determined by applying physical reasoning and curve fitting techniques to the calibration curve. The correlation can then be used in later measurements to ascertain the unknown

FIGURE 1.4   Representative calibration curve.

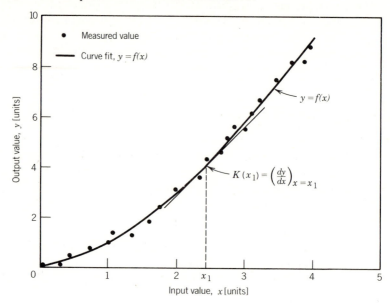

input value based on the output value, the value indicated by the measurement system.

The most common type of calibration is known as a *static calibration*. The term static refers to a procedure in which time is not a factor in the variable being measured. In static calibrations, only the magnitude of the known input is important. Accordingly, measurement systems that have been calibrated using a static calibration procedure are to be used to infer static variables or, in the most extreme instance, variables that vary relatively slowly with time. When time-dependent variables are to be measured, a procedure known as a *dynamic calibration* is also performed. In the broad sense, dynamic variables are time dependent in both their magnitude and their frequency content. The magnitude of input indicated by any measurement system will depend on the frequency content of the input value. The specific mechanism for this behavior will be introduced later, but is related to the ability of a particular system to react to a change in input value with time. Dynamic calibrations then determine the relationship between an input, of known magnitude and frequency, and the measurement system output. Usually in such calibrations a step change in input is applied or a sinusoidal input over a range of known frequencies is applied to the measurement system under calibration and the resulting system behavior is quantified with respect to time.

Static Sensitivity

The slope of a static calibration curve, evaluated at the input value, yields the *static sensitivity*[2] of the measurement system. As depicted graphically in the

---

[2]Some texts refer to this as the static gain or the steady sensitivity.

calibration curve of Figure 1.4, the static sensitivity, K, at any particular static input value, say x_1, is evaluated by

$$K = K(x_1) = \left(\frac{dy}{dx}\right)_{x=x_1} \tag{1.1}$$

where K is a function of x. The static sensitivity is a measure that relates the change in the indicated output associated with a given change in a static input. Since calibration curves can be linear or nonlinear depending on the measurement system and on the variable being measured, K may or may not be constant over a range of input values. Periodic calibrations of all measurement systems are the norm to ensure that the proper relationship between the applied and indicated system output is maintained.

Range

The proper procedure for calibration is to apply known inputs ranging from the minimum and to the maximum values for which the measurement system is to be used. This defines the operating *range* of the system. The input operating range is defined by

$$r_i = x(\text{maximum}) - x(\text{minimum}) \tag{1.2}$$

and the output range by

$$r_o = y(\text{maximum}) - y(\text{minimum}) \tag{1.3}$$

It is important to avoid extrapolation beyond the range of known calibration during measurement since the behavior of the measurement system is uncharted in these regions. As such, the range of calibration should be carefully selected.

Accuracy

The accuracy of a system can be estimated during calibration. If we assume for the moment that the input value is known exactly, then the known input value can be called the true value. The *accuracy* of a measurement system refers to its ability to indicate a true value exactly. Accuracy is related to absolute error. *Absolute error*, ϵ, is defined as the difference between the true value applied to a measurement system and the indicated value of the system:

$$\epsilon = \text{true value} - \text{indicated value} \tag{1.4}$$

from which the relative accuracy is found from

$$A = 1 - \frac{\epsilon}{\text{true value}} \tag{1.5}$$

Based on this definition, accuracy can be determined only when the true value is known, such as during a calibration.

An alternative form of calibration curve is the deviation plot shown in Figure 1.5. Such a curve plots the difference or deviation between the true or expected value, y', and the indicated value, y, versus the indicated value. Deviation curves are extremely useful when the differences between the input and the indicated value are too small to suggest possible trends in the differences between them on direct calibration plots. They are often required in situations that require errors to be reduced to the minimums possible. In Figure 1.5, the voltage output during calibration from a temperature-sensing thermocouple is compared with the nominal expected values obtained from reference tables.

The repeatability or precision of a measurement system refers to the ability of the system to indicate a particular value upon repeated but independent applications of a specific value of input. *Precision error* is a measure of the random variation to be expected during repeatability trials. Precision can be estimated at any time during a measurement provided that the input condition can be held constant. An estimate of a measurement system precision does not require a calibration, per se. But note that a system that repeatedly indicates the same wrong value upon repeated application of a particular input would be considered to be very precise regardless of its known accuracy. The average error in a series of repeated calibration measurements defines the error measure known as *bias*. Both precision and bias errors affect the measure of a system's accuracy.

The concepts of accuracy, error, and precision in measurements can be illustrated by the throw of darts. Consider the dart board of Figure 1.6 where the goal will be to throw the darts into the bull's-eye. For this analogy, the bull's-eye can represent the true value and each throw can represent a measurement value. In Figure 1.6*a*, the thrower displays good precision in that each throw repeatedly hits the same spot on the board, but the thrower is not accurate in

---

FIGURE 1.5   Calibration deviation plot for temperature sensor output.

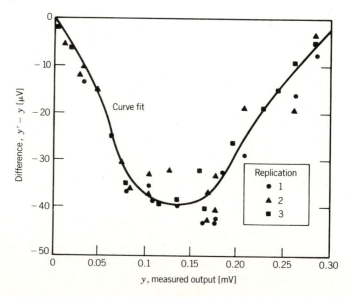

FIGURE 1.6   Throws on a dart board: illustration of precision and accuracy.

(*a*) High repeatability gives low precision error but gives no direct indication of accuracy

(*b*) High accuracy means low precision and bias errors

(*c*) Bias and precision errors lead to poor accuracy

that the dart misses the bull's-eye each time. This thrower is precise, but we see that precision alone is not a measure of accuracy. The error in each throw can be computed from the distance between the bull's-eye and each dart. The average value of the error yields the bias. This thrower has a bias to the left of the target. If the bias could be reduced, then this thrower's accuracy would improve. In Figure 1.6*b*, the thrower displays high accuracy and high repeatability, hitting the bull's-eye on each throw. Both throw scatter and bias are near zero. High accuracy must imply high precision and low bias as shown. In Figure 1.6*c*, the thrower displays neither high precision nor accuracy with the errant throws scattered around the board. Each throw contains a different amount of error. While the bias is the average of the errors in each throw, precision is related to the varying amount of error in the throws. The accuracy of this thrower's technique appears to be biased and lacking precision. The precision and bias of the thrower can be computed using the statistical methods that are discussed in Chapter 4. Both precision and bias quantify the error in any set of measurements and are used to estimate accuracy.

Suppose a measurement system was used to measure a variable whose value was kept constant and known exactly, as in a calibration. Ten independent measurements are made with the results, as shown in Figure 1.7. The variations in the measurements, the observed scatter in the data, would be related to the system precision error associated with the measurement of the variable. That is, the scatter is mainly due to (1) the measurement system and (2) the method of its use, since the value of the variable is essentially constant. However, the offset between the apparent average of the readings and the actual known value would provide a measure of the bias error to be expected from this measurement system.

In any measurement other than a calibration the error cannot be known exactly since the true value is not known. But based on the results of a calibration, the operator might feel confident that the error is within certain bounds, a plus or minus range of the indicated reading. Since the magnitude of the error in any measurement can only be estimated, one refers to an estimate of the error in

FIGURE 1.7   Effects of precision and bias error during calibration.

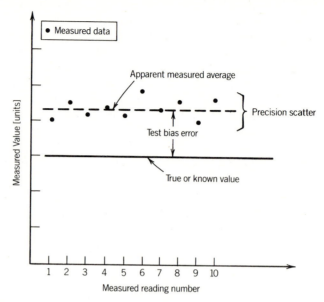

the measurement as the *uncertainty* present in the measured value. Uncertainty is brought about by errors that are present in the measurement system, its calibration, and measurement technique, and is manifested by measurement system bias and precision. A method of estimating the uncertainty in a measurement is treated in detail in a later chapter.

The precision and bias errors of a measurement system are the result of several interacting errors inherent to the measurement system, the calibration procedure, and the standard used to provide the known value. These errors can be delineated and quantified as elemental errors through the use of particular calibration procedures and data reduction techniques. An example of such elemental error quantification for a particular device is given for a typical pressure transducer in Table 1.1.

TABLE 1.1   Manufacturer's Specifications: Typical Pressure Transducer

Operation	
Input range	0 to 1000 cm H_2O
Excitation	±15 V dc
Output range	0 to 5 V
Performance	
Linearity error	±0.5% full scale
Hysteresis error	Less than ±0.15% full scale
Sensitivity error	±0.25% of reading
Thermal sensitivity error	±0.02%/°C of reading
Thermal zero drift	0.02%/°C full scale
Temperature range	0 to 50 °C

Sequence Calibration

A *sequence calibration* applies a sequential variation in the input value over the desired input range. This may be accomplished by increasing the input value (upscale direction) or by decreasing the input value (downscale direction) over the full input range. A sequence calibration can refer to either a static or a dynamic calibration as appropriate.

Hysteresis

The sequence calibration is an effective diagnostic technique for identifying and quantifying hysteresis error in a measurement system. *Hysteresis error* refers to differences between an upscale sequence calibration and a downscale sequence calibration. The hysteresis error of the system is given by $e_h = (y)_{\text{upscale}} - (y)_{\text{downscale}}$. The effect of hysteresis in a sequence calibration curve is illustrated in Figure 1.8a. Hysteresis is usually specified for a measurement system in terms of the maximum hysteresis error as a percentage of full-scale output range (FSR):

$$\%(e_h)_{\text{max}} = \frac{[e_h(x)]_{\text{max}}}{r_o} \times 100 \tag{1.6}$$

such as that indicated in Table 1.1. Hysteresis occurs when the output of a measurement system is somehow dependent on the previous value indicated by the system. Such dependencies can be brought about through some realistic system limitations such as friction or viscous damping in moving parts or residual charge in electrical components. Some hysteresis is normal for any system and affects the precision of the system.

Random Calibration

A *random calibration* applies the known input over the intended input range in a random manner. The application of random values of input tends to minimize the impact of any existing trends in the variation of uncontrolled variables on the measurement system, as well as hysteresis effects and human observation errors. It ensures that each application of input value is independent of the previous. This reduces calibration bias error. Generally, such a random variation in input value will more closely simulate the actual measurement situation. Random test methods can be used in either a static or a dynamic calibration.

A random calibration provides an important diagnostic test for the delineation of several measurement system performance characteristics based on a set of random calibration test data. In particular, linearity error, sensitivity error, zero error, and instrument repeatability error, as illustrated in Figure 1.8b–e, can be quantified from a static random calibration.

Linearity Error

Many instruments are designed to achieve a linear relation between an applied static input and indicated output values. Such a linear static calibration curve would have the general form

$$y_L(x) = a_0 + a_1 x \tag{1.7}$$

FIGURE 1.8   Examples of elements of instrument error.

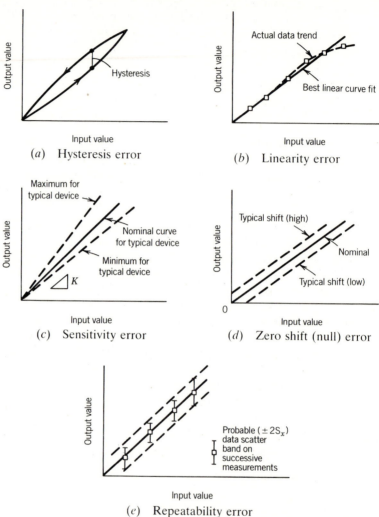

(a)   Hysteresis error

(b)   Linearity error

(c)   Sensitivity error

(d)   Zero shift (null) error

(e)   Repeatability error

where the curve fit $y_L(x)$ provides a predicted output value based on a linear relation between x and y. However, in real systems, truly linear behavior is an ideal goal only approximately achieved in practice. As a result, measurement device specifications usually provide a statement as to the expected linearity of the static calibration curve for the device. The relation between $y_L(x)$ and measured value $y(x)$ is a measure of the nonlinear behavior of a real system:

$$e_L(x) = y(x) - y_L(x) \qquad (1.8)$$

where $e_L(x)$ is the *linearity error* that arises in describing the actual system behavior by a linear polynomial. Such behavior is illustrated in Figure 1.8b in which a linear curve has been fit through a calibration data set. For a measurement system that is essentially linear in behavior, the extent of possible nonlinearity in a measurement device is often specified in terms of the maximum expected linearity error as a percentage of full-scale output range:

$$\%(e_L)_{max} = \frac{[e_L(x)]_{max}}{r_o} \times 100 \tag{1.9}$$

This value is listed as the linearity error expected from the pressure transducer in Table 1.1. Statistical methods of quantifying the error introduced in attempting to fit real data to a desired order polynomial are discussed in Chapter 4.

Sensitivity and Zero Errors

The expected scatter in data found during a calibration affects the precision in the estimate of the static sensitivity of the measurement system. As shown for the linear calibration curve in Figure 1.8c, if we fix the zero intercept at zero (a zero output from the system for zero input), then the scatter in the data leads to precision error in estimating the slope of the calibration curve. The *sensitivity error*, e_K, is a statistical measure of the precision error in the estimate of the slope of the calibration curve (as discussed in Chapter 4). The static sensitivity of a device is also temperature dependent and this is often specified. In Table 1.1, the listed sensitivity error reflects calibration results at a constant reference ambient temperature, whereas the thermal sensitivity error was found by calibration at different temperatures.

If the zero intercept is not fixed but the sensitivity is held fixed, then drifting of the zero intercept introduces a vertical shift of the calibration curve, as shown in Figure 1.8d. This shift of the zero intercept of the calibration curve is known as the *zero error*, e_z, of the measurement system. Zero error can usually be reduced by the user by periodically adjusting the output from the measurement system under a zero input condition. However, some random variation in the zero intercept is common, particularly with electronic and digital equipment subjected to temperature variations (e.g., thermal zero drift in Table 1.1).

Instrument Repeatability

Manufacturer specifications will sometimes refer to a measure called instrument repeatability. The ability of a measurement system to indicate the same value upon repeated but independent application of the same input value is known as the *instrument repeatability*. Such a measure is inclusive of the linearity and sensitivity errors. Repeatability is best quantified using statistical methods (Chapter 4) and is based on a measure of the variation in the measured data known as the standard deviation, S_x. This reflects only a measure of the device repeatability in an indicated value for a given input value found under controlled calibration conditions, such as shown in Figure 1.8e. The value normally claimed is in terms of the maximum expected instrument repeatability error as a percentage of full scale output range:

$$\%(e_R)_{max} = \frac{2(S_x)}{r_o} \times 100 \tag{1.10}$$

The instrument repeatability error reflects only that error to be expected from that instrument under controlled calibration conditions. The precision error (repeatability) in an actual measurement made using such an instrument would necessarily be greater.

1.5 STANDARDS

When a measurement system is calibrated it is compared with some standard whose value is presumably known. This standard may be a piece of equipment that the operator trusts. Perhaps it is an object having a well-defined physical attribute to be used as a comparison. Or it may even be a well-accepted technique known to produce a reliable value. Regardless, the quality of the standard is judged by how well its value is known. But the suitability of the standard for use in a calibration depends only on the intended application and required accuracy of the measurement system calibrated. Let us explore how certain standards come to be and how these standards are the foundation of all measurements.

A *dimension* defines a physical variable that is used to describe some aspect of a physical system. The fundamental value associated with any dimension is given by a unit. A *unit* defines a measure of a dimension. For example, terms such as mass, length, and time describe basic dimensions to which we associate the units of kilogram, meter, and second. A *primary standard* defines the value of a unit. It provides the means to describe the unit with a unique number that can be understood throughout the world. The primary standard then assigns a unique value to a unit by definition! As such it must define the unit exactly.

Primary standards are necessary, because the value assigned to a unit is actually arbitrary. Whether a meter is the length of a king's arm or the distance light travels in a fraction of a second depends only on how we want to define it. To avoid confusion, units are defined by international agreement through the use of primary standards. Once agreed upon, the primary standard forms the exact definition of the unit until it is changed by some later agreement. Important features sought in any standard should include:

- Global availability
- Continued reliability
- Stability

with minimal sensitivity to external environmental sources. Next we examine some basic dimensions and the primary standards that form the definition of the units that describe them.

Mass

The dimension of mass is defined in the International System of Units (SI) by the kilogram. Originally, the unit of the kilogram was defined by the mass of one liter of water at room temperature. But today an equivalent yet more consistent definition defines the kilogram exactly as the mass of a particular platinum–iridium bar which is maintained under very specific conditions at the International Bureau of Weights and Measures located in Sèvres, France. This particular bar forms the primary standard for the kilogram.

In the United States, an alternative system of units, the U.S. engineering standard unit system, has become so ingrained in the engineering industry that it is still used. In the U.S. engineering unit system, mass is defined by the pound-mass, lb_m, which can be derived from the definition of the kilogram:

$$1 \ lb_m = 0.45359237 \ kg$$

Equivalent standards for the kilogram and the pound-mass are maintained by the U.S. National Institute for Science and Technology (NIST, formerly the National Bureau of Standards, or NBS) in Gaithersburg, Maryland.

Time

The dimension of time is defined by the unit of a second in both SI and U.S. engineering units. One second has been defined [7] as the time elapsed during 9 192 631 770 periods of the radiation emitted between two excitation levels of the fundamental state of cesium-133. Despite this seemingly unusual definition, this primary standard can be reliably reproduced at suitably equipped laboratories throughout the world to within 2 parts in one billion.

The Bureau Internationale de l'Heure (BIH) in Paris maintains the primary standard for clock time. Periodically, adjustments to clocks around the world are made relative to the BIH clock so as to keep time synchronous.

The standard for cyclical frequency is based on the time standard. The standard unit is the hertz (1 Hz = 1 cycle/second). The cyclical frequency is related to the circular frequency (radians/second) by

$$1 \ hertz = \frac{2\pi \ radians}{1 \ second}$$

Both time and frequency standard signals are broadcast worldwide over designated radio stations for use in navigation and as a source of a standard for these dimensions.

Length

New primary standards are defined when the ability to determine the new standard becomes more accurate than the existing standard. In 1982, a new primary standard used to define the unit of a meter, the basic SI unit of the dimension of length, was adopted. The primary standard for the fundamental unit for the dimension of length is now based on the dimension of time. One meter is exactly defined as the length traveled by light in 3.335641×10^{-9} seconds, a number derived from the known velocity of light in a vacuum.

The U.S. engineering standard unit of the foot and related unit of the inch can be derived from the meter.

$$1 \ foot = 0.30480060 \ meters$$

$$1 \ inch = 0.02540005 \ meters$$

In the United States, primary standards for the meter are maintained at NIST.

Temperature

A temperature scale, based on thermodynamic principles, was devised by William Thomson, Lord Kelvin (1824–1907), and forms the basis for the absolute practical temperature scale in common use. This practical scale closely follows the thermodynamic scale from which it is derived. The absolute practical scale is defined by the basic SI unit of a kelvin, K. It is based on polynomial interpolation between the equilibrium phase change points of a number of common pure substances from the triple point of equilibrium hydrogen (13.81 K) to the freezing point of pure gold (1337.58 K). Above 1337.58 K, the scale is based on Plank's law of radiant emissions. The details of the standard have been modified over the years but are governed by the International Temperature Scale–1990 [8], which superseded the International Practical Temperature Scale–1968 [9].

The U.S. engineering unit system uses the absolute scale of Rankine (°R). This and the common scales of Celsius (°C) and Fahrenheit (°F) are related to the Kelvin scale by the following:

$$(°C) = (K) - 273.15$$

$$(°F) = (°R) - 459.67$$

$$(°F) = 1.8 \times (°C) + 32.0$$

Electrical Dimensions

Fundamental dimensions of electrical potential, current, and resistance are based on the definitions of the absolute units of volt (V), ampere (A), and ohm (Ω), respectively. These units are derivable from the basic units of mass, length, and time. The unit of one ampere absolute is defined by 1.00165 times the current in a water-based solution of AuN_2 that deposits Au at an electrode at a rate of 1.118×10^{-5} kg/s. One ohm absolute is defined by 0.9995 times the resistance to current flow of a column of mercury that is 1.063 m in length and has a mass of 0.0144521 kg at 273.15 K.

On a practical level, working standards for resistance take the form of certified standard resistors or resistance boxes and are used as standards for comparison in the calibration of resistance measuring devices. The practical potential standard makes use of a standard cell consisting of a saturated solution of cadmium sulfate. The potential difference of two conductors connected across such a solution is set at 1.0183 V at 293 K. The standard cell maintains a constant electromotive force over very long periods of time, provided that it is not subjected to a current drain exceeding 100 μA for more than a few minutes. The standard cell is typically used as a standard for comparison for voltage measurement devices.

Derived Units

Although the above discussion does not cover all of the basic dimensions, it does discuss those most commonly referred to in this text. Most other dimensions and their associated units can usually be defined in terms of the basic dimensions and units already described [10]. For example, force is defined through Newton's second law:

$$\text{force} = \frac{\text{mass} \times \text{acceleration}}{g_c}$$

In the SI system of units, force is defined by a newton (N) and acceleration by one meter per second squared, so that

$$1 \text{ newton} = 1 \frac{\text{kg} - \text{m}}{\text{s}^2 - g_c}$$

where $g_c \equiv 1 \text{ kg} - \text{m/s}^2 - \text{N}$. So that the newton is defined by the primary standards for mass, length, and time. In U.S. engineering units, force is defined by the unit of the pound force (lb) where

$$1 \text{ lb} = 1 \text{ lb}_m \times 32.174 \frac{\text{ft}}{\text{s}^2 - g_c}$$

where $g_c \equiv 32.174 \text{ lb}_m - \text{ft/s}^2 - \text{lb}$. Along similar lines, energy is defined as force times distance with the standard SI unit of the joule (J):

$$1 \text{ joule} = 1 \frac{\text{kg} - \text{m}^2}{\text{s}^2}$$

and power is defined as energy per unit time in terms of the watt (W):

$$1 \text{ watt} = 1 \frac{\text{joule}}{\text{second}}$$

A chart for converting between units is included inside the text cover. Table 1.2 lists the standard units used in the SI and U.S. engineering systems. (See page 24)

Hierarchy of Standards

The known value input to a measurement system during a calibration is called the standard. This value becomes the standard upon which the calibration is based. Obviously, the actual primary standards can be impractical as standards for normal calibration use. But they serve as a reference for exactness. It would not be reasonable to travel to France to calibrate an ordinary laboratory scale using the primary standard for mass (nor would it be permitted!). So for practical reasons, there exists a hierarchy of secondary standards that attempt to duplicate the primary standards. *Secondary standards* provide reasonable approximations to the primary standards but can be accessed more readily for calibration purposes. However, an amount of uncertainty must be accepted with the use of standards that are replicas of primary standards. At the top of the standards hierarchy, just below the primary standard, are the national reference standards maintained by designated standards laboratories throughout the world. In the

TABLE 1.2   Standard Dimensions and Units

Dimension	Unit SI	Unit US
Primary		
Length	meter (m)	foot (ft)
Mass	kilogram (kg)	pound-mass (lb_m)
Time	second (s)	second (s)
Force	newton (N)	pound-force (lb)
Temperature	kelvin (K)	rankine (°R)
Derived		
Voltage	volt (V)	volt (V)
Current	ampere (A)	ampere (A)
Resistance	ohm (Ω)	ohm (Ω)
Capacitance	farad (F)	farad (F)
Inductance	henry (H)	henry (H)
Pressure	pascal (Pa)	pound/foot^2 (psf)
Energy	joule (J)	British thermal unit (BTU)
Power	watt (W)	foot-pound (ft-lb)

United States, NIST maintains primary and secondary standards and recommends standard procedures for the calibration of measurement systems.

Each subsequent level of the hierarchy is derived by calibration against the standard at the previous higher level. Table 1.3 lists an example of such a lineage for standards from a primary standard at a national lab down to a working standard, such as might be used in a typical laboratory to calibrate a measurement system. As a common example, an institution might maintain its own local standard which it uses to calibrate the everyday measurement devices found in the individual laboratories throughout the institution. Periodically this institution would send its local standard to NIST for calibration against an NIST transfer standard (for a fee). NIST will periodically calibrate its own transfer standard. As one moves down through this lineage, the degree of exactness by which a standard approximates the primary standard from which it is derived deteriorates. That is, increasing elements of error are introduced into the standard as one goes from one generation of standard to the next. This is illustrated for a temperature standard traceability hierarchy in Table 1.4. Therefore, the uncertainty in the approximation of the known value increases. It follows, then, that since the calibration determines the relationship between the input value and the output value, the accuracy of the calibration will depend in part on the accuracy of the standard. But if typical working standards contain some error, how is accuracy ever determined? At best, accuracy can only be estimated. And the confidence in an estimate will depend on the quality of the standard and calibration techniques used.

TABLE 1.3   Hierarchy of Standards

Primary standard
Interlaboratory transfer standard
Local standard
Working instrument

TABLE 1.4 Temperature Standard Traceability

Standard		Error[a]
Level	Method	[°C]
Primary	Fixed thermodynamic points	0
Transfer	Platinum resistance thermometer	±0.005
Local	Platinum resistance thermometer	±0.01
Working instrument	Glass bulb thermometer	±0.1

[a]Typical instrument bias and precision errors.

The term "standard" can also be used in a slightly different sense. For the calibration of measurement systems used to measure certain process variables, the calibration may make use of a test standard procedure. A *test standard* refers to a particular testing method that aims to provide consistency in the conduct of a measurement from one test facility to another. Test standards often make available tabulated coefficients, the values of which have been accepted by the national or international community [11, 12]. Examples of test standards are illustrated in readily available documents such as the Power Test Codes of the American Society of Mechanical Engineers (ASME) or the standards of the American National Standards Institute (ANSI) [11, 13]. Test standards often instruct the user to place the measurement system into a reference configuration such that the outcome of a measurement in this configuration can be compared to data already well established and accepted by the measurement community. The accuracy of the measurement system can thus be estimated.

1.6 MEASUREMENT OVERVIEW

The design of a measurement system itself consists of the selection of an appropriate technique, acceptable instruments, appropriate calibration or system verification tests, and a well-thought-out plan on how the measurement data are to be handled, reduced, and presented. The overall planning procedure can be divided into several individual but interrelated steps that are performed **prior** to the actual execution of the measurement.

Collectively, the experimental design can be summarized as follows:

1. *Objective*. Identify the variables to be measured. Identify all pertinent related variables and control parameters.
2. *Plan*. Develop a strategy for variation of the independent variables and randomization of the extraneous variables. Can the acquired information meet the objective? The plan must use concomitant methods to check results for bias in the results.
3. *Methodology Assessment*. Identify one or more candidate measurement methods based on a literature survey of available techniques and instruments. A large body of information exists in trade journals, scientific journals, test codes, and vendor technical bulletins. The procedure should allow for a reference case to check system operation.
4. *Uncertainty Analysis*. Decide the minimum level of accuracy that can be tolerated in the final data results and still meet the objective. Use

uncertainty analysis methods to select a measurement method and instruments (methods to perform this step are detailed in Chapter 5). One method will stand out as the better choice for the measurement.

5. *Costs*. Reevaluate each candidate procedure in light of (4). Can the measurement be done within budget? Do the accuracy constraints need to be relaxed to get the costs down? A decision may require input from other engineering and budget management groups associated with the test program.

6. *Calibration*. Calibrations can reduce error in a measurement, but take time and money. Manufacturer specifications may be sufficient. The decision will depend on the accuracy levels determined necessary as per (4) above.

7. *Data Acquisition*. Choosing the method by which the variable is sensed, the signal conditioned, and the output recorded are obvious steps in designing a measurement process. The amount of data and frequency at which data need to be taken, as well as the range of operating conditions, will also need to be estimated.

8. *Data Reduction*. A means to manipulate the data into a final form should be conceived, including the form for final presentation. Large amounts of data or data requiring considerable mathematical manipulation may require computer processing. Statistical methods and curve-fitting regression analyses are discussed in Chapter 4.

Several important statements can be applied to a measurement design which are related to controlling a measurement test:

1. A sufficient number of variables should be measured so that at least two concomitant methods for estimating a result can be used.

2. If several parameters must be controlled in a measurement, the measurement procedure should be designed such that a change in one parameter does not affect the values of the other parameters. When this is not physically possible, the number of parameters affected should be minimized.

3. The measurement should be designed such that the measured variable is the only dependent process variable during each measurement. Other variables should be controlled.

4. The sensor should be sensitive to the variable to be measured but not sensitive to other variables.

5. The signal path of the measurement system should be designed to minimize the effects of extraneous variables.

6. The test plan should randomize the effects of extraneous variables.

1.7   SUMMARY

During a measurement the input signal is not known but is inferred from the value of the output signal from the measurement system. However, the output signal can be affected by many variables which will introduce variation and trends. Careful test planning is required to reduce such effects. Only during calibration is the input value actually known and this information is used to develop the relation between input and output necessary to interpret the output signal during later measurements. An important step in the design of a measurement system is the inclusion of a means for a reproducible calibration that closely simulates the type of signal to be input during actual measurements.

REFERENCES

1. Bridgeman, P. W., *Dimensional Analysis*, 2d ed., Yale University Press, New Haven, CN, 1931; paperback Y-82, 1963.
2. Duncan, W. J., *Physical Similarity and Dimensional Analysis*, Arnold, London, 1953.
3. Massey, B. S., *Units, Dimensions and Physical Similarity*, Van Nostrand-Reinhold, New York, 1971.
4. Lipsen, C., and N. J. Sheth, *Statistical Design and Analysis of Engineering Experimentation*, McGraw-Hill, New York, 1973.
5. Peterson, R. G., *Design and Analysis of Experiments*, Marcel-Dekker, New York, 1985.
6. Mead, R., *The Design of Experiments: Statistical Principles for Practical Application*, Cambridge Press, New York, 1988.
7. *NBS Technical News Bulletin* 52:1, January 1968.
8. International Temperature Scale—1990, *Metrologia* 27:3, 1990.
9. International Practical Temperature Scale—1968 (English version), *Metrologia* 4, October 1968. Also see the amended version EPT-76, *Metrologia* 12, 1976.
10. Kverneland, K. O., *World Metric Standards for Engineering*, Industrial Press, New York, 1978.
11. *ASME Power Test Codes*, American Society of Mechanical Engineers, New York.
12. *ASHRAE Handbook of Fundamentals*, American Society of Heating, Refrigeration and Air Conditioning Engineers, New York, 1982.
13. *American National Standard*, American National Standards Institute, New York.

NOMENCLATURE

e_h hysteresis error
e_K sensitivity error
e_L linearity error
e_R repeatability error
e_z zero error
p pressure $[m-l^{-1}-t^{-2}]$
r_i input range
r_o output range
x independent variable; input value

y dependent variable; output value
y_L linear polynomial
A accuracy
T temperature [°]
V volume $[l^3]$
ϵ absolute error

PROBLEMS

1.1 Discuss what is meant by a primary standard and by a secondary standard. In general, what is meant by the measurement term *standard*?

1.2 When should a calibration be performed?

1.3 Suppose you found an old dial thermometer in a stockroom. Discuss several methods by which you might estimate its precision? Its accuracy?

1.4 Under which situations would a dynamic calibration of a measurement system be required?

1.5 How would the resolution of the display scale of an instrument affect its accuracy? Explain in terms of precision and bias.

1.6 How would the hysteresis of an instrument affect its accuracy? Explain in terms of precision and bias.

1.7 Select three different types of measurement systems with which you have experience and identify which attributes of the system comprise the measurement system stages of Figure 1.1.

1.8 Identify the measurement system stages for the following systems (refer to Figure 1.1):
a. Thermostat
b. Automobile speedometer
c. Microphone/public address system
d. Stethoscope
e. Audio speaker
f. Telephone receiver

1.9 Specify the range of the calibration data of Table 1.5?

1.10 For the calibration data of Table 1.5, plot the results on rectangular and on log–log paper. Discuss the apparent advantages of either presentation.

1.11 For the calibration data of Table 1.5, determine the static sensitivity of the system at (a) $X = 5$; (b) $X = 10$; (c) $X = 20$. For which input values is the system more sensitive?

TABLE 1.5   Calibration Data

X [cm]	Y [V]	X [cm]	Y [V]
0.5	0.4	10.0	15.8
1.0	1.0	20.0	36.4
2.0	2.3	50.0	110.1
5.0	6.9	100.0	253.2

1.12 Consider the voltmeter calibration data in Table 1.6. Plot the data on suitable graph paper. Specify the percent maximum hysteresis based on full-scale range. Input X is based on a standard known to be accurate to better than 0.05 mV.

TABLE 1.6   Voltmeter Calibration Data

Increasing Input [mV]		Decreasing Input [mV]	
X	Y	X	Y
0.0	0.1	5.0	5.0
1.0	1.1	4.0	4.2
2.0	2.1	3.0	3.2
3.0	3.0	2.0	2.2
4.0	4.1	1.0	1.2
5.0	5.0	0.0	0.2

1.13 Three clocks are compared to a time standard at three successive hours. The data are given in Table 1.7. Arrange these clocks in order of estimated accuracy. Justify your choices.

TABLE 1.7   Clock Calibration Data

Clock	Standard Time		
	1:00:00	2:00:00	3:00:00
	Indicated Time		
A	1:02:23	2:02:24	3:02:25
B	1:00:05	2:00:05	3:00:05
C	1:00:01	1:59:58	3:00:01

1.14 Plot the following equations on appropriate graph scales (for log scales use at least 3 cycles). Variable y has units of volts. Variable x has units of meters (use a range of $0.1 \leq x \leq 10.0$).
a. $y = x^3$
b. $y = 0.1x$
c. $y = x^{0.6}$
d. $y = 20x^3 + 2.6$
e. $y = 5x^2 + 0.5$
f. $y = 1.1x + 3.5$

1.15 Plot $y = 10e^{-5x}$ volts on semilog paper (3 cycles). Determine the slope of the equation at $x = 0$; $x = 2$; and $x = 20$.

1.16 Plot the following data on appropriate graph paper. Estimate the static sensitivity at each X.

Y [V]	X [V]
2.9	0.5
3.5	1.0
4.7	2.0
9.0	5.0

1.17 The following data have the form $y = ax^b$. Plot the data on appropriate paper and estimate the coefficients a and b. Estimate the static sensitivity at each value of X. How is K affected by X?

Y [cm]	X [m]
0.14	0.5
2.51	2.0
15.30	5.0
63.71	10.0

1.18 For the calibration data given, plot the calibration curve on suitable paper. Estimate the static sensitivity of the system at each X and then plot this against X. Comment on the behavior of the static sensitivity with static input magnitude for this system.

Y [cm]	X [psi]
4.76	0.05
4.52	0.1
3.03	0.5
1.84	1.0

1.19 A bulb thermometer is used to measure outside temperature. Comment on the extraneous variables that might affect the difference between actual outside temperature and indicated temperature.

1.20 A synchronous electric motor test stand permits either the variation of input voltage or output shaft load and the subsequent measurement of motor efficiency, winding temperature, and input current. Comment on the independent, dependent, and extraneous variables for a motor test.

1.21 The transducer specified in Table 1.1 is chosen to measure a nominal pressure of 500 cm H_2O. The ambient temperature is expected to vary between 18 and 25 °C during tests. Estimate the magnitude of each elemental error affecting the measured pressure.

1.22 Develop a test plan that might be used to estimate the average temperature that could be maintained in a heated room as a function of the heater thermostat setting.

1.23 Develop a test plan that might be used to evaluate the fuel efficiency of a production model automobile. Explain your reasoning.

1.24 A large batch of carefully made machine shafts can be manufactured on one of four lathes by one of 12 quality machinists. Set up a test plan pattern to estimate the tolerances that can be held within a production batch. Explain your reasoning.

CHAPTER 2
STATIC AND DYNAMIC CHARACTERISTICS OF SIGNALS

2.1   INTRODUCTION

A measurement system transforms an *input* quantity into an *output* quantity that can be observed or recorded, such as the movement of a pointer on a dial or the magnitude of a digital display. The ultimate goal of any measurement system is to provide an exact indication of the value of the input, that is, the measured quantity. We use the output of a measurement system to provide information concerning the input variable. However, no real system measures exactly the value of a specific input variable, and the form or characteristics of the input quantity affect the error in the measurement. But the quality of the information that results from a measurement depends on proper selection of a measurement system, and proper interpretation of the measured results. An understanding of the nature and form of the input to a measurement system is necessary for the selection of the measurement system for a specific task.

Our goal is to understand the characteristics of both input signals to a measurement system and the resulting output signals. Information about the *magnitude*, which indicates the size of the input quantity, and the *frequency*, which indicates the way the signal changes in time, is required for the evaluation of measurement systems. The shape and form of a signal is often referred to as its *waveform*.

2.2   INPUT/OUTPUT SIGNAL CONCEPTS

Two important tasks that engineers face in the measurement of physical variables are

1.   Selecting a measurement system.
2.   Evaluating the output from a measurement system.

A simple example of selecting a measurement system might be the selection of a tire gauge for measuring the air pressure in a bicycle tire or in a car tire. The gauge for the car tire would need to be able to indicate pressures up to 40 lb/in.^{2}, but the bicycle tire gauge would be required to indicate higher pressures, maybe up to 100 lb/in.^{2}. This idea of the range of an instrument is basic to all measurement systems, and demonstrates that some basic understanding of the nature of the input signal, in this case the magnitude, is necessary in evaluating or selecting a measurement system for a particular application. A much more difficult task is the evaluation of the output of a measurement system when the behavior of the input is not known. The pressure in a tire does not change while we are trying to measure it, but what if we somehow connected the tire gauge to a cylinder in an automobile engine? We know that the pressure in the cylinder varies with time. If our task was to select a measurement system to determine this time-varying pressure, information about the pressure variations in the cylinder would be necessary. From thermodynamics and the speed range of the engine it may be possible to estimate the range of pressures to be expected, and the speed with which they change. Quantitative means of expressing these ideas of the magnitude and rate of change of a measured variable must be developed.

Many measurement systems exhibit similar responses under a variety of conditions, which suggests that the performance and capabilities of measuring systems may be described in a generalized way. The word generalized implies that the relationship between input and output for various measurement systems can be described using a common set of variables. To examine further the generalized behavior of measurement systems, we need first to examine the possible forms of the input and output signals.

The term *signal* is used in association with the transmission of information. In the present context, it is used to describe a physical variable that is a function of time and that interacts with a measurement system, either as an input to the measurement system or as an output from the measurement system. Thus, a signal is information about a measured variable being transmitted between a process and the measurement system, or between the stages of a measurement system.

Classification of Waveforms

Signals may be classified as either analog, discrete time, or digital. The term *analog* describes a signal that is continuous in time. Since physical variables tend to be continuous in nature, an analog signal provides a ready representation of their time-dependent behavior. In addition, the magnitude of the signal is continuous and thus can have any value within the operating range. An analog signal is shown in Figure 2.1a; a similar continuous signal would result from a recording of the pointer rotation with time for the output display shown in Figure 2.1b. Contrast this continuous signal with the signal shown in Figure 2.2a. This format represents a *discrete time* signal, where information about the magnitude of the signal is available only at discrete points in time. A discrete time signal usually results from the sampling of a continuous variable at finite time intervals.

Since information in the signal shown in Figure 2.2a is available only at discrete times, some assumption must be made about the behavior of the variable during

FIGURE 2.1 Analog signal concepts.

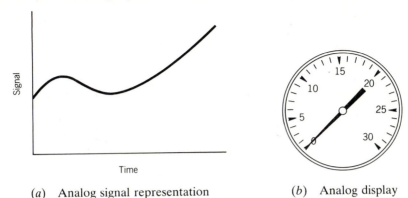

(*a*)   Analog signal representation   (*b*)   Analog display

the times when it is not being sampled. Generally, the sampling rate should be sufficiently high that the signal may be assumed constant over the period of time between samples. The waveform that results from this assumption is shown in Figure 2.2*b*. Clearly, as the time between samples is reduced, the difference between the discrete variable and the continuous signal it represents decreases.

Digital signals are particularly useful when data acquisition and processing are performed using a digital computer. A *digital signal* has two important characteristics. First, a digital signal exists at discrete values in time, like a discrete time signal, and the magnitude of a digital signal is determined by a process known as quantization, at each point in time. Quantization assigns a single number to represent a range of magnitudes of a continuous signal or variable. Figure 2.3*a* shows digital and analog forms of the same signal where the magnitude of the digital signal can have only certain discrete values. Thus, a digital signal provides a quantized magnitude at discrete times. The waveform that would result from assuming that the signal is constant between sampled points in time is shown in Figure 2.3*b*. As an example of quantization, consider a digital watch which displays time in hours and minutes. For the entire duration of one minute a single numerical value is displayed. As such the continuous physical variable time is quantized in its conversion to a digital display.

FIGURE 2.2 Discrete time signal concepts.

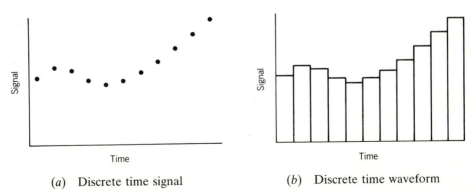

(*a*)   Discrete time signal   (*b*)   Discrete time waveform

FIGURE 2.3   Digital signal representation and
waveform.

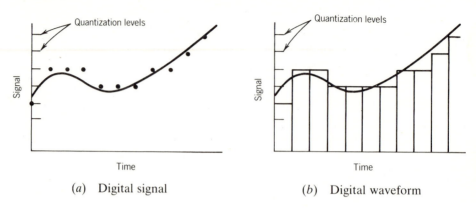

(a)   Digital signal   (b)   Digital waveform

Sampling of an analog signal to produce a digital signal can be accomplished
by using an analog-to-digital (A/D) converter, a type of solid-state device that
converts an analog voltage signal to a binary number system representation. The
limited resolution of the binary number that corresponds to a range of voltages
creates the quantization levels and ranges.

The technology of a compact disc player is an example of the conversion of
a continuously available signal, from a microphone, into a digital form. The
digital information is stored on a compact disc, and later read in digital form
by a laser playback system [1]. However, to serve as input to a traditional stereo
amplifier, and since speakers and the human ear are analog devices, the digital
information is again transformed into a continuous voltage signal for playback.

Signal Waveforms

In addition to classifying signals as analog, discrete time, or digital, some
description of the waveform associated with a signal is useful. Signals may be
characterized as either static or dynamic. A *static* signal does not vary with time.
This idea of a completely steady signal is an idealization and does not occur in
any real situation. The distinction is still a useful one since we are often concerned
with how quickly an input signal changes in time compared to the time required
for our measurement system output to change. Many physical variables change
slowly enough in time, compared to the response time of the measurement system
with which they interact, that for all practical purposes these signals may be
considered steady in time. For example, the voltage across the terminals of a
battery is approximately constant over its useful life. Or consider measuring
temperature using an outdoor thermometer; since the outdoor temperature does
not change significantly in a matter of minutes, this input signal is steady when
compared to our time period of interest. On the other hand, measurement
systems are often required to determine the measured variable as a continuous
function of time. In other words, the measurement system output must follow
in a predictable manner the changes in the input signal. This leads us to consider
further time-varying signals.

A *dynamic signal* is defined as a time-dependent signal. In general, dynamic
signal waveforms, $y(t)$, may be classified as shown in Table 2.1. A *deterministic*

TABLE 2.1   Classification of Waveforms

I. Static

$$y(t) = A_0$$

II. Dynamic

Periodic waveforms

Simple periodic waveform

$$y(t) = A_0 + C \sin(\omega t + \phi)$$

Complex periodic waveform

$$y(t) = A_0 + \sum_{n=1}^{\infty} C_n \sin\left(\frac{2n\pi t}{T} + \phi_n\right)$$

Aperiodic waveforms

Step[a]

$$\begin{aligned} y(t) &= A_0 U(t) \\ &= A_0 \quad \text{for } t > 0 \end{aligned}$$

Ramp

$$y(t) = Kt \quad \text{for } 0 < t < t_f$$

Pulse[b]

$$y(t) = A_0 U(t) - A_0 U(t - t_1)$$

III. Nondeterministic waveform

$$y(t) \approx A_0 + \sum_{n=1}^{\infty} C_n \sin\left(\frac{2n\pi t}{T} + \phi_n\right)$$

[a] $U(t)$ represents the unit step function, which is zero for $t < 0$ and 1 for $t \geq 0$.
[b] t_1 represents the pulse width.

signal is one that varies in time in a predictable manner, such as a sine wave, a step input, or a ramp input, as shown in Figure 2.4. A signal is *steady periodic* if the variation of the magnitude of the signal repeats at regular intervals in time. Examples of steady periodic behaviors would include the motion of an ideal pendulum, or the temperature variations in the cylinder of an internal combustion engine, under steady operating conditions. Periodic waveforms may be classified as simple or complex; a *complex periodic waveform* contains multiple frequencies. *Aperiodic* is used to describe deterministic signals that do not repeat at regular intervals, such as a step function.

Also described in Figure 2.4 is a *nondeterministic signal* which has no discernible pattern of repetition. A nondeterministic signal cannot be prescribed before it occurs, although certain characteristics of the signal may be known in advance. As an example, consider the transmission of data files from one computer to another. Signal characteristics such as the rate of data transmission and the possible range of signal magnitude are known for any signal in this system. But it would not be possible to predict future signal characteristics based on

FIGURE 2.4   Examples of dynamic signals.

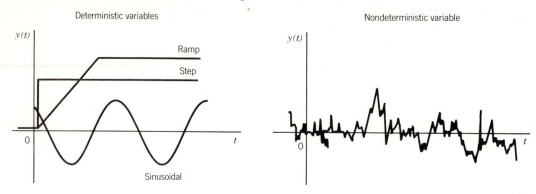

existing information in such a signal. Such a signal is properly characterized as being nondeterministic.

A key factor in measurement system behavior is the nature of the input to the system. A means is needed to classify waveforms for both the input signal and the resulting output signal relative to their magnitude and frequency. It would be very helpful if the behavior of measurement systems could be defined in terms of their response to a limited number and type of input signals. This is, in fact, exactly the case: A very complex signal, even one that is nondeterministic in nature, can be approximated as an infinite series of sine and cosine functions, as suggested in Table 2.1. The method of expressing such a complex signal as a series of sines and cosines is called *Fourier analysis*. Nature provides some experiences that support our contention that complex signals can be represented by the addition of a number of simpler periodic functions. For example, rich musical tones can be generated by combining a number of different pure tones. And an excellent physical analogy for Fourier analysis is provided by a prism. Figure 2.5 illustrates the transformation of a complex waveform, repre-

FIGURE 2.5   Separation of white light into its color spectrum. Color corresponds to a particular frequency or wavelength, while light intensity corresponds to varying amplitudes.

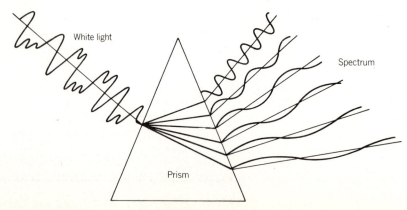

sented by white light, into its simpler components, represented by the colors in the spectrum. In this example, the colors in the spectrum are represented as simple periodic functions that combine to form white light. Fourier analysis is roughly the mathematical equivalent of a prism, and yields a representation of a complex signal in terms of simple periodic functions.

The representation of complex waveforms by simple periodic functions allows measurement system response to be reasonably well defined by examining the output resulting from a few specific input waveforms, one of which is a simple periodic. Before this generalized response of measurement systems can be determined, an understanding of the method of representing complex signals in terms of simpler functions is necessary.

2.3 WAVEFORMS AND FOURIER ANALYSIS

The goal of this section is to develop a basic understanding of the frequency and amplitude characteristics of both deterministic and nondeterministic signals. A simple, periodic waveform has a single, well-defined amplitude and a definite frequency. On the other hand, nondeterministic signals have multiple frequency components of differing amplitudes, and require analysis to determine their frequency content.

The fundamental concepts of frequency and amplitude result from the observation and analysis of periodic motions. Although sines and cosines are by definition geometric quantities related to the lengths of the sides of a right triangle, for our purposes sines and cosines are best thought of as mathematical functions that describe specific physical behaviors of systems. These behaviors are described by differential equations which have sines and cosines as their solutions. As an example, consider a mechanical vibration of a mass attached to a linear spring, as shown in Figure 2.6. A linear spring is one where the spring

FIGURE 2.6 Spring–mass system.

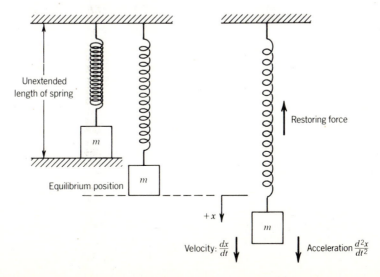

force, F, and displacement, x, are related by $F = kx$, where k is the constant of proportionality, called the spring constant. Application of Newton's second law to this system yields a governing equation for the displacement, x, as

$$m \frac{d^2x}{dt^2} + kx = 0 \qquad (2.1)$$

This linear, second-order differential equation with constant coefficients describes the motion of the idealized spring mass system when there is no external force applied. The general form of the solution to this equation is

$$x = A \sin \omega t + B \cos \omega t \qquad (2.2)$$

where $\omega = \sqrt{k/m}$. Equation 2.2 indicates that the mass will oscillate. Physically we know that if the mass is displaced from the equilibrium point and released, it will oscillate about the equilibrium point. The time required for the mass to finish one complete cycle of the motion is called the *period*. Frequency is related to the period and is defined as the number of complete cycles of the motion per unit time. This frequency, f, is measured in cycles per second, hertz (Hz). The term ω is also a frequency, but instead of having units of cycles per second, it has units of radians per second. This frequency, ω, is called the circular frequency since it relates directly to cycles on the unit circle, as illustrated in Figure 2.7. The relationship between ω, f, and the period, T, is

$$T = \frac{2\pi}{\omega} = \frac{1}{f} \qquad (2.3)$$

In equation 2.2, the sine and cosine terms can be combined if a phase angle is introduced such that

$$x = C \cos(\omega t - \phi) \qquad (2.4a)$$

or

$$x = C \sin(\omega t + \phi^*) \qquad (2.4b)$$

The values of C, ϕ, and ϕ^* are found from the following trigonometric identities:

$$A \cos \omega t + B \sin \omega t = \sqrt{A^2 + B^2} \cos(\omega t - \phi)$$
$$A \cos \omega t + B \sin \omega t = \sqrt{A^2 + B^2} \sin(\omega t + \phi^*) \qquad (2.5)$$

$$\phi = \tan^{-1} \frac{B}{A} \qquad \phi^* = \tan^{-1} \frac{A}{B} \qquad \phi^* = \frac{\pi}{2} - \phi$$

The size of the maximum and minimum displacements from the equilibrium position, or the value C, is the amplitude of the oscillation. The concepts of amplitude and frequency are essential for the description of time-dependent signals.

FIGURE 2.7   Relationship between cycles on the unit circle and circular frequency.

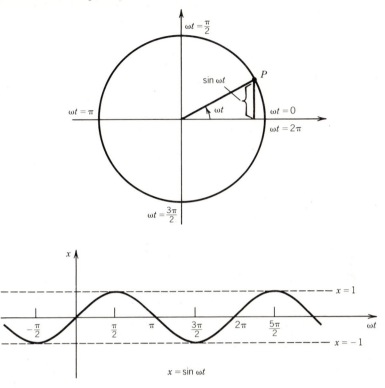

Many such signals, which result from the measurement of dynamic variables, are nondeterministic in nature and have a continuously varying rate of change. These signals, having *complex waveforms*, present difficulties in the selection of a measurement system and in the interpretation of an output signal. However, it is possible to separate a complex signal, or any signal for that matter, into a number of sine and cosine functions. In other words, any complex signal can be thought of as made up of sines and cosines of differing periods and amplitudes which are added together in an infinite series. This representation of a signal as a series of sines and cosines is called a *Fourier series*. Once a signal is broken down into a series of periodic functions, the importance of each frequency can be easily determined. This information about frequency content allows proper choice of a measurement system, and precise interpretation of output signals.

In theory, Fourier analysis allows essentially all mathematical functions of practical interest to be represented by an infinite series of sines and cosines.[1]

[1]A periodic function may be represented as a Fourier series if the function is piecewise continuous over the limits of integration and the function has a left- and right-hand derivative at each point in the interval. The sum of the resulting series is equal to the function at each point in the interval except points where the function is discontinuous.

The following definitions relate to Fourier analysis:

1. A function $y(x)$ is a *periodic function* if there is some positive number T such that

$$y(x + T) = y(x)$$

The *period* for $y(x)$ is T. If both $y_1(x)$ and $y_2(x)$ have period T, then

$$ay_1 + by_2$$

also has a period of T (a and b are constants).

2. A *trigonometric series* is given by

$$A_0 + A_1 \cos x + B_1 \sin x + A_2 \cos 2x + B_2 \sin 2x + \cdots$$

where A_n and B_n are the coefficients of the series.

Fourier Series and Coefficients (optional*)

A periodic function $y(x)$ with a period $T = 2\pi$ is to be represented by a trigonometric series, such that for any x,

$$y(x) = A_0 + \sum_{n=1}^{\infty} (A_n \cos nx + B_n \sin nx) \qquad (2.6)$$

With $y(x)$ known, the coefficients A_n and B_n are to be determined. To determine A_0, equation 2.6 is integrated from $-\pi$ to $+\pi$:

$$\int_{-\pi}^{\pi} y(x)\, dx = A_0 \int_{-\pi}^{\pi} dx + \sum_{n=1}^{\infty} \left(A_n \int_{-\pi}^{\pi} \cos nx\, dx + B_n \int_{-\pi}^{\pi} \sin nx\, dx \right)$$

$$(2.7)$$

Since

$$\int_{-\pi}^{\pi} \cos nx\, dx = 0 \quad \text{and} \quad \int_{-\pi}^{\pi} \sin nx\, dx = 0$$

equation 2.7 yields

$$A_0 = \frac{1}{2\pi} \int_{-\pi}^{\pi} y(x)\, dx \qquad (2.8)$$

---

*This development of the Euler formulas may be omitted without loss of continuity of the Chapter material.

The coefficient A_m may be determined by multiplying equation 2.6 by cos mx (m = a fixed positive integer), and integrating from $-\pi$ to $+\pi$:

$$\int_{-\pi}^{\pi} y(x) \cos mx \, dx$$

$$= \int_{-\pi}^{\pi} \left[A_0 + \sum_{n=1}^{\infty} (A_n \cos nx + B_n \sin nx) \right] \cos mx \, dx \quad (2.9)$$

The term associated with A_0 is zero:

$$A_0 \int_{-\pi}^{\pi} \cos mx \, dx = 0 \quad (2.10)$$

and the terms that remain to be evaluated are

$$\sum_{n=1}^{\infty} \left(A_n \int_{-\pi}^{\pi} \cos nx \cos mx \, dx + B_n \int_{-\pi}^{\pi} \sin nx \cos mx \, dx \right) \quad (2.11)$$

Using the following trigonometric identities,

$$\cos \alpha \cos \beta = \tfrac{1}{2} \cos(\alpha - \beta) + \tfrac{1}{2} \cos(\alpha + \beta)$$
$$\sin \alpha \cos \beta = \tfrac{1}{2} \sin(\alpha + \beta) + \tfrac{1}{2} \sin(\alpha - \beta)$$

yields

$$\int_{-\pi}^{\pi} \cos nx \cos mx \, dx = \frac{1}{2} \int_{-\pi}^{\pi} \cos(n + m)x \, dx + \frac{1}{2} \int_{-\pi}^{\pi} \cos(n - m)x \, dx$$

$$\int_{-\pi}^{\pi} \sin nx \cos mx \, dx = \frac{1}{2} \int_{-\pi}^{\pi} \sin(n + m)x \, dx + \frac{1}{2} \int_{-\pi}^{\pi} \sin(n - m)x \, dx$$

With this substitution, we see that all the terms in the series are zero except in the case when $n = m$. When $n = m$, the term

$$\frac{1}{2} \int_{-\pi}^{\pi} \cos(n - m)x \, dx$$

yields

$$\frac{1}{2} \int_{-\pi}^{\pi} dx = \pi$$

Then from equation 2.9

$$\int_{-\pi}^{\pi} y(x) \cos mx \, dx = A_m \pi$$

and

$$A_m = \frac{1}{\pi} \int_{-\pi}^{\pi} y(x) \cos mx \, dx \tag{2.12}$$

Similarly, multiplying equation 2.6 by sin mx and integrating from $-\pi$ to $+\pi$ yields B_m. Thus, for a function $y(x)$ with a period 2π, the coefficients of the trigonometric series representing $y(x)$ are given by the Euler formulas:

$$A_0 = \frac{1}{2\pi} \int_{-\pi}^{\pi} y(x) \, dx$$

$$A_n = \frac{1}{\pi} \int_{-\pi}^{\pi} y(x) \cos nx \, dx \tag{2.13}$$

$$B_n = \frac{1}{\pi} \int_{-\pi}^{\pi} y(x) \sin nx \, dx$$

The trigonometric series corresponding to $y(x)$ is called the Fourier series for $y(x)$, and the coefficients A_n and B_n are called the Fourier coefficients of $y(x)$. Functions represented by a Fourier series normally do not have a period of 2π. However, the transition to an arbitrary period can be affected by a change of scale, yielding a new set of Euler formulas, described below.

Fourier Coefficients for Functions Having Arbitrary Periods

The coefficients of a trigonometric series representing a function of period T are given by the Euler formulas,

$$A_0 = \frac{1}{T} \int_{-T/2}^{T/2} y(t) \, dt$$

$$A_n = \frac{2}{T} \int_{-T/2}^{T/2} y(t) \cos \frac{2n\pi t}{T} \, dt \tag{2.14}$$

$$B_n = \frac{2}{T} \int_{-T/2}^{T/2} y(t) \sin \frac{2n\pi t}{T} \, dt$$

where $n = 1, 2, 3, \ldots$, and T is the period of $y(t)$. The trigonometric series that results from these coefficients is a Fourier series, and may be written as

$$y(t) = A_0 + \sum_{n=1}^{\infty} \left(A_n \cos \frac{2n\pi t}{T} + B_n \sin \frac{2n\pi t}{T} \right) \tag{2.15}$$

Even and Odd Functions

A function $g(x)$ is even if it is symmetric about the origin, which may be stated, for all x,

$$g(-x) = g(x)$$

A function $h(x)$ is odd if, for all x,

$$h(-x) = -h(x)$$

For example, cos nx is even, while sin nx is odd. A particular function or waveform may be even, odd, or neither even nor odd.

Fourier Cosine Series

If $y(t)$ is even, its Fourier series will contain only cosine terms:

$$y(t) = A_0 + \sum_{n=1}^{\infty} A_n \cos \frac{2n\pi t}{T} \tag{2.16}$$

Fourier Sine Series

If $y(t)$ is odd, its Fourier series will contain only sine terms

$$y(t) = \sum_{n=1}^{\infty} B_n \sin \frac{2n\pi t}{T} \tag{2.17}$$

Functions that are neither even nor odd result in Fourier series that contain both sine and cosine terms.

EXAMPLE 2.1

Determine the Fourier series that represents the function shown in Figure 2.8.

KNOWN

$T = 10$

FIND

The Fourier coefficients A_1, A_2, . . . and B_1, B_2, ...

FIGURE 2.8   Function represented by a Fourier series in Example 2.1.

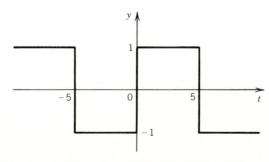

SOLUTION

Since the function shown in Figure 2.8 is odd, the Fourier series will contain only sine terms (see equation 2.17):

$$y(t) = \sum_{n=1}^{\infty} B_n \sin \frac{2n\pi t}{T}$$

where

$$B_n = \frac{2}{10} \left[\int_{-5}^{0} (-1) \sin \left(\frac{2n\pi t}{10} \right) dt + \int_{0}^{5} (1) \sin \left(\frac{2n\pi t}{10} \right) dt \right]$$

$$B_n = \frac{2}{10} \left\{ \left[\frac{10}{2n\pi} \cos \left(\frac{2n\pi t}{10} \right) \right]_{-5}^{0} + \left[\frac{-10}{2n\pi} \cos \left(\frac{2n\pi t}{10} \right) \right]_{0}^{5} \right\}$$

$$B_n = \frac{2}{10} \left\{ \frac{10}{2n\pi} [1 - \cos(-n\pi) - \cos(n\pi) + 1] \right\}$$

For even values of n, B_n is identically zero and for odd values of n,

$$B_n = \frac{4}{n\pi}$$

The resulting Fourier series is then

$$y(t) = \frac{4}{\pi} \sin \frac{2\pi}{10} t + \frac{4}{3\pi} \sin \frac{6\pi}{10} t + \frac{4}{5\pi} \sin \frac{10\pi}{10} t + \cdots$$

COMMENT

Consider the function given by

$$y(t) = 1.0 \qquad 0 < t < 5$$

We may represent $y(t)$ by a Fourier series if we extend the function beyond the specified range (0–5), either as an even periodic extension or as an odd periodic extension. (Because we are interested only in the Fourier series representing the function over the range $0 < t < 5$, we can impose any behavior outside of that domain that helps to generate the Fourier coefficients!) Let's choose an odd periodic extension of $y(t) = 1.0$; the resulting function is identical to the function shown in Figure 2.8.

In the series for $y(t)$ in example 2.1, when $n = 1$ the corresponding terms in the Fourier series are called *fundamental*, and have the lowest frequency in the series. The coefficient $2\pi/10$ is the *fundamental frequency* for this Fourier series. Frequencies corresponding to $n = 2, 3, 4, \ldots$ are known as harmonics, with, for example, $n = 2$ representing the second harmonic.

A series of sines and cosines may be written as a series of either sines or cosines through the introduction of a phase angle, so that the Fourier series

$$y(t) = A_0 + \sum_{n=1}^{\infty} \left(A_n \cos \frac{2n\pi t}{T} + B_n \sin \frac{2n\pi t}{T} \right)$$

may be written as

$$y(t) = A_0 + \sum_{n=1}^{\infty} C_n \cos \left(\frac{2n\pi t}{T} - \phi_n \right) \qquad (2.18)$$

or

$$y(t) = A_0 + \sum_{n=1}^{\infty} C_n \sin \left(\frac{2n\pi t}{T} + \phi_n^* \right) \qquad (2.19)$$

where

$$C_n = \sqrt{A_n^2 + B_n^2}$$

$$\tan \phi_n = \frac{B_n}{A_n} \qquad \tan \phi_n^* = \frac{A_n}{B_n} \qquad (2.20)$$

EXAMPLE 2.2

Find the Fourier coefficients of the periodic function

$$y(x) = \begin{array}{ll} -5 & \text{when } -\pi < x < 0 \\ +5 & \text{when } \quad 0 < x < \pi \end{array}$$

and $y(x + 2\pi) = y(x)$. Plot the resulting first four partial sums for the Fourier series.

KNOWN

Function $y(x)$ over the range $-\pi$ to π

FIND

Coefficients A_n and B_n

SOLUTION

The function as stated is periodic, having a period of 2π, and is identical in form to the odd function examined in Example 2.1. Since this function is also odd, the Fourier series contains only sine terms, and

$$B_n = \frac{1}{\pi} \int_{-\pi}^{\pi} y(x) \sin nx \, dx = \frac{1}{\pi} \left[\int_{-\pi}^{0} (-5) \sin nx \, dx + \int_{0}^{\pi} (5) \sin nx \, dx \right]$$

which yields upon integration

$$\frac{1}{\pi} \left\{ \left[(5) \frac{\cos nx}{n} \right]_{-\pi}^{0} - \left[(5) \frac{\cos nx}{n} \right]_{0}^{\pi} \right\}$$

Thus,

$$B_n = \frac{10}{n\pi} (1 - \cos n\pi)$$

which is zero for even n, and $20/n\pi$ for odd values of n. The Fourier series can then be written[2]

$$y(x) = \frac{20}{\pi} \left(\sin x + \frac{1}{3} \sin 3x + \frac{1}{5} \sin 5x + \cdots \right)$$

Figure 2.9 shows the first four partial sums of this function, as they compare to the function they represent. Note that at the points of discontinuity of the function, the Fourier series takes on the arithmetic mean of the two function values.

COMMENT

As the number of terms in the partial sum increases, the Fourier series approximation of the square wave function becomes even better. For each additional term in the series, the number of "humps" in the approximate function in each half cycle corresponds to the number of terms included in the partial sum.

EXAMPLE 2.3

As an example of interpreting the frequency content of a given signal, consider the output voltage from a rectifier. A rectifier functions to "flip" the negative half of an alternating current (ac) into the positive half plane, resulting in a

---

[2]If we assume that the sum of this series most accurately represents the function y at $x = \pi/2$, then

$$y \left(\frac{\pi}{2} \right) = \frac{20}{\pi} \left(1 - \frac{1}{3} + \frac{1}{5} - + \cdots \right)$$

or

$$\frac{\pi}{4} = 1 - \frac{1}{3} + \frac{1}{5} - \frac{1}{7} + - \cdots$$

This series approximation of π was first obtained by Gottfried Wilhelm Leibniz (1646–1716) in 1673 from geometrical reasoning.

FIGURE 2.9 First four partial sums of the Fourier series
$(20/\pi)(\sin x + \frac{1}{3}\sin 3x + \frac{1}{5}\sin 5x + \cdots)$ in comparison
with the exact waveform.

(*a*) First partial sum

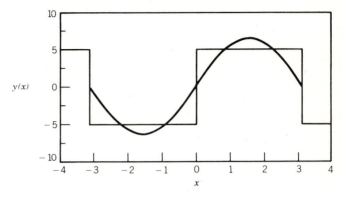

(*b*) Second partial sum

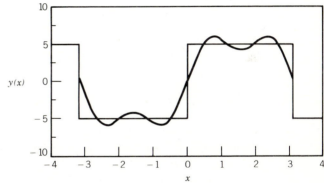

(*c*) Third partial sum

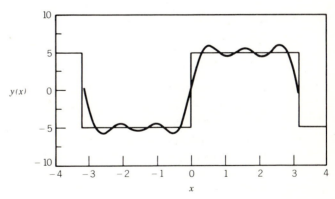

(*d*) Fourth partial sum

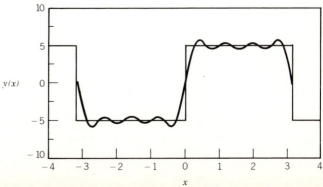

FIGURE 2.10 Rectified sine wave.

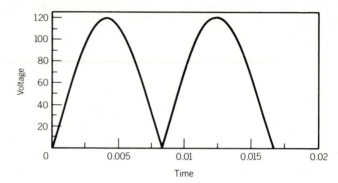

signal that appears as shown in Figure 2.10. For the ac signal the voltage is given by

$$E(t) = 120 \sin 120\pi t$$

The period of the signal is 1/60 of a second, and the frequency 60 Hz.

KNOWN

The rectified signal can be expressed as

$$E(t) = |120 \sin 120\pi t|$$

FIND

The frequency content of this signal as determined from a Fourier series analysis

SOLUTION

The frequency content of this signal can be determined by expanding the function in a Fourier series. The coefficients may be determined using the Euler formulas, keeping in mind that the rectified sine wave is an even function. The coefficient A_0 is determined from

$$A_0 = 2\left[\frac{1}{T}\int_0^{T/2} y(t)\, dt\right] = \frac{2}{1/60}\int_0^{1/120} 120 \sin 120\pi t\, dt$$

with the result that

$$A_0 = \frac{2 \times 60 \times 2}{\pi} = 76.4$$

The remaining coefficients in the Fourier series may be expressed as

$$A_n = \frac{4}{T} \int_0^{T/2} y(t) \cos \frac{2n\pi t}{T} \, dt$$

$$= \frac{4}{1/60} \int_0^{1/120} 120 \sin 120\pi t \cos 120 n\pi t \, dt$$

For values of n that are odd, the coefficient A_n is identically zero. For values of n that are even, the result is

$$A_n = \frac{120}{\pi} \left(\frac{-2}{n-1} + \frac{2}{n+1} \right) \tag{2.21}$$

The Fourier series for the function $|120 \sin 120\pi t|$ is

$$76.4 - 50.93 \cos 240\pi t$$

$$- 10.19 \cos 480\pi t - 4.37 \cos 720\pi t \ldots$$

Figure 2.11 shows the frequency content of the rectified signal, based on a Fourier series expansion.

COMMENT

The frequency content of a signal is determined by examining the amplitude of the various frequency components that are present in the signal. For a periodic mathematical function, these frequency contributions can be illustrated by expanding the function in a Fourier series and plotting the amplitudes of the contributing sine and cosine terms.

FIGURE 2.11   Frequency content of the function $y(t) = |120 \sin 120\pi t|$.

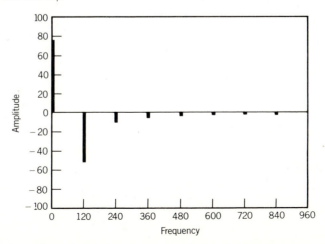

2.4 FOURIER TRANSFORM AND THE FREQUENCY SPECTRUM

The previous discussion of Fourier analysis demonstrates that an arbitrary, but known, function can be expressed as a series of sines and cosines known as a Fourier series. The coefficients of the Fourier series specify the amplitudes of the sines and cosines, each having a specific frequency. Unfortunately, in most practical measurement applications the input signal may not be known in functional form. So that while the theory of Fourier analysis demonstrates that any function can be expressed as a Fourier series, the analysis presented so far has not provided a specific technique for analyzing measured signals. Such a technique for the decomposition of a measured dynamic signal in terms of amplitude and frequency is now described.

Recall that the dynamic portion of a signal of arbitrary period can be described from equations 2.14 by

$$A_n = \frac{2}{T} \int_{-T/2}^{T/2} y(t) \cos 2\pi n f t \, dt$$

$$B_n = \frac{2}{T} \int_{-T/2}^{T/2} y(t) \sin 2\pi n f t \, dt$$

where the amplitudes A_n and B_n correspond to the nth frequency of a Fourier series. Suppose the signal $y(t)$ is an electric current and is represented as a Fourier series. If the nth harmonic of the Fourier series

$$A_n \cos 2\pi n f t + B_n \sin 2\pi n f t$$

is applied across a 1-Ω resistor, the average power dissipated over a time t_f is given by

$$\frac{A_n^2}{t_f} \int_0^{t_f} \cos^2 2\pi n f t \, dt + \frac{B_n^2}{t_f} \int_0^{t_f} \sin^2 2\pi n f t \, dt = \frac{1}{2} (A_n^2 + B_n^2)$$

So the total average power contained at the nth harmonic of the fundamental frequency $2\pi f = 2\pi / T$ is

$$\frac{C_n^2}{2} = \frac{1}{2} (A_n^2 + B_n^2) \qquad (2.22)$$

If we consider the period of the function to approach infinity, we can eliminate the constraint on Fourier analysis that the signal be a periodic waveform. In the limit as T approaches infinity the Fourier series becomes an integral, and the resulting coefficients may be expressed (with $\omega = 2\pi f$)

$$A(f) = \int_{-\infty}^{\infty} y(t) \cos 2\pi f t \, dt \qquad (2.23)$$

and

$$B(f) = \int_{-\infty}^{\infty} y(t) \sin 2\pi ft \, dt \qquad (2.24)$$

The coefficients in a Fourier series that represents a periodic function were associated with a particular frequency, but in the integral representation for an infinite period the values of A and B become continuous functions of frequency.

The amplitude information contained in $A(f)$ and $B(f)$ can also be described by a single amplitude, $C(f)$, where

$$C(f) = \sqrt{A(f)^2 + B(f)^2} \qquad (2.25)$$

Now consider the term defined by $Y(f) = A(f) - iB(f)$ where $i = \sqrt{-1}$. It follows directly that

$$Y(f) = \int_{-\infty}^{\infty} y(t)(\cos 2\pi ft - i \sin 2\pi ft) \, dt$$

Introducing the identity

$$e^{-i\theta} = \cos \theta - i \sin \theta$$

leads to

$$Y(f) = \int_{-\infty}^{\infty} y(t)e^{-i2\pi ft} \, dt \qquad (2.26)$$

Alternatively, we can recover $y(t)$ from

$$y(t) = \frac{1}{2\pi} \int_{-\infty}^{\infty} Y(f)e^{i2\pi ft} \, dt \qquad (2.27)$$

The term $Y(f)$ as described by equation 2.26 is known as the *Fourier transform* of $y(t)$. Similarly, equation 2.27 describes the *inverse Fourier transform* of $y(t)$. The significance of $Y(f)$ is that it describes $y(t)$ in terms related to amplitude and frequency content. For real test data, which typically begins at time zero, the limits of integration of equation 2.26 can be changed to extend from 0 to infinity.

The Fourier transform is a complex function (containing real and imaginary parts). The square of the magnitude of $Y(f)$

$$|Y(f)|^2 = \text{Re}[Y(f)]^2 + \text{Im}[Y(f)]^2$$

yields a measure of the power contained within the dynamic portion of $y(t)$, where $\text{Re}[Y(f)]$ and $\text{Im}[Y(f)]$ refer to the real and imaginary parts of $Y(f)$, respectively. It is seen that the reconstruction of $y(t)$ can be accomplished directly from the Fourier transform. The amplitude and phase shift relations are given by

$$C(f) = \sqrt{2|Y(f)|^2} \qquad (2.28)$$

and

$$\phi(f) = \tan^{-1} \frac{\text{Im } Y(f)}{\text{Re } Y(f)} \qquad (2.29)$$

Further details concerning the Fourier transform may be found in [2,3]. An excellent historical account and discussion of the wide ranging applications is found in [4].

A time-domain plot, such as $y(t)$ versus t, provides information as to how signal amplitude varies with time. The value of $C(f)$ plotted as a function of frequency is called an amplitude frequency spectrum of $y(t)$. Such a frequency-domain plot yields information that describes how signal amplitude varies with frequency. A more common method of displaying this information is the power frequency spectrum which plots the signal power variation with frequency. From (2.25) and (2.28) the power contained at any frequency is given by

$$|Y(f)|^2 = \tfrac{1}{2}[A(f)^2 + B(f)^2]$$

For a dynamic signal, $y(t)$, from which the static portion, A_0, has been removed, the area under a plot of signal power versus frequency, referred to as a power spectrum plot,

$$\text{Area} = \frac{[y(t)]^2}{2} = \int_0^\infty Y(f) \, df \qquad (2.30)$$

provides a measure of the power contained in the dynamic portion of $y(t)$. Furthermore, the area under the curve between any frequency interval defined by $f \pm \delta f/2$, where f is the center frequency of the interval,

$$\frac{[y(t)]^2}{2} = \int_{f-(\delta f/2)}^{f+(\delta f/2)} Y(f) \, df \qquad (2.31)$$

provides a measure of the power contained in the signal within that frequency interval. The amplitude and power frequency spectrum presentation is a powerful diagnostic and analytical tool for interpreting signal content.

It should now be clear that $C(f)$ and $\phi(f)$ can be found by application of the Fourier transform to a signal $y(t)$. The required integration for the Fourier transform given by equation 2.26 can be performed numerically and computer software packages are readily available for this purpose.

Consider the time-dependent portion of the signal $y(t)$ which is measured N times at equally spaced time intervals, t. In this case, the continuous signal $y(t)$ will be replaced by the discrete time signal given by $y(r \, \delta t)$ $r = 0, 1, \ldots,$ $N - 1$. In effect, the relationship between $y(t)$ and $y(r \, \delta t)$ is described by a set of impulses of an amplitude determined by the value of $y(t)$ at each time step $r\,\delta t$. This transformation from a continuous to discrete signal is described by

$$y(r\,\delta t) = y(t) \sum_{r=0}^{N-1} \delta(t - r\,\delta t)$$

where $\delta(t - r\,\delta t)$ is the delayed unit impulse function.

To analyze a measured signal, the signal will normally be converted to a digital form, often through data acquisition using a digital computer. Such data will not extend from zero to infinity, but over some finite time interval 0 to t_f. There exists a length of this observation period for which statistical parameters describe with reasonable accuracy the behavior of the signal, and longer observation times would not change the descriptive parameters such as the Fourier transform or the power spectral density. For all practical purposes, a discrete data set containing N values representing a time interval 0 to t_f will represent the parameters of the signal, provided the sample time t_f is sufficiently long.

An approximation to the Fourier transform integral of equation 2.26 for use on a discrete data set is the discrete Fourier transform (DFT). Using the relation $\omega = 2\pi f$, the DFT is given by

$$Y(f_k) = \frac{1}{N} \sum_{r=0}^{N-1} y(r\,\delta t)e^{-i2\pi rk/N} \qquad k = 0, 1, 2, \ldots, \frac{N}{2} - 1 \qquad (2.32)$$

where the values of $Y(f_k)$ represent $k = N/2$ discrete power values spaced at discrete frequency intervals having a resolution $1/(N\,\delta t)$. Equation 2.32 performs the numerical integration required by the Fourier integral. Equations 2.25, 2.28, and 2.32 demonstrate that the application of the DFT on the discrete series of data, $y(r\,\delta t)$, permits the decomposition of the discrete data in terms of frequency and amplitude content. Hence, using this method, a measured discrete signal of unknown functional form can be reconstructed as a trigonometric series using Fourier transform techniques.

An algorithm for computing the discrete Fourier transform of a discrete signal is provided in Appendix A. This particular algorithm was chosen for its simplicity for demonstration purposes. The computational time of this DFT algorithm increases at a rate that is proportional to N^2. This makes it too inefficient for use with data sets of large N. A faster algorithm for computing the discrete Fourier transform, known as the fast Fourier transform (FFT), was developed by Cooley and Tukey [5]. This method is widely available and is the basis of most Fourier analysis software packages. The FFT is discussed in most advanced texts on signal analysis (e.g., [6,7]). The accuracy of discrete Fourier analysis depends on the frequency content[3] of $y(t)$ and on the DFT resolution. An extensive discussion of these interrelated parameters is given in Chapter 6.

EXAMPLE 2.4

Convert the continuous signal described by $y(t) = 10 \sin 2\pi t$ into a discrete set of 128 numbers using a time increment of 0.1 s between each discrete measurement.

---

[3]The value of $1/\delta t$ must be more than twice the highest frequency contained in $y(t)$.

KNOWN

$C(f_1 = 1) = 10$

$f_1 = \omega_1/2\pi = 1$ Hz

$\phi(f_1 = 1) = 0$

$\delta t = 0.1$ s

$N = 128$

FIND

$y(r\, \delta t)$

SOLUTION

The measurement of $y(t)$ every 0.1 s produces the discrete data set $y(r\, \delta t)$. One period of the original signal, $1/f_1 = 1$ s, and its discrete representation as

FIGURE 2.12   Representation of a simple periodic function as a discrete signal.

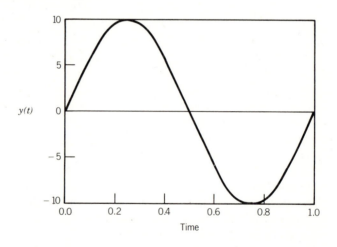

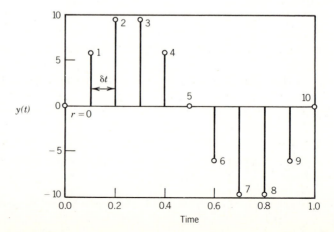

TABLE 2.2   Discrete Data Set for $10 \sin 2\pi t$

r	$y(r\,\delta t)$	r	$y(r\,\delta t)$	r	$y(r\,\delta t)$	r	$y(r\,\delta t)$
1	0.0000	33	9.5105	65	5.8783	97	−5.8775
2	5.8778	34	9.5106	66	0.0006	98	−9.5104
3	9.5106	35	5.8780	67	−5.8774	99	−9.5107
4	9.5106	36	0.0003	68	−9.5104	100	−5.8782
5	5.8779	37	−5.8776	69	−9.5107	101	−0.0004
6	0.0000	38	−9.5105	70	−5.8784	102	5.8776
7	−5.8778	39	−9.5107	71	−0.0006	103	9.5105
8	−9.5106	40	−5.8781	72	5.8774	104	9.5107
9	−9.5106	41	−0.0003	73	9.5104	105	5.8781
10	−5.8779	42	5.8776	74	9.5108	106	0.0003
11	0.0000	43	9.5105	75	5.8784	107	−5.8776
12	5.8778	44	9.5107	76	0.0007	108	−9.5105
13	9.5105	45	5.8781	77	−5.8773	109	−9.5106
14	9.5106	46	0.0003	78	−9.5103	110	−5.8781
15	5.8779	47	−5.8776	79	−9.5108	111	−0.0002
16	0.0001	48	−9.5104	80	−5.8784	112	5.8777
17	−5.8778	49	−9.5107	81	−0.0007	113	9.5105
18	−9.5105	50	−5.8782	82	5.8772	114	9.5106
19	−9.5106	51	−0.0004	83	9.5103	115	5.8779
20	−5.8779	52	5.8775	84	9.5108	116	0.0001
21	−0.0001	53	9.5104	85	5.8784	117	−5.8778
22	5.8778	54	9.5107	86	0.0007	118	−9.5105
23	9.5105	55	5.8782	87	−5.8773	119	−9.5106
24	9.5106	56	0.0005	88	−9.5104	120	−5.8779
25	5.8780	57	−5.8775	89	−9.5108	121	0.0000
26	0.0001	58	−9.5104	90	−5.8783	122	5.8779
27	−5.8777	59	−9.5107	91	−0.0006	123	9.5106
28	−9.5105	60	−5.8783	92	5.8774	124	9.5106
29	−9.5106	61	−0.0005	93	9.5104	125	5.8778
30	−5.8780	62	5.8774	94	9.5107	126	−0.0001
31	−0.0002	63	9.5104	95	5.8782	127	−5.8779
32	5.8777	64	9.5107	96	0.0005	128	−9.5106

a series of impulses in a time domain are shown in Figure 2.12. The first 128 terms in the data set are given in Table 2.2. Note that 128 data points represent 12.8 periods of the signal. The duration of this measurement of $y(t)$ lasts for $(N - 1)\,\delta t = 12.7$ s. Close inspection of the discrete data reveals the effects of roundoff (noise).

EXAMPLE 2.5

Use the algorithm of the discrete Fourier transform in Appendix A to estimate the amplitude and power frequency spectrum of the discrete data set of Example 2.4.

KNOWN

$y(r\,\delta t)$ of Table 2.2

$\delta t = 0.1$ s

$N = 128$

FIND

$C(f_k)$

SOLUTION

We should note that the use of a time increment of 0.1 s conforms to the guidelines provided in Appendix A, which state that the time increment used must be at least twice the highest frequency in the data set. For $N = 128$ data points in the discrete set, the DFT will return $N/2$ values for $C(f_k)$, each spaced at a frequency interval of $1/(N\,\delta t) = 0.078$ Hz. The results of the numerical analysis are plotted in the amplitude frequency spectrum in Figure 2.13 and in the power spectrum of Figure 2.14. In each figure the data points shown are located at the center of the $N/2$ frequency intervals returned by the DFT.

COMMENT

The data of Figure 2.13 clearly indicate an amplitude spike of approximately 10 over the range 1 ± 0.078 Hz. It is interesting to note that the rise toward the spike, while easily distinguishable and not ambiguous, is gradual beginning at frequencies below 1 Hz with a decay continuing through frequencies above 1 Hz. This is a normal tendency found when using discrete Fourier analysis methods. Advanced signal analysis techniques [6] can be used when necessary to apply certain "window" functions to the data set to minimize the artificial manifestations introduced by discrete methods. Experimenting with the param-

FIGURE 2.13   Amplitude as a function of frequency for a discrete representation of $10 \sin 2\pi t$ resulting from a discrete Fourier transform algorithm.

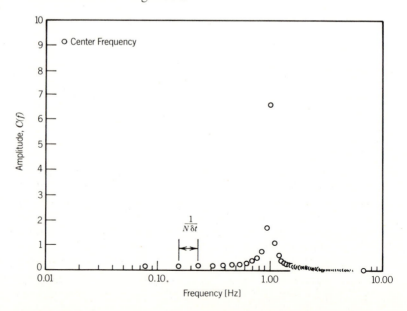

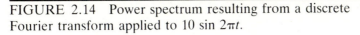
FIGURE 2.14   Power spectrum resulting from a discrete
Fourier transform applied to $10 \sin 2\pi t$.

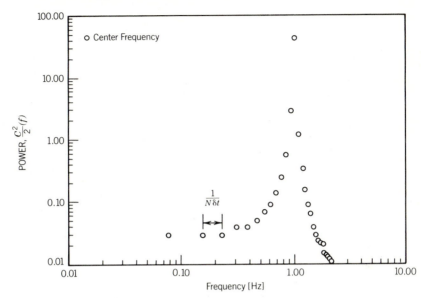

eters of resolution and N for a function of known frequency content, such as in
this example, will lend further insight into the behavior of the Fourier transform
algorithm for signals of unknown frequency content.

2.5   SUMMARY

This chapter has provided a fundamental basis for the description of signals
associated with measurements and measurement systems. The capabilities of a
measurement system can be properly specified only if the nature of the input
signal is known. Such a signal may be classified as static or dynamic. Dynamic
signals may be further classified according to the nature of their waveform, either
as deterministic or nondeterministic. Such descriptions of general classes of input
signals will be seen, in Chapter 3, to allow universal descriptions of measurement
system dynamic behavior.

Any signal is appropriately thought of and modeled as being composed of
many individual frequencies or harmonics. Mathematically, any signal can be
represented by the amplitudes associated with a particular frequency range. As
such, measurement system selection and design must consider the frequency
content of the input signals the system is intended to measure. In addition,
Fourier analysis allows a precise definition of phase relationships among various
frequency components within a particular signal.

In conclusion, signal characteristics form an important basis for the selection
of measurement systems and the interpretation of measurement system output.
In Chapter 3 these ideas are combined with the concept of a generalized set of

measurement system behaviors. The combination of generalized measurement system behavior and generalized descriptions of input waveforms provides for an understanding of a wide range of instruments and measurement systems.

REFERENCES

1. Monforte, J., The digital reproduction of sound, *Scientific American*, 251(6):78, 1984.
2. Bracewell, R. N., *The Fourier Transform and Its Applications*, 2d ed., rev., McGraw Hill, New York, 1986.
3. Champeney, D. C., *Fourier Transforms and Their Physical Applications*, Academic, London, 1973.
4. Bracewell, R. N., The Fourier transform, *Scientific American*, 260(6):86, 1989.
5. Cooley, J. W., and Tukey, J. W., *Math. Comput.* 19:207, April 1965 (see also Special Issue on the fast Fourier transform, *IEEE Transactions on Audio and Electroacoustics* AU-2, June 1967).
6. Bendat, J. S., and A. G. Piersol, *Measurement and Analysis of Random Data*, Wiley, New York, 1966 (see also Bendat, J. S., and A. G. Piersol, *Engineering Applications of Correlation and Spectral Analysis*, Wiley, New York, 1980).
7. Cochran, W. T., et al., What is the fast Fourier transform?, *Proc. IEEE* 55(10):1664, 1967.

Suggested Reading

Halliday, D., and R. Resnick, *Fundamentals of Physics*, 2d ed., Wiley, New York, 1981.

Kreyszig, E., *Advanced Engineering Mathematics*, 4th ed., Wiley, New York, 1979.

NOMENCLATURE

f	frequency in Hz $[t^{-1}]$		F	force $[m-l-t^{-2}]$
k	spring constant $[m-t^{-2}]$		N	total number of discrete data points
m	mass $[m]$			
t	time $[t]$		T	period $[t]$
x	displacement or independent variable		U	unit step function
			Y	Fourier transform
y	dependent variable		α	angle [radians]
y_m	discrete data points		β	angle [radians]
A	amplitude		ϕ	phase angle [radians]
B	amplitude		ω	circular frequency in rad/s
C	amplitude			

PROBLEMS

2.1 Define the term *signal* as it relates to measurement systems. Provide two examples of static and dynamic input signals to measurement systems.

2.2 List the important characteristics of input and output signals, and define each.

2.3 Determine the value of the spring constant that would result in a spring–mass system that would execute one complete cycle of oscillation every second, for a mass of 0.5 kg. What natural frequency does this system exhibit in radians/second?

2.4 For the following sine and cosine functions determine the period, the frequency in hertz and the circular frequency in rad/s. (*Note: t* represents time in seconds.)
  a. $\sin 10t$
  b. $\cos \dfrac{2\pi t}{3}$
  c. $\sin 3n\pi t$ for $n = 1$ to ∞

2.5 Express the following function in terms of a cosine term only:

$$y(x) = 10 \sin 2t + 3 \cos 2t$$

2.6 If T is a period of $y(x)$, show that nT for $n = 2, 3, \ldots$ is a period of that function.

2.7 The nth partial sum of a Fourier series is defined as

$$A_0 + A_1 \cos \omega_1 t + B_1 \sin \omega_1 t + \cdots$$
$$+ A_n \cos \omega_n t + B_n \sin \omega_n t$$

For the third partial sum of the Fourier series given by

$$y(t) = \sum_{n=1}^{\infty} \frac{3n}{2} \sin nt + \frac{5n}{3} \cos nt$$

  a. What is the fundamental frequency and the associated period?
  b. Express this partial sum as cosine terms only.

2.8 For the Fourier series given by

$$y(t) = 4 + \sum_{n=1}^{\infty} \frac{2n\pi}{10} \cos \frac{n\pi}{4} t + \frac{120n\pi}{30} \sin \frac{n\pi}{4} t$$

where t is time in seconds:
  a. What is the fundamental frequency in hertz and radians/second?
  b. What is the period T associated with the fundamental frequency?

    c.   Express this Fourier series as an infinite series containing sine terms only.

2.9   Find the Fourier series of the function shown in Figure 2.15, assuming the function has a period of 2π. Plot an accurate graph of the first three partial sums of the resulting Fourier series.

FIGURE 2.15   Function to be expanded in a Fourier series in Problem 2.9.

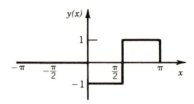

2.10   Determine the Fourier series for the function,

$$y(x) = x \qquad -5 < x < 5$$

by expanding the function as an odd periodic function with a period of 10 units, as shown in Figure 2.16. Plot the first, second, and third partial sums of this Fourier series.

FIGURE 2.16   Sketch for Problem 2.10.

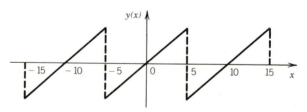

2.11   a. Show that $y(x) = x^2\,(-\pi < x < \pi)$, $y(x + 2\pi) = y(x)$ has the Fourier series

$$y(x) = \frac{\pi^2}{3} - 4\left(\cos x - \frac{1}{4}\cos 2x + \frac{1}{9}\cos 3x - + \cdots\right)$$

    b. By setting $x = \pi$ in the above series, show that a series approximation for π, first discovered by Euler, results as

$$\sum_{n=1}^{\infty} \frac{1}{n^2} = 1 + \frac{1}{4} + \frac{1}{9} + \frac{1}{16} + \cdots = \frac{\pi^2}{6}$$

2.12 Classify the following signals as static or dynamic, and identify any that may be classified as periodic:
a.   sin 10t
b.   Figure 2.17

FIGURE 2.17   Sketch for Problem 2.12.

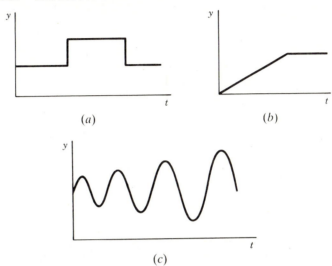

(a)   (b)

(c)

2.13 A particle executes linear harmonic motion around the point $x = 0$. At time zero the particle is at the point $x = 0$ and has a velocity of 5 cm/s. The frequency of the motion is 1.0 Hz. Determine:
a.   The period
b.   The amplitude of the motion
c.   The displacement as a function of time
d.   The maximum speed

2.14 Define the following characteristics of signals:
a.   Frequency content   c.   Magnitude
b.   Amplitude   d.   Period

2.15 Construct a frequency spectrum plot for the Fourier series in Problem 2.11 for $y(x) = x^2$. Discuss the significance of this frequency spectrum for selecting a measurement system.

2.16 Sketch representative waveforms of the following signals, and represent them as mathematical functions (if possible):
a.   The output signal from the thermostat on a refrigerator
b.   The electrical signal to a spark plug in a car engine
c.   The input to a cruise control for an automobile
d.   A pure musical tone (e.g., 440 Hz is an A)
e.   The note produced by a guitar string
f.   AM and FM radio signals

2.17 Represent the function

$$e(t) = 5 \sin 31.4t + 2 \sin 44t$$

as a discrete set of $N = 128$ numbers separated by a time increment of $(1/N)$. Use the DFT algorithm to construct an amplitude spectrum from this data set.

2.18 Repeat Problem 2.17 using a data set of 256 numbers at $\delta t = (1/N)$ s and $\delta t = (1/2N)$ s. Compare and discuss the results.

2.19 A particular strain sensor is mounted to an aircraft wing that is subjected to periodic wind gusts. The strain measurement system indicates a periodic strain that ranges from 3250×10^{-6} in./in. to 4150×10^{-6} in./in. at a frequency of 1 Hz. Determine:
a. The average value of this signal
b. The amplitude and the frequency of this output signal when expressed as a simple periodic function
c. A Fourier series that represents this signal
Construct an amplitude spectrum plot for the output signal.

2.20 For a dynamic calibration involving a force measurement system, a known force is applied to a sensor. The force varies between 100 and 170 N at a frequency of 10 rad/s. State the average (static) value of the input signal, its amplitude, and its frequency. Assuming that the signal may be represented by a simple periodic waveform, express the signal as a Fourier series.

2.21 A displacement sensor is placed on a dynamic calibration rig known as a shaker. This device produces a known periodic displacement which serves as the input to the sensor. If the known displacement is set to vary between 2 and 5 mm at a frequency of 100 Hz, express the input signal as a Fourier series. Plot the signal in the time domain, and construct an amplitude spectrum plot.

2.22 Classify the following signals as completely as possible:
a. Clock face having hands
b. Morse code
c. Musical score, as input to a musician
d. Flashing neon sign
e. Telephone conversation
f. Fax transmission

CHAPTER 3
MEASUREMENT SYSTEM BEHAVIOR

3.1 INTRODUCTION

The time-dependent behavior of a measurement system subjected to a particular input signal defines the system response to that input. Each measurement system will respond differently to different types of input signals. A particular system may not be suitable for the measurement of certain input signals. Yet a measurement system will always output information regardless of how well this information might reflect the actual input signal. To acquire accurate information about the input signal based on the information contained in the output signal, the response of the system to a variety of input signals needs to be known. In this chapter, the concept of simulating the measurement system behavior through mathematical modeling is introduced. From this modeling, the important aspects of measurement system response pertinent to system design and specification will be introduced and investigated.

Throughout this chapter, we will use the term "measurement system" in a generic sense. When we discuss the response of a measurement system, we refer either to the response of the measurement system as a whole, or to the response of any component or instrument that makes up that system. Both are important and are interpreted in similar ways. Each individual stage of the measurement system will have its own response to a given input. The overall system response will be governed by the response of each stage that makes up the complete system.

3.2 GENERAL MODEL FOR A MEASUREMENT SYSTEM

As pointed out in Chapter 2, all input and ouput signals can be broadly classified as being static, dynamic, or some combination of the two. For a static signal, only the signal magnitude is needed to reconstruct the input signal based on the indicated output signal. Consider the measurement of the length of a board using a ruler. Once the ruler is positioned, the indication (output) of the magnitude of length is immediately displayed since the board length (input) does not change over the time required to make the measurement. Thus, the board length represents a static input signal which is interpreted through the static magnitude output by the ruler.

For dynamic signals, signal amplitude, frequency, and/or general waveform become necessary to reconstruct the input signal. Since dynamic signals vary with time, the measurement system must be able to respond fast enough literally to keep up with the input signal. Consider the time response of a typical bulb thermometer during the act of measuring body temperature. The thermometer, initially at room temperature or so, is placed under the tongue. But even after several seconds, the thermometer does not indicate the expected value of body temperature and its display continually changes. What has happened? Surely body temperature does not change in a short period of time. Use of the magnitude of the output signal after several seconds might lead to false conclusions about one's health. Experience shows that within a few minutes, the correct body temperature will be indicated; so we wait. In this example, body temperature itself is a constant during the measurement, but the input signal to the thermometer is suddenly changed from room temperature to body temperature. This is a dynamic event as the thermometer (the measurement system) sees it! The thermometer must gain energy from its new environment to reach thermal equilibrium and this takes a finite amount of time. The ability of any measuring system to follow dynamic signals is a characteristic of the design of the measuring system components. We recognize the importance of measurement system response even when using the most common of measurement systems.

The response of a particular measurement system to different dynamic inputs could be ascertained during exhaustive static and dynamic calibrations. But in general, the behavior of measurement systems to a few special inputs will, for the most part, define the input–output signal relationships necessary to enable the correct interpretation of the measured signal. It will be shown that only a few measurement system characteristics define the system response, so that dynamic calibrations can be restricted to a few specific tests. The results of these tests can be used to judge the suitability of a particular system to interpret the signal information.

With the previous discussion in mind, consider that the primary task of a measurement system is to sense an input signal and to translate that information into a readily understandable and quantifiable output form. Then one can assume that a measurement system performs some mathematical operation on a sensed input. In fact, a general measurement system can be represented by a differential equation which describes the operation that a measurement system performs on the input signal.

This concept of an input signal being operated on by the measurement system that provides an output signal is illustrated in Figure 3.1 with the mathematical operation that the measurement system performs represented within the box. For some input signal, $F(t)$, the system performs some operation that yields the output signal, $y(t)$. Then we must use $y(t)$ to infer $F(t)$. Therefore, at least a qualitative understanding of the operation that the measurement system performs is imperative to correctly interpret the input signal. We will propose a general mathematical model for a measurement system. Then, by representing a typical input signal as some function that acts as an input to the model, we can study just how the measurement system would behave by solving the model

FIGURE 3.1   Measurement system operation on an input
signal.

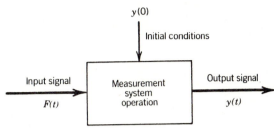

equation. In essence, we will perform the analytical equivalent of a system
calibration. This information can then be used to determine those input signals
for which a particular measurement system is best suited.

Consider, then, the following model of a measurement system which consists
of a general nth-order linear ordinary differential equation in terms of a general
output signal, represented by variable $y(t)$, and subject to a general input signal,
represented by the forcing function, $F(t)$:

$$a_n\frac{d^n y}{dt^n} + a_{n-1}\frac{d^{n-1}y}{dt^{n-1}} + \cdots + a_1\frac{dy}{dt} + a_0 y = F(t) \qquad (3.1)$$

The coefficients $a_0, a_1, a_2, \ldots, a_n$ represent physical system parameters whose
properties and values will depend on the measurement system itself. The forcing
function can be generalized into the mth-order form

$$F(t) = b_m\frac{d^m x}{dt} + b_{m-1}\frac{d^{m-1}x}{dt^{m-1}} + \cdots + b_0 x \qquad m \le n$$

where $b_0, b_1, \ldots, b_m$ also represent physical system parameters. Real measurement systems can be modeled in this manner by considering their governing
system equations. These equations are generated by application of pertinent
fundamental physical laws of nature to the measurement system. Our discussion
will be limited to measurement system concepts, but a general treatment of
systems can be found in text books dedicated to that topic (e.g., [1–3]).

EXAMPLE 3.1

As an illustration, consider the seismic accelerometer depicted in Figure 3.2a.
Various configurations of this instrument are used in seismic and vibration engineering to determine the motion of large bodies to which the accelerometer
is attached. Basically, as the small accelerometer mass reacts to motion it places
the piezoelectric crystal into compression or tension, which causes a surface
charge to develop on the crystal. The charge is proportional to the motion. As
the large body moves, the mass of the accelerometer will move with an inertial
response. The stiffness of the spring, k, will provide a restoring force to move

FIGURE 3.2   Accelerometer of Example 3.1.

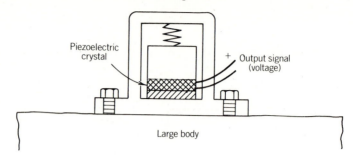

(a)   Piezoelectric accelerometer attached to large body

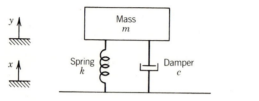

(b)   Representation using mass,
spring, and damper

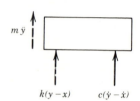

(c)   Free-body diagram

the accelerometer mass back to equilibrium while internal frictional damping, c, will oppose any displacement away from equilibrium. A model of this measurement device in terms of ideal elements of stiffness, mass, and damping is given in Figure 3.2b and the corresponding free-body diagram in Figure 3.2c. Let y denote the position of the small mass within the accelerometer and x denote the displacement of the body. Solving Newton's second law for the free-body yields the second-order linear, ordinary differential equation,

$$m \frac{d^2y}{dt^2} + c \frac{dy}{dt} + ky = c \frac{dx}{dt} + kx$$

Since the displacement y is the pertinent output from the accelerometer due to displacement x, the equation has been written such that all output terms, that is, all the y terms, are on the left side. All other terms are considered to be input signals and are placed on the right side. Comparing this to the general form for a second-order equation from (3.1),

$$a_2 \frac{d^2y}{dt^2} + a_1 \frac{dy}{dt} + a_0 y = b_1 \frac{dx}{dt} + b_0 x$$

we observe that $a_2 = m$, $a_1 = b_1 = c$, $a_0 = b_0 = k$, and that the forces developed due to the velocity and displacement of the body become the inputs to the accelerometer. If we could anticipate the waveform of x (e.g., $x(t) = x_0 \sin \omega t$), we could simulate the measurement system response to this input by solving the equation.

Fortunately, many measurement systems can be modeled by zero-, first-, or second-order linear, ordinary differential equations. More complex systems can usually be simplified to a lower order or linearized for the intention of simply studying system response to specific types of input. Our intention here is to attempt to understand how systems behave and how response is closely related to the design features of a measurement system; it is not to simulate the exact system behavior. The exact input–output relationship is **always** found from calibration. But modeling will guide us in narrowing down our choice to specific instruments and determining the type, range, and specifics of calibration. Next, we will examine several special cases of equation 3.1 that identify the most important concepts of measurement system behavior.

3.3 SPECIAL CASES OF THE GENERAL SYSTEM MODEL

Zero-Order Systems

The simplest model for a measurement system is that of a zero-order differential equation:

$$a_0 y = F(t)$$

with $F(t) = b_0 x$, or dividing through by a_0,

$$y = KF(t) \tag{3.2}$$

where $K = 1/a_0$. K is called the *static sensitivity* or *steady gain* of the system. This system property was introduced in Chapter 1 as the relation between the static output change associated with a change in input. In zero-order behavior the system output is considered to respond to the input signal instantly. If an input signal of magnitude $F(t) = A$ were applied, the instrument would indicate KA, as specified by (3.2). Often the scale of the measuring device is modified to accommodate K so as to indicate A directly.

For real systems, the zero-order system concept is used to model measurement system response to static inputs. In fact, the zero-order concept appropriately models any system during a static calibration. When dynamic input signals are involved, the output signal can be considered valid only when it is at static equilibrium. This is because most real measurement systems possess inertial or storage capabilities that require higher-order differential equations to correctly model their time-dependent behavior to dynamic input signals.

Determination of K.   The static sensitivity is found from the static calibration of the measurement system. It is the slope of the calibration curve.

EXAMPLE 3.2

A pencil-type pressure gauge commonly used to measure tire pressure can be modeled at static equilibrium by considering the force balance on the gauge sensor, a piston that slides up and down a cylinder. In Figure 3.3a, the piston motion is restrained by an internal spring of stiffness, k, so that at static equilibrium the absolute pressure force, F, bearing on the piston equals the force exerted on the piston by the spring, F_s, plus the atmospheric pressure force, F_{atm}. In this manner, the transduction of pressure into displacement occurs. Considering the piston free-body at static equilibrium in Figure 3.3b, the static force balance, $\Sigma \mathbf{F} = 0$, yields

$$ky = F - F_{atm}$$

where y is measured relative to some static reference position marked as zero on the output display. Pressure is simply the force acting inward over the piston surface area, A. So dividing through by area provides the zero-order response equation between output displacement and input pressure

$$y = \frac{A}{k}(p - p_{atm})$$

The term $p - p_{atm}$ represents the pressure relative to atmospheric pressure. It is the pressure indicated by this gauge. Direct comparison with (3.2) implies that the input pressure is translated into piston displacement through the static sensitivity, $K = A/k$. The system operates on pressure so as to bring about a relative displacement of the piston, the magnitude of which is used to indicate the pressure. The spring stiffness and piston area affect the magnitude of this displacement. The exact static input–output relationship is found through calibration of the gauge. Since such elements as piston inertia and frictional dissi-

FIGURE 3.3   Pressure gauge of example 3.2.

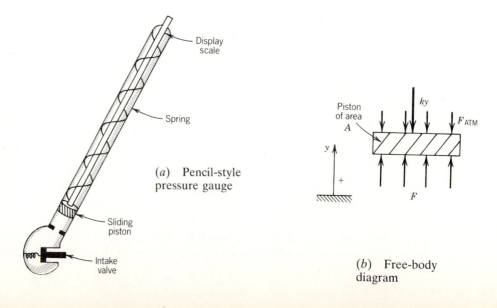

(a)   Pencil-style pressure gauge

(b)   Free-body diagram

pation were not considered, this model would not be appropriate for studying the dynamic response of this system.

First-Order Systems

Measurement systems that contain storage elements cannot respond instantaneously to changes in input. The bulb thermometer considered earlier in this chapter is a good example. The bulb exchanges energy with its environment until the two are at the same temperature, storing energy during the exchange. The temperature of the bulb sensor will change with time until this equilibrium is reached, which accounts physically for its less than immediate response. The rate at which temperature changes with time can be modeled with a first-order derivative and the thermometer behavior modeled as a first-order equation. In general, systems with a storage or dissipative capability may be modeled using a first-order differential equation of the form

$$a_1 \dot{y} + a_0 y = F(t) \tag{3.3}$$

with $\dot{y} = dy/dt$. Dividing through by a_0 gives

$$\tau \dot{y} + y = KF(t) \tag{3.4}$$

where $\tau = a_1/a_0$. τ is called the *time constant* of the system. Regardless of the physical dimensions of a_0 and a_1, their ratio will always have the dimensions of time. The time constant provides a measure of the speed of system response, and as such is an important specification in measuring dynamic input signals. To explore this concept more fully, consider the response of the general first-order system to the following two forms of an input signal: the step function and the simple periodic function.

Step Function Input

With reference to Figure 3.4, the step function, $AU(t)$, is defined as

$$AU(t) = 0 \qquad t \leq 0^-$$
$$AU(t) = A \qquad t \geq 0^+$$

FIGURE 3.4   The unit step function.

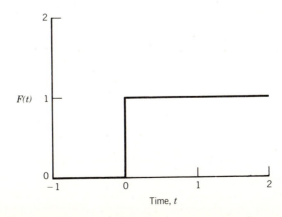

Time, t

where A is the amplitude of the step function and $U(t)$ is defined as the unit step function. Physically, this function describes a sudden change in the input signal from a constant value of one magnitude to a constant value of some other magnitude, such as a sudden change in temperature, pressure, or loading. Application of a step function input to a measurement system provides information about the speed at which a system responds to a change in input signal. To illustrate this, let us apply a step function as an input to the general first-order system. Setting $F(t) = AU(t)$ in (3.4) gives

$$\tau \dot{y} + y = KAU(t)$$

with an arbitrary initial condition, $y(0) = y_0$. The solution of this differential equation yields for $t \geq 0^+$

$$y(t) = KA + (y_0 - KA)e^{-t/\tau} \tag{3.5}$$

The solution of the differential equation, $y(t)$, is the *time response* (or simply the *response*) of the system. Equation 3.5 describes the behavior of the system to a step change in input. This means that $y(t)$ is in fact the output indicated by the display stage of the system. It should represent the time variation of the output display of the measurement system if an actual step change were applied to the system. We have simply used mathematics to simulate this response.

The second term on the right side of (3.5) is known as the *transient response* of $y(t)$ since, as $t \rightarrow \infty$, the magnitude of this term eventually reduces to zero. The first term is known as the steady response since, as $t \rightarrow \infty$, the response of $y(t)$ approaches this steady value. The *steady response* is that portion of the output signal that remains after the transient response has decayed to zero. In light of this, we will define a *steady signal* as a signal that consistently repeats itself with time (such as a static or a periodic signal).

For illustrative purposes only, let $y_0 < A$ so that the time response becomes as shown in Figure 3.5. Over time, the indicated output value rises from its initial value, at the instant the change in input is applied, to an eventual constant

FIGURE 3.5   First-order system response to a step function input.

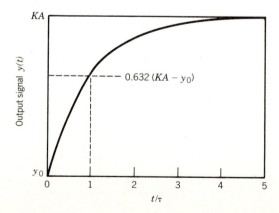

value, $y_\infty = KA$, at steady response. Compare this general time response to the recognized behavior of the bulb thermometer when measuring body temperature as discussed earlier. We see a qualitative similarity. In fact, the measurement of body temperature represents a real step function input to the thermometer, itself a real first-order measuring system.

Suppose we rewrite the response equation 3.5 in a form

$$\Gamma(t) = \frac{y(t) - y_\infty}{y_0 - y_\infty} = e^{-t/\tau} \qquad (3.6)$$

where the term Γ will be called the *error fraction*. Equation 3.6 is plotted in Figure 3.6 where the time axis has been nondimensionalized by the time constant. It can be seen that the error fraction decreases from a value of 1 and approaches a value of 0 with increasing t/τ. At the instant just after the step change in input is introduced, $\Gamma = 1.0$ so that the indicated output from the measurement system is 100% in error. This implies that the system has responded 0% to the input change. From Figure 3.6, it is apparent that as time moves forward the system responds with decreasing error in its indicated value. Let the percent response of the system to a step change be given as $(1 - \Gamma) \times 100$. Then by $t = \tau$, where $\Gamma = 0.368$, the system will have responded to 63.2% of the step change. Further, when $t = 2.3\tau$ the system will have responded to 90% of the step change; by $t = 5\tau$, 99.9%. The time required for a system to respond to a value that is 90% of the step input, $y_\infty - y_0$, is called the *rise time* of the system.

Based on this behavior, one can see that the time constant is in fact a measure of the speed of first-order measurement system response to a change in input value. A smaller time constant indicates a shorter time between the instant that an input is applied and when the system reaches an essentially steady output. The time constant will be defined as the time required for a first-order system to achieve 63.2% of the step change magnitude, $y_\infty - y_0$. The time constant is a system property, as indicated by equation 3.4. Its value will depend on the measurement system considered, ranging from extremely short to very long values of time.

FIGURE 3.6   First-order system response to a step function input: the error fraction.

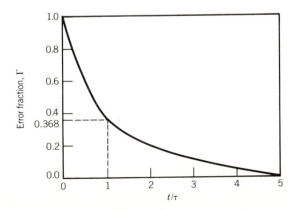

Determination of τ.   According to the above development, the time constant of a first-order measurement system can be experimentally determined by recording the system's response to a step function input of a known magnitude. In practice, it is best to record that response from $t = 0$ until steady response is achieved. The data can then be plotted as error fraction versus time on a semilog plot, such as in Figure 3.7. This type of plot is equivalent to the transformation

$$\ln \Gamma = 2.3 \log \Gamma = -\frac{1}{\tau} t \qquad (3.7)$$

which is of the linear form, $Y = mX + B$. A curve fit through the data will provide a good estimate of the slope, m, of the resulting plot. From equation 3.7, we see that $m = -1/\tau$, which yields the time constant estimate.

The above method offers several advantages over attempting to compute τ directly from the time required to achieve 63.2% of the step-change magnitude. First, real systems may deviate somewhat from perfect first-order behavior. In itself, this is fine, since our intention is to study the system behavior not to simulate it exactly. On rectangular grid plots, such as Figure 3.6, deviations from perfect first-order behavior will not be easy to detect. But on a semilog plot such deviations are readily apparent as clear trends away from a linear curve. Modest deviations do not pose any problem and can be ignored. But strong deviations provide a good indication that the system is not behaving as expected, thus requiring a closer examination of either the system operation, the step function experiment or the assumed form of the system model. Second, the acquisition of data during the step function experiment is prone to some error (precision error) in each data point. The use of a data curve fit to determine τ utilizes all of the data over time so as to minimize the influence of an error in any one data point. Third, the method eliminates the need to accurately determine the $\Gamma = 1.0$ and 0.368 points, which are difficult to establish in practice.

FIGURE 3.7   The error fraction plotted on semilog coordinates.

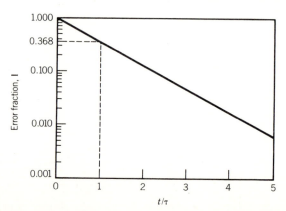

EXAMPLE 3.3

Suppose a bulb thermometer, originally indicating 68 °F, is suddenly exposed to a fluid temperature of 98 °F. Develop a simple model to simulate the thermometer output response.

KNOWN

$T(0) = 68 \ °F$
$T_\infty = 98 \ °F$
$F(t) = [T_\infty - T(0)]U(t)$

ASSUMPTIONS

To keep things simple, assume:

No installation effects (neglect conduction and radiation effects)
Sensor mass is mass of liquid in bulb only
Uniform temperature within bulb (lumped mass)
Thermometer scale is calibrated to indicate temperature

FIND

$T(t)$

SOLUTION

Consider the energy balance developed in Figure 3.8. According to the first law of thermodynamics, the rate at which energy is exchanged between the sensor and its environment through convection, $\dot{Q}$, must be balanced by the storage of energy within the thermometer, dE/dt. This conservation of energy is written as

$$\frac{dE}{dt} = \dot{Q}$$

FIGURE 3.8 Thermometer and energy balance of Example 3.3.

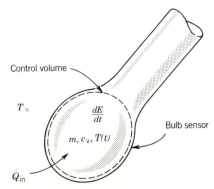

Control volume

T_∞

$\dfrac{dE}{dt}$

$m, c_V, T(t)$

Bulb sensor

$\dot{Q}_{in}$

Energy storage in the bulb is manifested by a change in bulb temperature so that for a constant mass bulb, $dE(t)/dt = mc\,dT(t)/dt$. Energy exchange by convection between the bulb at $T(t)$ and an environment at T_∞ has the form $\dot{Q} = hA\,\Delta T$. The first law can be written as

$$mc_v\,\frac{dT(t)}{dt} = hA_s[T_\infty - T(t)]$$

This equation can be written in the form

$$mc_v\,\frac{dT(t)}{dt} + hA_s[T(t) - T(0)] = hA_s[T_\infty - T(0)]U(t)$$

$$= hA_sF(t)$$

with initial condition $T(0)$ and where

m = mass of liquid within thermometer
c_v = specific heat of liquid within mass
h = convection heat transfer coefficient between bulb and environment
A_s = thermometer surface area

The term hA_s controls the rate at which energy can be transferred between a fluid and a body; it is analogous to an electrical conductance. By comparison with equation 3.3, $a_0 = hA_s$, $a_1 = mc_v$ and $b_0 = hA_s$. Rewriting for $t \geq 0^+$ and simplifying yields

$$\frac{mc_v}{hA_s}\,\frac{dT(t)}{dt} + T(t) = T_\infty$$

From equation 3.4, this implies

$$\tau = \frac{mc_v}{hA_s} \qquad K = \frac{hA_s}{hA_s} = 1$$

Solving (or direct comparison with equation 3.5) yields the thermometer response:

$$T(t) = T_\infty + [T(0) - T_\infty]e^{-t/\tau}$$

$$= 98 - 30e^{-t/\tau}\ [°F]$$

COMMENT

It is clear that the time constant, τ, of the thermometer can be reduced by decreasing its mass to area ratio or by increasing h (for example, this can be accomplished by stirring, shaking, or other methods that increase the velocity of the fluid around the sensor). Without modeling, such information could be ascertained only by trial and error, a time-consuming and costly method with no assurance of success. Also, it is significant that we found that the response of the temperature measurement system in this case will depend on the environmental conditions of the measurement that control h, since the magnitude of h affects the magnitude of τ.

Review of Example 3.3 should make it apparent that the results of a well-executed step calibration may not be indicative of an instrument's performance during a measurement if the measurement conditions differ from those existing during the step calibration.

EXAMPLE 3.4

For the thermometer of Example 3.3 subjected to a step change in input, calculate the rise time in terms of t/τ.

KNOWN

Same as Example 3.3

Sine

ASSUMPTIONS

Same as Example 3.3

FIND

90% response time in terms of t/τ

SOLUTION

The percent response of the system is given by $(1 - \Gamma) \times 100$ with the error fraction, Γ, defined by (3.6). From (3.8), we note that at $t = \tau$, the thermometer will indicate $T(\tau) = 79$ °F, which represents only 63.2% of the step change from 68 to 98 °F. The time required to reach 79 °F would be called the 63.2% response time of the measurement system. The rise time represents the time required for Γ to drop to a value of 0.10. Then

$$\Gamma = 0.10 = e^{-t/\tau}$$

or

$$\frac{t}{\tau} = 2.3$$

COMMENT

In general, a time equivalent to 2.3τ is required to achieve 90% of the applied step input for a first-order system.

EXAMPLE 3.5

A particular thermometer is subjected to a step change, such as in Example 3.3, in an experimental exercise to determine its time constant. The temperature

Equation 3.8 can be rewritten in a general form

$$y(t) = Ce^{-t/\tau} + B \sin[\omega t + \phi(\omega)] \qquad (3.9)$$

$$B = \frac{KA}{[1 + (\omega\tau)^2]^{1/2}}$$

$$\phi(\omega) = -\tan^{-1} \omega\tau$$

where B represents the amplitude of the steady response and the angle $\phi(\omega)$ is known as the *phase shift*. A relative illustration between the input signal and the system output response is given in Figure 3.10 for an arbitrary frequency and system time constant. From equation 3.9, it is seen that both B and ϕ are frequency dependent. Hence, the exact form of the output response will depend on the value of the frequency of the input signal. The steady response of any system to which a periodic input of frequency, ω, is applied is known as the *frequency response* of that system. The frequency dependence of a system affects the magnitude of amplitude B and brings about a time delay between when the input is applied and when the measuring system responds to the input signal. This time delay, β_1, is seen in the phase shift, $\phi(\omega)$, of the steady response. For a phase shift given in radians, the time delay in units of time is

$$\beta_1 = \frac{\phi(\omega)}{\omega}$$

Since equation 3.9 applies to all first-order measuring systems, the magnitude and phase shift by which the output signal differs from the input signal are

FIGURE 3.10   Relationship between sinusoidal input and output: amplitude, frequency, and time lag.

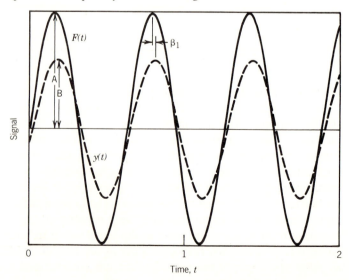

FIGURE 3.11   First-order system frequency response:
magnitude ratio.

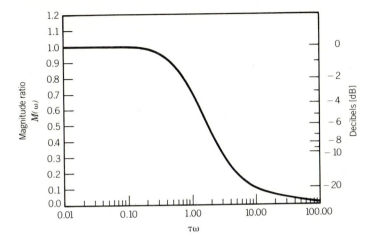

predictable. Here we define the magnitude ratio, $M(\omega)$, for a first-order system
subjected to a simple periodic input as

$$M(\omega) = \frac{B}{KA} = \frac{1}{[1 + (\omega\tau)^2]^{1/2}} \tag{3.10}$$

The magnitude ratio for a first-order system is plotted in Figure 3.11 and the
corresponding phase shift is plotted in Figure 3.12. The effects of both system
time constant and input signal frequency on frequency response are apparent in
both figures. This behavior can be interpreted in the following manner. For those
values of $\omega\tau$ for which the system can respond with values of $M(\omega)$ near unity,
the measurement system will transfer all or nearly all of the input signal amplitude
to the output and with very little time delay; that is, B will be close to KA in

FIGURE 3.12   First-order system frequency response:
phase shift.

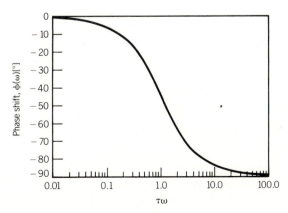

magnitude and $\phi(\omega)$ will be near $0°$. At large values of $\omega\tau$ the measurement system will essentially remove, that is, filter out, the frequency information of the input signal by responding with very small amplitudes, shown by small $M(\omega)$, and larger time delays.

Any combination of ω and τ will produce the same results. If one wants to measure signals with high frequency content and transfer the information about the signal to the output stage of the measurement system, then a system having a small τ will be necessary. A large τ system will result in the removal of the high-frequency information from the output signal. On the other hand, systems of large τ may be adequate to measure signals of very low-frequency content.

The *dynamic error*, $\delta(\omega)$, of a system can be defined as $\delta(\omega) = M(\omega) - 1$, and represents a measure of the inability of a system to adequately reconstruct the amplitude of the input signal for a particular input frequency. Measurement systems with a magnitude ratio at or near unity over the anticipated frequency band of the input signal are preferred to minimize $\delta(\omega)$. Perfect reproduction of the input signal is not possible, so some dynamic error is inevitable. For a first-order system, the *frequency bandwidth* is defined traditionally as the frequency band over which $M(\omega) \geq 0.707$; in terms of the decibel, defined as

$$dB = 20 \log M(\omega)$$

it is the band of frequencies within which $M(\omega)$ does not drop by more than -3 dB.

The functions $M(\omega)$ and $\phi(\omega)$ represent the frequency response of the measurement system to periodic inputs. These equations and universal curves are intended to be used for guidance in the selection of measurement systems and system components. They are not meant to be used to correct measured data. This can be seen by considering Figures 3.11 and 3.12. As $M(\omega)$ falls away from unity and as $\phi(\omega)$ falls away from zero, these curves take on rather steep slopes. In these regions of the curves, small errors in the prediction of τ and deviations of the real systems from ideal first-order behavior can lead to errors of unpredictable magnitude.

Determination of Frequency Response.

The frequency response of a measurement system is found by a dynamic calibration. In this case, the calibration would entail applying a simple periodic waveform of known amplitude to the system input stage and measuring the corresponding output stage amplitude and phase shift. In practice, however, developing a method to produce a periodic input signal in the form of a physical variable may demand considerable ingenuity and effort. Hence, in many situations an engineer will elect to rely on modeling to infer system frequency response behavior.

EXAMPLE 3.6

A temperature sensor is to be selected to measure temperature within a reaction vessel. It is suspected that the temperature will behave as a simple periodic waveform with a frequency somewhere between 1 and 5 Hz. Several size sensors are available, each with a known time constant. Based on time constant, select a suitable sensor, assuming that a dynamic error of $\pm 2\%$ is acceptable.

KNOWN

$1 \leq f \leq 5$ Hz $|\delta(\omega)| \leq 0.02$

ASSUMPTIONS

First-order system

$F(t) = A \sin \omega t$

FIND

τ

SOLUTION

A $|\delta(\omega)| \leq 0.02$ implies that a magnitude ratio of between $0.98 \leq M \leq 1.02$ is sought. From Figure 3.11, it is seen that first-order systems never exceed $M = 1$. So we need

$$0.98 \leq M(\omega) = \frac{1}{[1 + (\omega\tau)^2]^{1/2}}$$

From Figure 3.11, this constraint is maintained over the range $0 \leq \omega\tau \leq 0.2$. It is also seen in this figure that for a system of fixed time constant, the smallest value of $M(\omega)$ will occur at the largest frequency. So setting $\omega = 2\pi(5)$ rad/s and solving for $M(\omega) = 0.98$ yields, $\tau \leq 6.5$ ms. Accordingly, a sensor with a time constant of 6.5 ms or less will be suitable.

Second-Order Systems

Systems that possess inertia will contain a second-derivative term in their model equation (see Example 3.1). A system that is modeled by a second-order differential equation is called a second-order system. Examples of second-order measurement systems include accelerometers and diaphragm pressure transducers (including microphones).

In general, a second-order measurement system subjected to an arbitrary input, $F(t)$, can be described by an equation of the form

$$a_2\ddot{y} + a_1\dot{y} + a_0y = F(t) \tag{3.11}$$

where a_0, a_1, and a_2 are physical parameters used to describe the system and $y^n = d^ny/dt^n$. This equation can be rewritten as

$$\frac{1}{\omega_n^2}\ddot{y} + \frac{2\zeta}{\omega_n}\dot{y} + y = KF(t) \tag{3.12}$$

where

$$\omega_n = \sqrt{\frac{a_0}{a_2}} \equiv \text{natural frequency of the system}$$

$$\zeta = \frac{a_1}{2(a_0 a_2)^{1/2}} \equiv \text{damping ratio of the system}$$

Consider the homogeneous solution to (3.12). Independent of input signal, the homogeneous solution will describe the natural or intrinsic response of any system. Its form, therefore, depends only on the order of the equation and the system parameters a_0, a_1, and a_2. The behavior of (3.12) will depend on the roots of the characteristic equation to (3.12)

$$\frac{1}{\omega_n^2} \lambda^2 + \frac{2\zeta}{\omega_n} \lambda + 1 = 0$$

which yields the two roots,

$$\lambda_{1,2} = -\zeta\omega_n \pm \omega_n\sqrt{\zeta^2 - 1}$$

Depending on the value for ζ three forms of homogeneous solution are possible:

$0 < \zeta < 1$ (underdamped system solution)

$$y_h(t) = Ce^{-\zeta\omega_n t} \sin(\omega_n\sqrt{1 - \zeta^2}\, t + \Theta) \tag{3.13a}$$

$\zeta = 1$ (critically damped system solution)

$$y_h(t) = C_1 e^{\lambda_1 t} + C_2 t e^{\lambda_2 t} \tag{3.13b}$$

$\zeta > 1$ (overdamped system solution)

$$y_h(t) = C_1 e^{\lambda_1 t} + C_2 e^{\lambda_2 t} \tag{3.13c}$$

The homogeneous solution determines the transient response of a system. It is seen that for systems having $0 \leq \zeta \leq 1$, the transient response will be oscillatory, whereas for $\zeta \geq 1$, the transient response will not oscillate. The critically damped solution, $\zeta = 1$, denotes the demarcation between oscillatory and nonoscillatory behavior in the transient response. The damping ratio is a measure of system damping. Damping is a property of a system that enables the system to dissipate energy internally.

Step Function Input

Again, the step function input is applied to determine the general transient system behavior and speed at which the system will respond to a change in input.

The response of a second-order measurement system to a step function input is found from the solution of (3.12), with $F(t) = AU(t)$, to be

$0 \leq \zeta \leq 1$

$$y(t) = KA - KAe^{-\zeta\omega_n t}\left[\frac{\zeta}{(1 - \zeta^2)^{1/2}}\sin(\omega_n\sqrt{1 - \zeta^2}\,t) + \cos(\omega_n\sqrt{1 - \zeta^2}\,t)\right]$$

$$(3.14a)$$

$\zeta = 1$

$$y(t) = KA - KA(1 + \omega_n t)e^{-\omega_n t} \qquad\qquad (3.14b)$$

$\zeta > 1$

$$y(t) = KA - KA$$
$$\times\left[\frac{\zeta + \sqrt{\zeta^2 - 1}}{2\sqrt{\zeta^2 - 1}}e^{(-\zeta + \sqrt{\zeta^2 - 1})\omega_n t} + \frac{\zeta - \sqrt{\zeta^2 - 1}}{2\sqrt{\zeta^2 - 1}}e^{(-\zeta - \sqrt{\zeta^2 - 1})\omega_n t}\right]$$

$$(3.14c)$$

where the initial conditions, $y(0)$ and $\dot{y}(0)$, have been set to zero for convenience.

Equations 3.14a–c are plotted in Figure 3.13 for several values of ζ. The interesting feature is the transient response. For underdamped systems, the transient response is oscillatory about the steady value and occurs with a period

$$T_d = \frac{2\pi}{\omega_d}$$

$$\omega_d = \omega_n\sqrt{1 - \zeta^2}$$

where ω_d is called the *ringing frequency*. In instruments, this oscillatory behavior is called "ringing." The duration of the transient response is controlled by the

FIGURE 3.13   Second-order system response to a step function input.

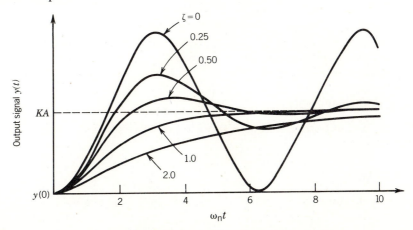

$\zeta\omega_n$ term. The system settles to KA more quickly when it is designed with a larger $\zeta\omega_n$. Nevertheless, for all systems with $\zeta > 0$, the response will eventually indicate the steady value of $y_\infty = KA$ as $t \to \infty$.

The time required for a second-order system to first achieve a value within 90% of the step input, $KA - y_0$, is defined as its *rise time*. Rise time can be decreased by decreasing the damping ratio, as seen in Figure 3.13. But the severe ringing associated with very lightly damped systems can delay the time to achieve a steady value relative to systems of higher damping. This is demonstrated by comparing the response at $\zeta = 0.3$ with the response at $\zeta = 1$ in Figure 3.13. The time required for a measurement system's oscillations to settle to within $\pm 10\%$ of the steady value, KA, is defined as the *settling time* or response time. The settling time is an approximate measure of the time to steady response. A damping ratio of about 0.7 appears to offer a good compromise between ringing and settling time. If an error fraction of a few percent is acceptable, then a system with $\zeta = 0.7$ will achieve steady response in about one-half the time of a system having $\zeta = 1$. For this reason, most measurement systems intended to measure sudden changes in input signal will typically be designed such that parameters a_0, a_1, and a_2 provide a damping ratio between 0.6 and 0.8.

Determination of Ringing Frequency and Rise and Settling Times. The experimental determination of the ringing frequency associated with underdamped systems is performed by applying a step input to the second-order measurement system and recording the response with time. This type of calibration will also yield information concerning the time to steady response of the system, which includes rise and settling times. Typically, measurement systems suitable for dynamic signal measurements are rated according to their 90% rise time and settling time.

EXAMPLE 3.7

Determine the physical parameters that affect the natural frequency and damping ratio of the accelerometer of Example 3.1.

KNOWN

Accelerometer shown in Figure 3.2

ASSUMPTIONS

Second-order system as modeled in Example 3.1

FIND

ω_n and ζ

SOLUTION

Comparison between the governing equation for the accelerometer in Example 3.1 and equation 3.12 suggest

$$\omega_n = \sqrt{\frac{k}{m}} \quad \zeta = \frac{c}{2(km)^{1/2}}$$

Accordingly, the physical parameters of mass, spring stiffness, and frictional damping control the natural frequency and damping ratio of this measurement system.

EXAMPLE 3.8

The curve given in Figure 3.14 was reproduced from an oscilloscope trace of the voltage signal output of a diaphragm pressure transducer subjected to a step change in input. During a static calibration using a pressure standard, it is found that the pressure–voltage relationship was linear over the range 1 atmosphere (atm) to 4 atm and that the static sensitivity of the measurement system was 1 V/atm. For the step test, it is known that the initial pressure was atmospheric

---

FIGURE 3.14   Pressure transducer response to a step input for Example 3.8.

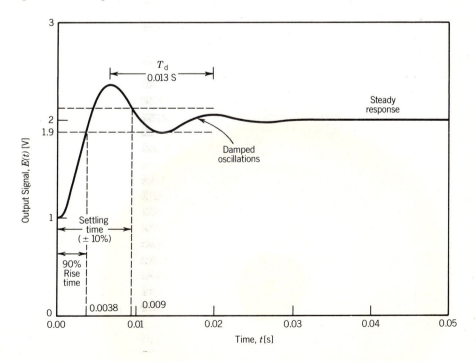

pressure, p_a, and the final pressure was $2p_a$. Estimate the 90% rise time, the 90% settling time, and the ringing frequency of the measurement system.

KNOWN

$p(0) = 1$ atm
$p_\infty = 2$ atm
$K = 1$ V/atm

ASSUMPTIONS

Oscilloscope does not affect transducer performance
Second-order system

FIND

90% rise and settling times; ω_d

SOLUTION

The ringing behavior of the system noted on the oscilloscope trace supports the assumption that the transducer can be described as having a second-order behavior. From the given information,

$$E(0) = Kp(0) = 1 \text{ V}$$

$$E_\infty = Kp_\infty = 2 \text{ V}$$

so that the step change observed on the oscilloscope trace should appear as a magnitude of 1 V. The 90% rise time will occur when the output first achieves a value of 1.9 V. The 90% settling time will occur when the output from the measurement system settles between $1.9 \text{ V} \leq E(t) \leq 2.1 \text{ V}$. From Figure 3.14, the 90% rise occurs in about 4 ms and the 90% settling time in about 9 ms. The period of the ringing behavior, T_d, is judged to be about 13 ms for an $\omega_d \approx 485$ rad/s.

Sine Function Input

The response of a second-order system to a sinusoidal input of the form, $F(t) = A \sin \omega t$, is given by

$$y(t) = y_h + \frac{KA \sin[\omega t + \phi(\omega)]}{\{(1 - (\omega/\omega_n)^2]^2 + (2\zeta \, \omega/\omega_n)^2\}^{1/2}} \qquad (3.15)$$

with frequency-dependent phase shift

$$\phi(\omega) = -\tan^{-1} \frac{2\zeta \, \omega/\omega_n}{1 - (\omega/\omega_n)^2} \qquad (3.16)$$

The exact form for y_h is found from equations 3.13a–c and depends on the value of ζ. The steady response, the second term on the right side, has the general form

$$y_{\text{steady}}(t) = B \sin[\omega t + \phi(\omega)] \tag{3.17}$$

with amplitude B. Comparison of equations 3.15 and 3.17 indicates that the amplitude of the steady response of a second-order system subjected to a sinusoidal input is dependent on the value of ω. Thus, the amplitude of the output signal from a second-order measurement system is frequency dependent. In general, we can define the magnitude ratio, $M(\omega)$, as

$$M(\omega) = \frac{B}{KA} = \frac{1}{\{[1 - (\omega/\omega_n)^2]^2 + (2\zeta\,\omega/\omega_n)^2\}^{1/2}} \tag{3.18}$$

So that the output signal of a second-order measurement system that is measuring a sine wave input may contain both a dynamic error and a phase shift.

The magnitude ratio for a second-order system is plotted in Figure 3.15 for several values of damping ratio. A corresponding plot of the phase-shift dependency on input frequency and damping ratio is shown in Figure 3.16. For an ideal measurement system, $M(\omega)$ would equal unity and $\phi(\omega)$ would equal zero for all values of frequency. Instead, real systems behave in a nearly ideal manner over only a portion of the frequency curve. In general, $M(\omega)$ will approach zero and $\phi(\omega)$ will approach $-\pi$ as ω/ω_n becomes large. Keep in mind that ω_n is a property of the measurement system, while ω is a property of the input signal.

FIGURE 3.15   Second-order system frequency response: magnitude ratio.

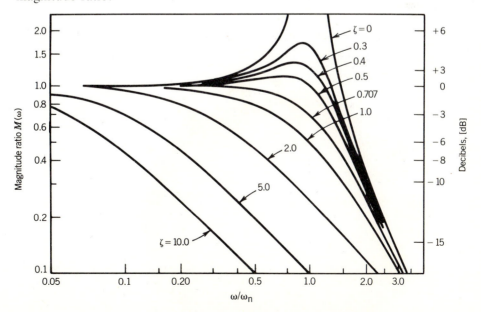

Several tendencies are apparent in Figures 3.15 and 3.16. For a system of zero damping, $\zeta = 0$, $M(\omega)$ will approach infinity and $\phi(\omega)$ jumps to $-\pi$ in the vicinity of $\omega = \omega_n$. This behavior is characteristic of system resonance. Real systems possess some amount of damping, which modifies the abruptness and magnitude of resonance, but underdamped systems may still achieve resonance. This region on Figures 3.15 and 3.16 is called the *resonance band* of the system, referring to the range of frequencies over which the system is in resonance. Resonance in underdamped systems occurs at the *resonance frequency*, $\omega_R = \omega_n \sqrt{1 - 2\zeta^2}$. Resonance behavior results in values of $M(\omega) > 1$ and considerable phase shift. For most applications, operation at frequencies within the resonance band is undesirable and could even be damaging to some delicate sensors. Resonance behavior is very nonlinear and results in distortion of the signal. On the other hand, systems having $\zeta > 0.707$ do not resonate.

At low values of ω/ω_n, $M(\omega)$ remains near unity and $\phi(\omega)$ near zero. This indicates that information concerning the input signal of frequency, ω, will be passed through to the output signal with little alteration. This region on the frequency response curves is called the *transmission band*. The actual extent of the frequency range for near unity gain depends on the system damping ratio. The transmission band of a system is either specified by its frequency bandwidth, typically defined for a second-order system as $3\,\text{dB} \geq M(\omega) \geq -3\,\text{dB}$, or specified explicitly by some other range of dynamic error. Operation of a measurement system within its transmission band is preferred when it is desirable to measure the dynamic content of the input signal.

FIGURE 3.16   Second-order system frequency response: phase shift.

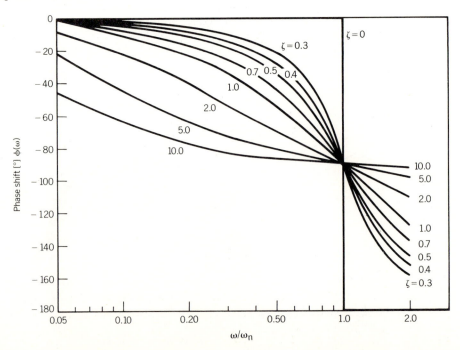

At large values of ω/ω_n, $M(\omega)$ approaches zero. In this region, the measurement system will attenuate amplitude information in the input signal. This region is known as the *filter band*, typically defined as the frequency range over which $M(\omega) \leq -3$ dB. Most readers are familiar with the use of a filter to remove undesirable features from desirable product. Operation within this band is useful to remove amplitude information at frequencies within the filter band.

EXAMPLE 3.9

Determine the frequency response of a pressure transducer that has a damping ratio of 0.5 and a ringing frequency (found by a step test) of 1200 Hz.

KNOWN

$\zeta = 0.5$

$\omega_d = 2\pi(1200 \text{ Hz}) = 7540 \text{ rad/s}$

ASSUMPTIONS

Second-order system

FIND

$M(\omega)$ and $\phi(\omega)$

SOLUTION

The frequency response of a measurement system is given by $M(\omega)$ and $\phi(\omega)$ as defined in equations 3.16 and 3.18. Since $\omega_d = \omega_n\sqrt{1 - \zeta^2}$, the natural frequency of the pressure transducer is known to be $\omega_n = 8706$ rad/s. The frequency response at selected frequencies is computed from (3.16) and (3.18):

ω [rad/s]	$M(\omega)$	$\phi(\omega)$ [°]
500	1.00	-3.3
2 600	1.04	-18.2
3 500	1.07	-25.6
6 155	1.15	-54.7
7 540	1.11	-73.9
8 706	1.00	-90.0
50 000	0.05	-170.2

COMMENT

Note the resonance behavior in the transducer response which peaks at $\omega_R = 6155$ rad/s. Resonance effects can be minimized by operating a measurement system at input frequencies less than about 30% of the system's natural frequency.

EXAMPLE 3.10

An accelerometer is to be selected to measure a time-dependent motion. In particular, input signal frequencies below 100 Hz are of prime interest. Select a set of acceptable parameter specifications for the instrument assuming a dynamic error of ±5%.

KNOWN

$f \leq 100$ Hz (i.e., $\omega \leq 628$ rad/s)

ASSUMPTIONS

Second-order system
Dynamic error of ± 5% acceptable

FIND

Select ω_n and ζ

SOLUTION

To meet a ±5% dynamic error constraint, we will want $0.95 \leq M(\omega) \leq 1.05$ over the frequency range $0 \leq \omega \leq 628$ rad/s. This problem is certainly open ended in that a number of instruments with different ω_n and will do the task. So as one solution, let us set $\zeta = 0.7$ and then solve for the required ω_n using equation 3.18:

$$1.05 \geq M(\omega) = \frac{1}{\{[1 - (\omega/\omega_n)^2]^2 + [2\zeta (\omega/\omega_n)]^2\}^{1/2}}$$

$$0.95 \leq M(\omega) = \frac{1}{\{[1 - (\omega/\omega_n)^2]^2 + [2\zeta(\omega/\omega_n)]^2\}^{1/2}}$$

With $\omega = 628$ rad/s, these equations yield $\omega_n \geq 1047$ rad/s. Alternatively, we could plot (3.18) with $\zeta = 0.7$, as shown in Figure 3.17. In Figure 3.17, we find that $0.95 \leq M(\omega) \leq 1.05$ over about the frequency range $0 \leq \omega/\omega_n \leq 0.6$. Again, this puts $\omega_n \geq 1047$ rad/s as being acceptable. So as one solution, an instrument having $\zeta = 0.7$ and $\omega_n \geq 1047$ rad/s will meet the problem constraints.

3.4 TRANSFER FUNCTIONS

Consider the schematic representation in Figure 3.18. The measurement system operates on the input signal, $F(t)$, by some function, $G(s)$, so as to indicate the output signal, $y(t)$. This operation can be explored by taking the Laplace

FIGURE 3.17   Frequency response at $\zeta = 0.7$ for
Example 3.10.

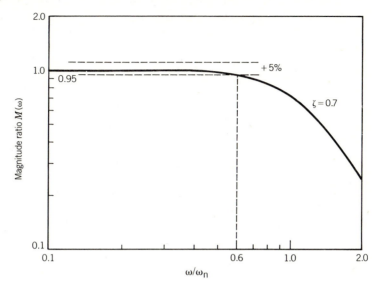

transform of both sides of the differential equation, (3.4), which describes the
general first-order measurement system. One obtains

$$Y(s) = K \frac{1}{\tau s + 1} F(s) + \frac{y_0}{\tau s + 1}$$

where $y_0 = y(0)$. This can be rewritten as

$$Y(s) = KG(s)F(s) + Q(s)G(s) \tag{3.19}$$

where $G(s)$ is the transfer function of the first-order system given by

$$G(s) = \frac{1}{\tau s + 1} \tag{3.20}$$

and $Q(s)$ is the system initial state function. The first term on the right side of
equation 3.19 contains the information that describes the steady response of the
measurement system to the input signal, whereas the second term describes its

FIGURE 3.18   Operation of the transfer function.

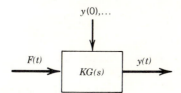

transient response. As indicated in Figure 3.18, the transfer function defines the mathematical operation that the measurement system performs on $F(t)$ to yield the time response of the system.

The system frequency response, which has been shown to be given by $M(\omega)$ and $\phi(\omega)$, can be found by finding the value of $G(s)$ at $s = i\omega$. This yields the complex number,

$$G(s = i\omega) = \frac{1}{\tau i\omega + 1}$$

$$= M(\omega)\, e^{i\phi(\omega)}$$

where $G(i\omega)$ is a vector on the real–imaginary plane having a magnitude, $M(\omega)$, and inclined at an angle, $\phi(\omega)$, relative to the real axis as indicated in Figure 3.19. For the first-order system, the magnitude of $G(i\omega)$ is simply equation 3.10 and the phase shift angle is given by equation 3.9.

For a second-order system, the governing equation is defined by equation 3.12, with initial conditions $y(0) = y_0$ and $\dot{y}(0) = \dot{y}_0$. The Laplace transform yields

$$Y(s) = \frac{1}{(1/\omega_n^2)\, s^2 + (2\zeta/\omega_n)\, s + 1}\, KF(s) + \frac{s\dot{y}_0 + y_0}{(1/\omega_n^2)\, s^2 + (2\zeta/\omega_n)\, s + 1}$$

$$= KG(s)F(s) + Q(s)G(s)$$

By inspection, the transfer function is given by

$$G(s) = \frac{1}{(1/\omega_n^2)\, s^2 + (2\zeta/\omega_n)s + 1}$$

FIGURE 3.19   Complex plane.

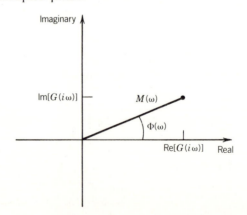

Solving for $G(s)$ at $s = i\omega$, one obtains

$$G(s = i\omega) = \frac{1}{(i\omega)^2/\omega_n^2 + 2\zeta i\omega/\omega_n + 1}$$

$$= M(\omega)\, e^{i\phi(\omega)}$$

(3.21)

which yields the same magnitude ratio and phase shift relations as given by equations 3.16 and 3.18.

3.5 PHASE LINEARITY

It can be noted from Figures 3.15 and 3.16 that systems having a damping ratio near 0.7 possess the broadest frequency range over which $M(\omega)$ will remain at or near unity and that over this same frequency range the phase shift will essentially vary in a linear manner with frequency. While it is not possible to design a measurement system without accepting some amount of phase shift, it is desirable to design a system such that the phase shift varies linearly with frequency. This is because measurement system operation in the regions of Figure 3.16 where phase shift varies nonlinearly with frequency will often bring about a significant distortion in the waveform of the output signal relative to the input signal. Distortion refers to a notable change in the shape of the waveform, as opposed to simply an amplitude alteration or relative phase shift. For these reasons, many measurement systems are designed with $0.6 \leq \zeta \leq 0.8$.

The distortion of a signal brought on by measurement system operation in the region where phase shift and frequency vary nonlinearly can be shown by considering a particular complex waveform represented by a general function, $u(t)$:

$$u(t) = \sum_{n=1}^{\infty} \sin n\omega t = \sin \omega t + \sin 2\omega t + \cdots$$

Suppose during measurement a phase shift of this signal were to occur such that the phase shift remained linearly proportional to the frequency; that is, the measured signal, $v(t)$, could be represented by

$$v(t) = \sin(\omega t - \phi) + \sin(2\omega t - 2\phi) + \cdots$$

Or, by setting

$$\theta = (\omega t - \phi)$$

we can obtain

$$v(t) = \sin \theta + \sin 2\theta + \cdots$$

which is equivalent to the original signal, $u(t)$. If the phase shift were not linearly related to the frequency, an output signal different in form from the input signal would result. This is demonstrated in Example 3.11.

EXAMPLE 3.11

Consider the effect of the variations in phase shift with frequency on a measured signal by examination of the signal defined by the function

$$u(t) = \sin t + \sin 5t$$

Suppose this signal is measured in such a way that a phase shift that is linearly proportional to the frequency occurs in the form

$$v(t) = \sin(t - 0.35) + \sin[5t - 5(0.35)]$$

Both $u(t)$ and $v(t)$ are plotted in Figure 3.20. It is apparent that the two waveforms have remained identical except that $v(t)$ lags $u(t)$ by some time increment. However, suppose $u(t)$ were measured in such a way that a nonlinear phase shift occurred, that is, the relation between phase shift and frequency was nonlinear, such as in the signal output form

$$w(t) = \sin(t - 0.35) + \sin(5t - 2)$$

FIGURE 3.20   Waveforms for Example 3.11.

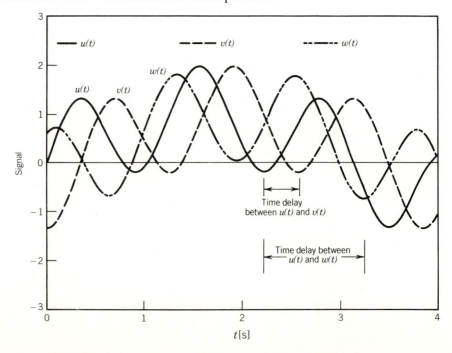

This is also plotted in Figure 3.20, which can be seen to produce a differently behaving waveform than $u(t)$. Such a difference is signal distortion. In comparing $u(t)$, $v(t)$, and $w(t)$, it should be apparent that distortion can be brought about by the nonlinear relation of phase shift with frequency.

3.6 MULTIPLE-FUNCTION INPUTS

So far we have discussed the response of a linear system to a signal containing only a single frequency. What about the response of a measurement system to multiple input frequencies? Or to an input that consists of both a static and a dynamic part, such as a periodic pressure, strain, or temperature signal? When using models that are linear, ordinary differential equations subjected to inputs that are linear in terms of the dependent variable, the principle of superposition of linear systems will apply to the solution of these equations. The *principle of superposition* states that a linear combination of input signals applied to a linear measurement system produces an output signal that is simply the linear addition of the separate output signals that would result if each input term had been applied separately. Multiple inputs do not change the form of the transient response (only its coefficients) so that, in general, we can write that if the forcing function of a form

$$F(t) = A_0 + \sum_{i=1}^{\infty} (A_i \sin \omega_i t) \tag{3.22}$$

is applied to a system, then the combined steady response will have the form

$$KA_0 + \sum_{n=1}^{\infty} B(\omega_i) \sin[\omega_i t + \phi(\omega_i)] \tag{3.23}$$

where $B(\omega_i) = KA_i M(\omega_i)$. The development of the superposition principle can be found in basic texts on dynamic systems (e.g., [4]).

EXAMPLE 3.12

A second-order instrument having a $K = 1$ unit/unit, $\zeta = 2$, and $\omega_n = 628$ rad/s is to be used to measure an input signal of the form

$$F(t) = 5 + 10 \sin 100t + 20 \sin 400t$$

Predict its steady output signal.

KNOWN

Second-order system
$K = 1$ unit/unit
$\zeta = 2.0$

ω_n = 628 rad/s

$F(t)$ = 5 + 10 sin 100t + sin 400t

ASSUMPTIONS

Linear system (superposition holds)

FIND

$y(t)$

SOLUTION

Since $F(t)$ has a form consisting of a linear addition of multiple input functions, the steady response signal will have the form of equation 3.22:

$$y(t) = 5K + 10KM(100 \text{ rad/s}) \sin[100t + \phi(100 \text{ rad/s})]$$
$$+ 20KM(400 \text{ rad/s}) \sin[400t + \phi(400 \text{ rad/s})]$$

Using equations 3.16 and 3.18, or, alternatively, Figures 3.15 and 3.16, with ω_n = 628 rad/s and ζ = 2.0, we find that

$$M(100 \text{ rad/s}) = 1.00 \qquad \phi(100 \text{ rad/s}) = -13°$$
$$M(400 \text{ rad/s}) = 0.39 \qquad \phi(400 \text{ rad/s}) = -56°$$

So that the output signal will have the form

$$y(t) = 5 + 10 \sin(100t - 13°) + 7.8 \sin(400t - 56°)$$

It can be seen that information concerning the average value and concerning 100-rad/s component of the input signal will be passed along to the output signal. However, information concerning the 400-rad/s component of the input signal has been severely reduced (down 61%) in the output signal. The output signal is plotted against the input signal in Figure 3.21. If information concerning the 400-rad/s component is important, then it would be better to select an instrument that has an improved frequency response in the 400-rad/s range.

3.7   COUPLED SYSTEMS

When a measurement system consists of more than one instrument, measurement system behavior can become more complicated. As instruments in each stage of the system are connected, the output from one stage becomes the input to the next stage and so forth. Such measurement systems will have an output response to the original input signal that is some combination of the individual instrument responses to the input. However, the system concepts of first- and second-order systems studied previously can be used for a case by case study of the coupled measurement system. This is done by considering the input to each stage of the measurement system as the output of the previous stage.

FIGURE 3.21   Input signal and output signal for
Example 3.12.

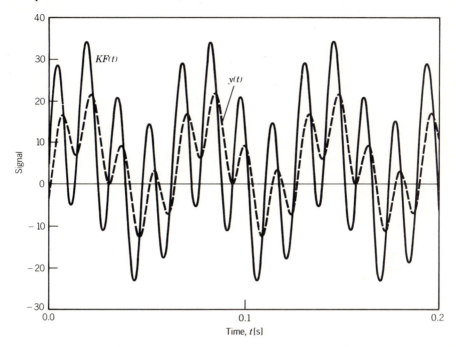

This concept is easily illustrated by considering a first-order sensor that may
be connected to a second-order output device (for example, a temperature
sensor–transducer connected to a strip chart recorder). This combination actually
forms a third-order measurement system (i.e., order one plus order two). Suppose the input to the sensor is a simple periodic waveform, $F(t) = A \sin \omega t$.
The transducer will respond with the output signal of the form of (3.8)

$$y_t(t) = Ce^{-t/\tau} + \frac{K_t A}{[1 + (\omega\tau)^2]^{1/2}} \sin(\omega t + \phi_t)$$

$$\phi_t = -\tan^{-1}(\omega\tau)$$

where subscript t refers to the transducer. However, the transducer output signal
now becomes the input signal, $F_2(t) = y_t$, to the second-order device. The output
from the second-order device, $y_s(t)$, will be a second-order response to the input
$F_2(t)$,

$$y_s(t) = y_h(t) + \frac{K_t K_s A \sin(\omega t + \phi_t + \phi_s)}{[1 + (\omega\tau)^2]^{1/2} \{[1 - (\omega/\omega_n)^2]^2 + (2\zeta \omega/\omega_n)^2\}^{1/2}} \quad (3.24)$$

$$\phi_s = -\tan^{-1} \frac{2\zeta \omega/\omega_n}{1 - (\omega/\omega_n)^2}$$

where subscript s refers to the strip chart recorder and $y_h(t)$ is the transient
response. The output signal displayed on the recorder, $y_s(t)$, is the measurement

system response to the original input signal to the transducer, $F(t) = A \sin \omega t$. The amplitude of the recorder output signal is the product of the static sensitivities and magnitude ratios of a first- and second-order system. The phase shift is the sum of the phase shifts of the two systems.

Based on (3.24) we can make a general observation. Consider the schematic representation in Figure 3.22, which depicts a measurement system consisting of H interconnected devices, $j = 1, 2, \ldots, H$, each device described by a linear system model. The overall transfer function of the combined system, $G(s)$, will be the product of the transfer functions of each of the individual devices, $G_j(s)$, such that

$$KG(s) = K_1 G_1(s) K_2 G_2(s) \cdots K_H G_H(s) \tag{3.25}$$

At $s = i\omega$, (3.25) gives

$$KG(i\omega) = (K_1 K_2 \cdots K_H)$$
$$\times [M_1(\omega) M_2(\omega) \cdots M_H(\omega)] \, e^{i[\phi_1(\omega) + \phi_2(\omega) + \cdots + \phi_H(\omega)]}$$

According to (3.25), given an input signal to device 1, the system steady output signal at device H will be described by the system frequency response $G(s = i\omega) = M(\omega)e^{i\phi(\omega)}$, with an overall system static sensitivity described by $K = K_1 K_2 \cdots K_H$. The overall system magnitude ratio will be the product $M(\omega) = M_1(\omega) M_2(\omega) \cdots M_H(\omega)$ and the overall system phase shift will be the sum $\phi(\omega) = \phi_1(\omega) + \phi_2(\omega) + \cdots + \phi_H(\omega)$. This will hold true provided that there do not exist any significant loading effects, a situation discussed in Chapter 6.

FIGURE 3.22   Coupled systems: system transfer function.

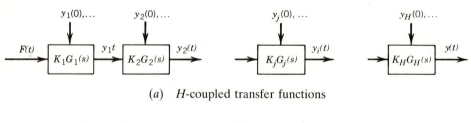

(a)   H-coupled transfer functions

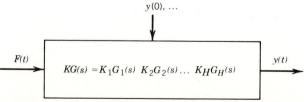

(b)   Equivalent system transfer function

3.8 SUMMARY

The response of a measurement system to a time-dependent input depends on several factors including the inherent speed of response of that system and the frequency response of that system. Modeling has enabled us to develop and to illustrate these concepts. Those system design parameters that affect system response are exposed through modeling and this assists in instrument selection. Modeling has also suggested the methods by which measurement system specifications such as time constant, response time, frequency response, damping ratio and resonance frequency can be determined both analytically and experimentally. Interpretation of these system properties and their effect on system performance was determined.

The speed of response of a system to a change in input is estimated by use of the step function input. The system parameters of time constant, for first-order systems, and natural frequency and damping ratio, for second-order systems, are used as indicators of system response speed. The magnitude ratio and phase shift define the frequency response of any system and are found by an input of a periodic waveform to a system. Figures 3.11, 3.12, 3.15, and 3.16 are universal frequency response curves for first- and second-order systems, respectively. These curves can be found in most engineering and mathematical handbooks and can be applied to any first- or second-order system, as appropriate.

REFERENCES

1. Close, C. M., and D. K. Frederick, *Modeling and Analysis of Dynamic Systems*, Houghton Mifflin, Boston, 1978.
2. Doebelin, E. O., *System Modeling and Response, Theoretical and Experimental Approaches*, Wiley, New York, 1980.
3. Ogata, K., *System Dynamics*, Prentice-Hall, Englewood Cliffs, NJ, 1978.
4. Palm, W. J., III, *Modeling, Analysis and Control of Dynamic Systems*, Wiley, New York, 1983.

SUGGESTED READING

Raven, F., *Automatic Control Engineering*, 2d ed., McGraw-Hill, New York, 1968.

NOMENCLATURE

$a_0, a_1, \ldots, a_n$	physical coefficients	f	cyclical frequency [Hz]
$b_0, b_1, \ldots, b_m$	physical coefficients	k	spring constant or stiffness $[m\text{-}t^{-2}]$
c	damping coefficient $[m\text{-}t\text{-}l]$	h	heat transfer coefficient $[m\text{-}l\text{-}t^{-3}/°]$

m	mass $[m]$	$G(s)$	transfer function
$p(t)$	pressure $[m\text{-}l^{-1}\text{-}t^{-2}]$	K	static sensitivity
t	time $[t]$	$M(\omega)$	magnitude ratio, B/KA
$x(t)$	independent variable	$T(t)$	temperature $[°]$
$y(t)$	dependent variable	T_d	ringing period $[t]$
y^n	nth time derivative of $y(t)$	$U(t)$	unit step function
		β_1	time lag $[t]$
$y^n(0)$	initial condition of y^n	$\delta(\omega)$	dynamic error
		τ	time constant $[t]$
A	input signal amplitude or magnitude	$\phi(\omega)$	phase shift
		ω	circular frequency $[t^{-1}]$
B	output signal amplitude	ω_n	natural frequency $[t^{-1}]$
C	constant	ω_d	ringing frequency $[t^{-1}]$
$E(t)$	energy $[m\text{-}l\text{-}t^{-3}]$ or voltage $[V]$	ω_R	resonance frequency $[t^{-1}]$
F	force $[m\text{-}l\text{-}t^{-2}]$	ζ	damping ratio
$F(t)$	forcing function	Γ	error fraction

Subscripts

0   initial value
h   homogeneous solution

∞ final or steady value

PROBLEMS

3.1 For each of the following models of measurement systems, state the time constant, resonance frequency, and damping ratio, as appropriate:
   a. $10\dot{T} + 5T = 5F(t)$
   b. $2\ddot{P} + 2\dot{P} + 3P = \sin 4t$
   c. $0.15\dot{y} + 14y = 6U(t)$
   d. $10y = 5F(t); F(t) = 2$
   e. $\ddot{z} + 4\dot{z} + 2z = 1 + 4 \sin 2t$
   f. $\dot{y} + 10y = 2 \sin 6t$

3.2 Determine the 75, 90, and 95% response time for each of the systems given (assume zero initial conditions):
   a. $0.4\dot{T} + T = 4U(t)$
   b. $\ddot{y} + 2\dot{y} + 4y = U(t)$
   c. $2\ddot{P} + 8\dot{P} + 8P = 2U(t)$
   d. $5\dot{y} + 5y = U(t)$

3.3 A measurement system is found to have a static sensitivity of 4 V/kg. If an input of 10 kg is measured, determine the expected value of the output signal. What would be the significance of increasing the static sensitivity?

3.4 A special sensor is designed to sense the percent vapor present in a liquid–vapor mixture. If during a static calibration the sensor indicates 80 units when in contact with 100% liquid, 0 units with 100% vapor, and 40 units with a 50–50% mixture, determine the static sensitivity of the sensor.

3.5 A system is modeled by the equation

$$0.5\dot{y} + y = F(t)$$

If the input signal is suddenly increased to 150 units at time $t = 0$ s and $y(0) = 100$ units.
a. Determine its response equation.
b. On the same graph, plot both the input signal and the system time response until steady response is reached.

3.6 Suppose a thermometer similar to that of Example 3.3 is known to have a time constant of 30 s in a particular application. Plot its time response to a step change from 32 to 120 °F. Determine its 90% rise time.

3.7 Determine the rise time of a first-order instrument having a time constant of 25 ms.

3.8 During a step function calibration, a first-order instrument is exposed to a step change of 100 units. If after 1.2 s the instrument indicates 80 units, estimate the instrument time constant. Estimate the error in the indicated value after 1.5 s. $y(0) = 0$ units.

3.9 Estimate the dynamic error that would result from measuring a 2-Hz periodic waveform using a first-order system with time constant of 0.7 s.

3.10 A first-order instrument with time constant of 1 s is used to measure a signal that can be represented by $F(t) = 10 \cos 2.5t$. Write the expected indicated steady response output signal. What is the expected time lag between input and output signal. $y(0) = 0$; $K = 1$

3.11 A first-order instrument with a time constant of 2 s is to be used to measure a periodic input. If a dynamic error of $\pm 2\%$ can be tolerated, determine the maximum frequency of a periodic input that can be measured. What is the associated time lag (in seconds) at this frequency?

3.12 For an instrument having a time constant of 0.01 s, determine its frequency response ($M(\omega)$ and $\phi(\omega)$). Plot your result. Determine the frequency range over which the dynamic error remains less than 10%. Indicate this range on your plot.

3.13 A temperature measuring device having a time constant of 0.15 s outputs a voltage that is linearly proportional to temperature. The device is used to measure an input signal of the form $T(t) = 115 + 12 \sin 2t$ °C. Plot the input signal and the predicted output signal with time assuming first-order behavior and a static sensitivity of 5 mV/°C. Determine the dynamic error and time lag in the steady response. $T(0) = 115$ °C

3.14 A first-order device is to be installed into a reactor vessel to monitor temperature. If a sudden rise in temperature greater than 100 °C should occur, shutdown of the reactor will need to begin within 5 s after reaching 100 °C. Determine the maximum allowable time constant for the device.

3.15 A single loop *LR* circuit having a resistance of 1 MΩ is to be used as a low-pass filter between an input signal and a voltage measurement device. To attenuate undesirable frequencies above 1000 Hz by at least 50% select a suitable inductor size if the time constant for this circuit is given by *L/R*.

3.16 Estimate the 90% rise time and settling time for each of the following systems if $F(t) = 2U(t)$ and all initial conditions are zero. Plot the response $y(t)$ and indicate transient and steady responses.
   a. $8\ddot{y} + 4\dot{y} + 2y = F(t)$
   b. $3\ddot{y} + 3\dot{y} + 3y = F(t)$
   c. $0.9\ddot{y} + 1.2\dot{y} + 0.4y = F(t)$
   d. $0.2\ddot{y} + 1.7\dot{y} + 0.9y = F(t)$

3.17 For an instrument having a damping ratio of 0.6, plot its frequency response using equations 3.16 and 3.18. Determine the frequency range over which the dynamic error remains within 5% and indicate this range on your plot. Repeat for a damping ratio of 0.9 and 2.0.

3.18 If an instrument has a known damping ratio of 0.8, ringing frequency of 1000 Hz and static sensitivity of 1.5 V/V, determine its expected steady response to an input signal that oscillates sinusoidally between 12 and 24 V at a frequency of 300 Hz.

3.19 An application demands that a sinusoidal pressure variation of 250 Hz be measured with no more than 2% dynamic error. In selecting a suitable pressure transducer from a vendor catalog, you note that a desirable line of transducers has a fixed natural frequency of 600 Hz but that you have a choice of transducer damping ratios of between 0.5 and 1.5 in increments of 0.05. Select a suitable transducer.

3.20 A record-playing turntable is to be isolated from room vibrations by placing it on an isolation pad. The isolation pad can be considered as a board of mass, *m*, a foam mat of stiffness, *k*, and a damping coefficient, *c*. For expected vibrations in the frequency range of between 2 and 40 Hz, select reasonable values for *m*, *k*, and *c* such that the room vibrations are attenuated by at least 50%. Assume that the only degree of freedom is in the vertical direction.

3.21 A single loop *R-C-L* electrical circuit can be modeled as a second-order system in terms of current. Show that the differential equation for such a circuit subjected to a forcing function potential $E(t)$ is given by

$$L\frac{d^2I}{dt^2} + R\frac{dI}{dt} + \frac{1}{C}I = \dot{E}(t)$$

Determine the natural frequency and damping ratio for this system. For a forcing potential, $E(t) = 1 + 0.5 \sin 2000t$ V, determine the system steady response when $L = 2$ H, $C = 1$ μF, and $R = 10000Ω$. Plot the steady output signal and input signal versus time. $I(0) = dI(0)/dt = 0$

3.22 A transducer that behaves as a second-order instrument has a damping ratio of 0.7 and a natural frequency of 1000 Hz. It is to be used to measure a signal containing frequencies as large as 750 Hz. If a dynamic error of $\pm 10\%$ can be tolerated, is this transducer a good choice?

3.23 A signal of the form $F(t) = \sin t + 0.3 \sin 20t$ lb is input to a first-order system that has a time constant of 0.2 s and a static sensitivity of 1 V/lb. Determine the steady response (steady output signal) from the system. Discuss the transfer of information from the input to the output. Can the input signal be resolved based on the output?

3.24 A force transducer having a damping ratio of 0.5 and a natural frequency of 4000 Hz is available for use to measure a periodic signal of 2000 Hz. Show that the transducer fails a $\pm 10\%$ dynamic error constraint. Estimate its resonance frequency.

3.25 An accelerometer, whose frequency response is defined by equations 3.16 and 3.18, has a damping ratio of 0.4 and a natural frequency of 18 000 Hz. It is used to sense the relative displacement of a beam to which it is attached. If an impact to the beam imparts a vibration at 4500 Hz, estimate the dynamic error and phase shift in the accelerometer output. Estimate its resonance frequency.

3.26 Derive the equation form for the magnitude ratio and phase shift of the seismic accelerometer of Example 3.1. Does its frequency response differ from that predicted by equations 3.16 and 3.18? For what type of measurement would you suppose this instrument would be best suited?

3.27 Suppose the pressure transducer of Example 3.9 had a damping ratio of 0.6. Plot its frequency response $M(\omega)$ and $\phi(\omega)$. At which frequency is $M(\omega)$ a maximum?

3.28 The output stage of a first-order transducer is to be connected to a second-order display stage device. The transducer has a known time constant of 1.4 ms and static sensitivity of 2 V/°C while the display device has values of sensitivity, damping ratio and natural frequency of 1 V/V, 0.9 and 5000 Hz, respectively. Determine the steady response of this measurement system to an input signal of the form, $10 + 50 \sin 628t$ °C.

3.29 The displacement of a solid body is to be monitored by a seismic transducer (second-order system) with signal output displayed on a recorder (second-order system). The displacement is expected to vary sinusoidally between 2 and 5 mm at a rate of 85 Hz. Select appropriate design specifications for the measurement system for no more than 5% dynamic error (i.e., specify an acceptable range for natural frequency and damping ratio for each device).

3.30 The displacement of a rail vehicle chassis as it rolls down a track is measured using a transducer ($K = 10$ mV/mm, $\omega_n = 10000$ rad/s, $\zeta = 0.6$) and a recorder ($K = 1$ mm/mV, $\omega_n = 700$ rad/s, $\zeta = 0.7$). The resulting amplitude spectrum of the output signal consists of a spike of 90 mV at 2 Hz and a spike of 50 mV at 40 Hz. Are the measurement system specifications suitable for the displacement signal? (If not, suggest changes.) Estimate the actual displacement of the chassis. State any assumptions.

3.31   It is expected that the amplitude spectrum of the time-varying displacement signal from a vibrating U-shaped tube will have a spike at 85, 147, 220, and 452 Hz. Each spike is related to an important vibrational mode of the tube. Displacement transducers available for this application have a range of natural frequencies from 500 Hz through 1000 Hz, with a fixed damping ratio of about 0.5. The transducer output is to be monitored on a DFT-based spectrum measurement device that has a frequency bandwidth extending from 0.1 to 250 kHz. Within the stated range of availability, select a suitable transducer for this measurement.

CHAPTER 4
PROBABILITY AND STATISTICS

4.1 INTRODUCTION

Engineering measurements repeatedly taken under seemingly identical conditions will normally show variations in measured values. Sources that contribute to these variations include:

Measurement System
- Resolution
- Repeatability

Measurement Procedure and Technique
- Repeatability

Measured Variable
- Temporal variation
- Spatial variation

For a given set of repeated measurements, one will want to be able to quantify, (1) some single representative value that best characterizes the average of the data set, as well as, (2) some representative value that provides a measure of the variation in the measured data set. Furthermore, an engineer will need to know how well this single average value represents the true average value of the measured variable. This is done by establishing, (3) an interval about the representative value in which the true value is expected to lie. Statistics and probability provide the tools needed to arrive at these three estimates.

This chapter presents an introduction to the concepts of probability and statistics at a level sufficient to provide information for a large class of engineering judgments. Such material allows for the reduction of raw data into results. Some of this material, such as the concepts of mean value and standard deviations, should already be familiar to the engineering student and are presented along with their relations with probability to provide for a correct interpretation of data.

4.2   STATISTICAL MEASUREMENT THEORY

A *sample* of data refers to a set of data obtained during repeated measurements of a variable under fixed operating conditions. This measured variable is known as the *measurand*. Fixed operating conditions imply that the external conditions that control the process from which the measurand is obtained are held at fixed values while obtaining the sample. In actual engineering practice, the ability to control the operating conditions at truly fixed conditions may be impossible and the term "fixed operating conditions" should be considered in a nominal sense. That is, the process conditions should be maintained as closely as possible.

In this chapter, we will consider only those situations where bias error is negligible.[1] Recall from Chapter 1 that this is the case where the average error in a data set is zero. We begin by considering the measurement problem of estimating the *true value*, x', based on the information derived from the repeated measurement of x. The true value of the measurand is the value we want to estimate from the measurement.

A sample of the variable x under controlled, fixed operating conditions renders a finite number of data points. We use this data to infer x'. We can imagine that if the number of data points is very small, then our estimation of x' from the data set could be heavily influenced by the value of any one data point. If this data point showed a large variation from x' relative to the other data, then the estimate could show a large error. If the data set were larger then the influence of any one data point would be offset by the larger influence of the other data. As $N \to \infty$, all the possible variations would be included in the data set. From a practical view, only finite sized data sets are possible and, hence, the measured data can provide only some estimate of the true value.

From a statistical analysis of the data set and an analysis of sources of error that influence this data, we can estimate x' as

$$x' = \bar{x} \pm u_x \, (P\%) \tag{4.1}$$

where $\bar{x}$ represents the most probable estimate of x' based on the available data and u_x the confidence interval or uncertainty in that estimate at some probability level, $P\%$. We seek methods that will estimate $\bar{x}$ and the precision in $\bar{x}$ based on the variation in the data set.

Probability Density Functions

Regardless of the care taken in obtaining a set of data for a measurement under identical conditions, random scatter in the data values will normally occur. As such, the measurand is also known as a *random variable*. If the variable is continuous in time or space, then it is said to be a continuous random variable. A variable represented only by discrete values while time is continuous, such as a discrete set of data points, is called a discrete random variable. During

---

[1] Bias error will not vary with repeated measurements and does not affect the statistics of the measurement. Bias is considered in Chapter 5.

TABLE 4.1   Sample of Variable x

i	x_i	i	x_i
1	0.98	11	1.02
2	1.07	12	1.26
3	0.86	13	1.08
4	1.16	14	1.02
5	0.96	15	0.94
6	0.68	16	1.11
7	1.34	17	0.99
8	1.04	18	0.78
9	1.21	19	1.06
10	0.86	20	0.96

repeated measurements of a variable under fixed operating conditions, each data point may tend to assume one particular value or lie within some interval about this value more often than not, when all the data are compared. This tendency toward one central value about which all the other values are scattered is known as a *central tendency* of a random variable.[2] *Probability* deals with the concept that a particular interval of values for a variable will be measured at some frequency relative to any other interval.

The central value and those values scattered about it can be determined from the probability density of the random variable. The frequency with which the random variable assumes a particular value or interval of values is described by its probability density. Consider a sample of the discrete random variable, x, shown in Table 4.1, which consists of N individual measurements, x_i, where $i = 1, 2, \ldots, N$, each measurement taken under identical test operating conditions. The measured values of this variable are plotted on a single axis as shown in Figure 4.1.

In Figure 4.1, there exists a region on the axis where the data points tend to clump; this region contains the central value. Such behavior is typical of most engineering measurands. We might expect that the true value of x is contained somewhere in this clump.

This description for variable x can be extended. Suppose we replot the data of Table 4.1. The abscissa will be divided between the maximum and minimum measured values of x into K small intervals of width, $\pm \delta x$. Let the number of times, n_j, that a measured value assumes a value within an interval $x \pm \delta x$ be plotted on the ordinate. For small N, K should be conveniently chosen but such that $n_j \geq 5$ for at least one interval. For $N > 40$, an estimate [1] of the number of intervals K required for a statistical analysis is found from

$$K = 1.87(N - 1)^{0.40} + 1 \tag{4.2}$$

The resulting plot is called a *histogram* of the variable. The histogram is just another way of viewing both the tendency and the density of a variable. If the ordinate were nondimensionalized by dividing n_j by the total number of measurements of the variable, N, a *frequency distribution* of the variable would

---

[2]Not all random variables display a central tendency; e.g., the value of a fair roll of the die could show values from 1 through 6 with equal probability of $\frac{1}{6}$.

FIGURE 4.1   Concept of density in reference to a
random variable (from Example 4.1).

result. For any value of the variable, the frequency, f_j, at which that value of
the variable occurred is found from its frequency distribution.

EXAMPLE 4.1

Compute the histogram and frequency distribution for the data of Table 4.1.

KNOWN

Data of Table 4.1

$N = 20$

ASSUMPTIONS

Fixed operating conditions

FIND

Histogram and frequency distribution

SOLUTION

To develop the histogram, compute a reasonable number of intervals for this
data set. For $N = 20$, a convenient estimate of K is found from (4.2) to be

$$K = 1.87(N - 1)^{0.40} + 1 = 7$$

Next, determine the maximum and minimum values of the data set and divide
this range into K intervals. For a minimum of 0.68 and a maximum of 1.34, an
interval increment of $\delta x = \pm 0.05$ centered at 1.00 is chosen.

j	Interval	n_j	$f_j = n_j/N$
1	$0.65 \leq x_i < 0.75$	1	0.05
2	$0.75 \leq x_i < 0.85$	1	0.05
3	$0.85 \leq x_i < 0.95$	3	0.15
4	$0.95 \leq x_i < 1.05$	7	0.35
5	$1.05 \leq x_i < 1.15$	4	0.20
6	$1.15 \leq x_i < 1.25$	2	0.10
7	$1.25 \leq x_i < 1.35$	2	0.10

Since at least one interval has an $n_j \geq 5$, the interval number is adequate.
The results are plotted in Figure 4.2. The plot displays a definite central tendency
at the maximum frequency of occurrence within the interval 0.95 to 1.05.

FIGURE 4.2   Histogram and frequency distribution for
data of Table 4.1.

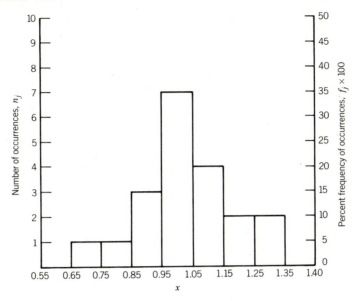

COMMENT

The sum of the number of occurrences, $\sum_{j=1}^{K} n_j$, must equal the total number
of measurements, N. Likewise, the area under the percent frequency distribution
curve will always equal the total frequency of occurrence of 100%, that is,
$100 \times \sum_{j=1}^{K} f_j = 100\%$.

The *probability density function*, $p(x)$, results from the frequency distribution,
in the limit as $N \to \infty$ and $\delta x \to 0$, by

$$p(x) = \lim_{N \to \infty, \delta x \to 0} \sum_{j=1}^{K} \frac{n_j}{N \delta x} \qquad (4.3)$$

The probability density function relates the measurand to its probability of
occurrence. The probability density function defines the probability that a measured variable might assume a particular value on any individual measurement
and it also defines the central tendency of the variable. This central tendency is
the desired representative value which gives the best estimate of the true value.
   The actual shape that the probability density function will assume will depend
on the random variable it represents and the circumstances affecting the process
from which the random variable is obtained. There are a number of standard
distribution shapes that suggest how a random variable will be distributed on
the probability density plot. The specific values of the random variable and the

width of the distribution depend on the actual process but the overall shape of the plot will most likely fit some standard distribution. A number of standard distributions that engineering data are likely to follow along with specific comments regarding the types of processes from which these are likely to be found are given in Table 4.2. Generally, the experimentally determined histograms are used to determine which standard distribution the random variable tends to follow. In turn, the standard distribution is used to interpret the data. Of course, the list in Table 4.2 is not all-inclusive and the reader is referred to more complete treatments of this subject [2, 4].

Regardless of the type of distribution assumed by a random variable, a variable that shows a central tendency can be described and quantified through its mean value and variance. The mean value or central tendency of a continuous random variable, $x(t)$, having a probability density function $p(x)$, is given by

$$x' = \lim_{T \to \infty} \frac{1}{T} \int_0^T x(t)\, dt \tag{4.4}$$

If the random variable is described by discrete data, the mean value of random variable, x_i, where $i = 1, 2, \ldots, N$, is given by

$$x' = \lim_{N \to \infty} \frac{1}{N} \sum_{i=1}^N x_i \tag{4.5}$$

Physically, the width of the density function reflects the data variation. For a continuous random variable, the variance is given by

$$\sigma^2 = \lim_{T \to \infty} \frac{1}{T} \int_0^T [x(t) - x']^2\, dt \tag{4.6}$$

or for discrete data, the variance is given by

$$\sigma^2 = \lim_{N \to \infty} \frac{1}{N} \sum_{i=1}^N (x_i - x')^2 \tag{4.7}$$

The standard deviation, σ, a commonly used statistical parameter, is defined as the square root of the variance.

A fundamental difficulty arises in the definitions given by equations 4.3–4.7 in that they assume an infinite number of measurements. Certainly no measured data set can be infinite. Real data set sizes may range from as few as one to some large but finite number. For now we will study "infinite statistics" to introduce the connection between probability and statistics. Then we will turn our attention to the practical treatment of finite data sets.

4.3   INFINITE STATISTICS

A common distribution found in measurements is the normal or Gaussian distribution.[3] Many measurands common to engineering measurements are de-

[3]Actually, this distribution was independently suggested in the 18th century by Gauss, Laplace, and DiMoivre. However, fate is fickle and Gauss retains the honor.

scribed by this distribution which predicts that the scatter seen in a measured data set will be distributed symmetrically about some central tendency. The measurands of length, temperature, pressure, and velocity will likely display such a distribution, provided the measurements are made repeatedly while holding the operating conditions fixed. Scatter brought about by process unsteadiness or by sources of precision error are "normally" distributed. The shape of the normal distribution is the familiar bell curve.

The probability density function for a random variable, x, having a normal distribution is defined as

$$p(x) = \frac{1}{\sigma(2\pi)^{1/2}} \exp\left[-\frac{1}{2}\frac{(x - x)^2}{\sigma^2} \right] \qquad (4.8)$$

where x' is defined as the true mean value of x and σ the true variance of x. Hence, the exact form of $p(x)$ depends on the specific values for x' and σ. Note that a maximum in $p(x)$ will occur at $x = x'$, the true mean value. This implies that in the absence of bias the central tendency of a random variable having a normal distribution is toward its true mean value. The most expected value from any single measurement, that is, the most probable value, would be the true mean value.

How can we predict the probability that any future measurement will fall within some stated interval? The probability, $P(x)$, that random variable, x, will assume a value within the interval $x' \pm \delta x = \pm x_1$ is given by the area under $p(x)$. This area is found by integration over the interval. Thus, this probability is given by

$$P(x' - \delta x \le x \le x' + \delta x) = \int_{x'-\delta x}^{x'+\delta x} p(x)\, dx \qquad (4.9)$$

Integration of (4.9) is made easier through the following transformations. Begin by defining the terms $\beta = (x - x')/\sigma$, as the standardized normal variate for any value x, and $z_1 = (x_1 - x')/\sigma$, as the z variable which specifies an interval on $p(x)$. It follows that

$$dx = \sigma\, d\beta \qquad (4.10)$$

so that equation 4.9 becomes

$$P(z_1 \le \beta \le z_1) = \frac{1}{(2\pi)^{1/2}} \int_{-z_1}^{z_1} e^{-\beta^2/2}\, d\beta \qquad (4.11)$$

Since for a normal distribution, $p(x)$ is symmetrical about x', one can write

$$\frac{1}{(2\pi)^{1/2}} \int_{-z_1}^{z_1} e^{-\beta^2/2}\, d\beta = 2\left[\frac{1}{(2\pi)^{1/2}} \int_{0}^{z_1} e^{-\beta^2/2}\, d\beta \right] \qquad (4.12)$$

Known as the *normal error function*, the value in brackets in equation 4.12 provides one-half of the probability sought from equation 4.9. This half value is tabulated in Table 4.3 for the interval defined by z_1 in Figure 4.3.

TABLE 4.2   Standard Statistical Distributions and
Relations to Measurements

Distribution	Applications	Mathematical Representation
Normal	Most physical properties that are continuous or regular in time or space. Variations due to precision error.	$p(x) = \dfrac{1}{\sigma(2\pi)^{1/2}} \exp\left[-\dfrac{1}{2} \dfrac{(x - x)^2}{\sigma^2} \right]$
Log normal	Failure or durability projections; events whose outcome tends to be skewed toward the extremity of the distribution.	$p(x) = \dfrac{1}{x\sigma(2\pi)^{1/2}} \exp\left[-\dfrac{1}{2} \ln \dfrac{(x - x)^2}{\sigma^2} \right]$
Poisson	Events randomly occurring in time; $p(x)$ refers to probability of observing x events in time t.	$p(x) = \dfrac{e^{-x} x^{xt}}{x!}$
Weibull	Fatigue tests; similar to log normal applications.	See [4]
Binomial	Situations describing the number of occurrences, n, of a particular outcome during N independent tests where the probability of any outcome, P, is the same.	$p(n) = \left[\dfrac{N!}{(N - n)!n!} \right] P^n (1 - P)^{N-n}$

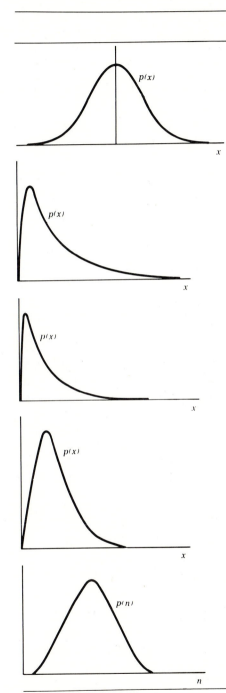

TABLE 4.3   Probability Values for Normal Error
Function

One-Sided Integral Solutions for $p(z_1) = \dfrac{1}{(2\pi)^{1/2}} \displaystyle\int_0^{z_1} e^{-\beta^2/2}\, d\beta$

$z_1 = \dfrac{x_1 - x'}{\sigma}$	0.00	0.01	0.02	0.03	0.04	0.05	0.06	0.07	0.08	0.09
0.0	.0000	.0040	.0080	.0120	.0160	.0199	.0239	.0279	.0319	.0359
0.1	.0398	.0438	.0478	.0517	.0557	.0596	.0636	.0675	.0714	.0753
0.2	.0793	.0832	.0871	.0910	.0948	.0987	.1026	.1064	.1103	.1141
0.3	.1179	.1217	.1255	.1293	.1331	.1368	.1406	.1443	.1480	.1517
0.4	.1554	.1591	.1628	.1664	.1700	.1736	.1772	.1808	.1844	.1879
0.5	.1915	.1950	.1985	.2019	.2054	.2088	.2123	.2157	.2190	.2224
0.6	.2257	.2291	.2324	.2357	.2389	.2422	.2454	.2486	.2517	.2549
0.7	.2580	.2611	.2642	.2673	.2704	.2734	.2764	.2794	.2823	.2852
0.8	.2881	.2910	.2939	.2967	.2995	.3023	.3051	.3078	.3106	.3233
0.9	.3159	.3186	.3212	.3238	.3264	.3289	.3315	.3340	.3365	.3389
1.0	.3413	.3438	.3461	.3485	.3508	.3531	.3554	.3577	.3599	.3621
1.1	.3643	.3665	.3686	.3708	.3729	.3749	.3770	.3790	.3810	.3830
1.2	.3849	.3869	.3888	.3907	.3925	.3944	.3962	.3980	.3997	.4015
1.3	.4032	.4049	.4066	.4082	.4099	.4115	.4131	.4147	.4162	.4177
1.4	.4192	.4207	.4222	.4236	.4251	.4265	.4279	.4292	.4306	.4319
1.5	.4332	.4345	.4357	.4370	.4382	.4394	.4406	.4418	.4429	.4441
1.6	.4452	.4463	.4474	.4484	.4495	.4505	.4515	.4525	.4535	.4545
1.7	.4554	.4564	.4573	.4582	.4591	.4599	.4608	.4616	.4625	.4633
1.8	.4641	.4649	.4656	.4664	.4671	.4678	.4686	.4693	.4699	.4706
1.9	.4713	.4719	.4726	.4732	.4738	.4744	.4750	.4758	.4761	.4767
2.0	.4772	.4778	.4783	.4788	.4793	.4799	.4803	.4808	.4812	.4817
2.1	.4821	.4826	.4830	.4834	.4838	.4842	.4846	.4850	.4854	.4857
2.2	.4861	.4864	.4868	.4871	.4875	.4878	.4881	.4884	.4887	.4890
2.3	.4893	.4896	.4898	.4901	.4904	.4906	.4909	.4911	.4913	.4916
2.4	.4918	.4920	.4922	.4925	.4927	.4929	.4931	.4932	.4934	.4936
2.5	.4938	.4940	.4941	.4943	.4945	.4946	.4948	.4949	.4951	.4952
2.6	.4953	.4955	.4956	.4957	.4959	.4960	.4961	.4962	.4963	.4964
2.7	.4965	.4966	.4967	.4968	.4969	.4970	.4971	.4972	.4973	.4974
2.8	.4974	.4975	.4976	.4977	.4977	.4978	.4979	.4979	.4980	.4981
2.9	.4981	.4982	.4982	.4983	.4984	.4984	.4985	.4985	.4986	.4986
3.0	.49865	.4987	.4987	.4988	.4988	.4988	.4989	.4989	.4989	.4990

It should now be clear that the statistical terms defined by equations 4.4–4.7 are actually statements of probability. The area under the portion of the probability density function curve, $p(x)$, defined by the interval $x' - z_1\sigma \le x' \le x' + z_1\sigma$ provides the probability that a measurement will assume a value within that interval. Direct integration of $p(x)$ for a normal distribution between the limits $x' \pm z_1\sigma$ yields that for $z_1 = 1.0$, 68.26% of the area under $p(x)$ lies within $\pm 1.0\sigma$ of x' or there is a 68.26% chance that a measurement of x will have a

FIGURE 4.3   Integration terminology relevant to the
normal error function.

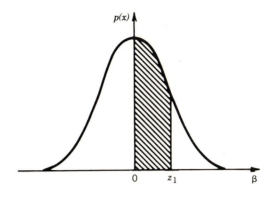

value within the interval $x' \pm 1.0\sigma$. As the interval defined by z_1 is increased,
the probability of occurrence increases. For

$z_1 = 2.0$:   95.45% of the area under $p(x)$ lies within $\pm z_1\sigma$.
$z_1 = 3.0$:   99.73% of the area under $p(x)$ lies within $\pm z_1\sigma$.

This concept is illustrated in Figure 4.4

EXAMPLE 4.2

Using the probability values in Table 4.3, show that the probability that a
measurement will yield a value within $x' \pm \sigma$ is 0.6826 or 68.26%.

FIGURE 4.4   Relation between the probability density
function and its statistical parameters, x' and σ.

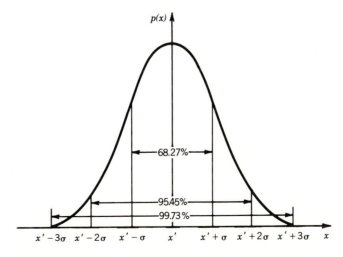

variable is described by its infinite statistics, finite statistics describe only the behavior of the finite data set.

Finite-sized data sets can provide the statistical estimates known as the *sample mean* value and the *sample variance* defined by

$$\bar{x} = \frac{1}{N} \sum_{i=1}^{N} x_i \qquad (4.14a)$$

$$S_x^2 = \frac{1}{N-1} \sum_{i=1}^{N} (x_i - \bar{x})^2 \qquad (4.14b)$$

where $(x_i - \bar{x})$ is called the deviation of x_i. The *sample standard deviation*, S_x, is defined as $\sqrt{S_x^2}$. The sample mean value provides a most probable estimate of the true mean value, x'. The sample variance represents a measure of the *precision* of a measurement. These equations are robust and provide reasonable statistical estimates regardless of the probability density function of the measurand.

In general, N independent measurements of a measurand provide for N degrees of freedom in that measurand. However, in N finite measurements of a measurand that has a central tendency, the data must be scattered about a mean value. The freedom of any data point in the measurement to assume any value, therefore, becomes restricted by this mean value. Hence, the degrees of freedom, v, in the measure of data scatter, the sample variance, is reduced to $N - 1$.

The predictive utility of infinite statistics can be extended to data sets of finite sample size with only some modification. When sample sizes are finite, the z variable described does not provide a reliable weight estimate of the true probability. However, the sample variance can be weighted in such a manner so as to compensate for the difference between the finite statistical estimates and the infinite statistics for a measurand. For a normal distribution of x about some sample mean value, x, one can state that statistically

$$x_i = \bar{x} + t_{v,P}S_x \quad (P\%) \qquad (4.15)$$

where the variable $t_{v,P}$ is obtained from a new weighting function for finite data sets which replaces the z variable. This new variable is referred to as the t estimator.

The value for the t estimator is a function of the probability, P, and the degrees of freedom, v, in the standard deviation. These t values can be obtained from Table 4.4, which is a tabulation of the *Student-t distribution* as developed by William S. Gosset[4] (1876–1937), who recognized that the use of the z variable with S_x in place of σ was not reliable. Careful inspection of the t chart shows that the t value inflates the size of the interval required to attain a percent probability, $P\%$, to describe x. That is, it has the effect of increasing the magnitude of $t_{v,P}S_x$ relative to $z_1\sigma$ ($P\%$) when N is finite. As N approaches infinity,

[4]At the time, Gosset was employed as a brewer and statistician by a well-known Irish brewery. Should you encounter the low melting point, small viscosity fluids of his trade you might pause to reflect on his multifarious contributions.

TABLE 4.4   Student t Distribution

ν	t_{50}	t_{90}	t_{95}	t_{99}
1	1.000	6.314	12.706	63.657
2	0.816	2.920	4.303	9.925
3	0.765	2.353	3.182	5.841
4	0.741	2.132	2.770	4.604
5	0.727	2.015	2.571	4.032
6	0.718	1.943	2.447	3.707
7	0.711	1.895	2.365	3.499
8	0.706	1.860	2.306	3.355
9	0.703	1.833	2.262	3.250
10	0.700	1.812	2.228	3.169
11	0.697	1.796	2.201	3.106
12	0.695	1.782	2.179	3.055
13	0.694	1.771	2.160	3.012
14	0.692	1.761	2.145	2.977
15	0.691	1.753	2.131	2.947
16	0.690	1.746	2.120	2.921
17	0.689	1.740	2.110	2.898
18	0.688	1.734	2.101	2.878
19	0.688	1.729	2.093	2.861
20	0.687	1.725	2.086	2.845
21	0.686	1.721	2.080	2.831
30	0.683	1.697	2.042	2.750
40	0.681	1.684	2.021	2.704
50	0.680	1.679	2.010	2.679
60	0.679	1.671	2.000	2.660
∞	0.674	1.645	1.960	2.576

t approaches those values given by the z variable just as S_x must approach σ. It should be understood that for very small sample sizes ($N \leq 10$), sample statistics can be misleading. In that situation other information regarding the measurement may be required, including additional measurements.

Standard Deviation of the Means

Suppose we were to measure a variable N times under fixed operating conditions. If we replicated this procedure M times, somewhat different estimates of the sample mean value and sample variance would be obtained for each of the M data sets. Why? The chance occurrence of events in any finite sample will affect the estimate of sample statistics. In fact, in experimental situations it is quite straightforward to demonstrate the variation of the sample mean value among different finite-sized samples of the same variable, even under identical operating conditions.[5] After M replications of the N measurements, a set of mean values that are themselves normally distributed about some central value would be obtained. In fact, regardless of the probability density function of the measurand, the mean values obtained from M replications will follow a normal distribution.[6] This process is visualized in Figure 4.5. The amount of variation

---

[5]This provides a compelling argument for providing replication in the measurement test plan.
[6]This is a consequence of the *central limit theorem* [2, 4].

FIGURE 4.5   The normal distribution tendency of
sample means about a true value in the absence of bias.

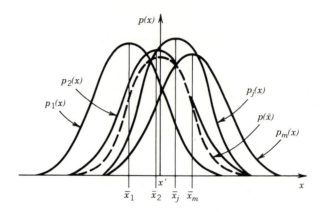

possible in the sample means would depend on two values: the sample variance, S_x^2, and sample size, N. The discrepancy tends to increase with variance and decrease with $N^{1/2}$.

This tendency for finite sample sets to have somewhat different statistics should not be surprising, for that is precisely the problem inherent to a finite data set. The variation in the sample statistics will be characterized by a normal distribution of the sample mean values about the true mean. The variance of the distribution of mean values that could be expected can be estimated from a single finite data set through the *standard deviation of the means*, $S_{\bar{x}}$

$$S_{\bar{x}} = \frac{S_x}{N^{1/2}} \qquad (4.16)$$

An illustration of the relation between the standard deviation of a data set and the standard deviation of the means is given in Figure 4.6. The standard deviation of the means is a property of a finite data set. It reflects an estimate of how the sample mean values may be distributed about a true mean value.

The standard deviation of the means represents a measure of the precision in a sample mean. The range over which the possible values of the true mean value might lie at some probability level, P, based on the information from a sample data set is given as

$$\bar{x} \pm t_{v,P} S_{\bar{x}} \quad (P\%)$$

where $t_{v,P} S_{\bar{x}}$ represents a *precision interval*, at the assigned probability, $P\%$, within which one should expect the true value of x to fall. As such, the precision interval is a quantified measure of the precision in the estimate of the true value of variable x. This estimate of the true mean value based on a finite data set is stated as

$$x' = \bar{x} \pm t_{v,P} S_{\bar{x}} \quad (P\%) \qquad (4.17)$$

FIGURE 4.6   Relation between S_x and a distribution and
the relation between $S_{\bar{x}}$ and the true value.

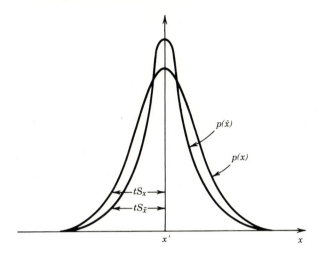

EXAMPLE 4.4

Consider the data of Table 4.1. (a) Compute the sample statistics for this data
set. (b) Estimate the interval of values over which 95% of the measurements of
the measurand should be expected to lie. (c) Estimate the true mean value of
the measurand at 95% probability based on this finite data set.

KNOWN

Table 4.1

$N = 20$

ASSUMPTIONS

Data set follows a normal distribution

FIND

$x, \bar{x} \pm tS_x$ and $\bar{x} \pm tS_{\bar{x}}$

SOLUTION

The sample mean value is computed for the $N = 20$ values by the relation

$$\bar{x} = \frac{1}{20} \sum_{i=1}^{20} x_i = 1.02$$

This, in turn, is used to compute the sample standard deviation

$$S_x = \sqrt{\frac{1}{19} \sum_{i=1}^{20} (x_i - 1.02)^2} = 0.16$$

The degrees of freedom in the standard deviation are $v = N - 1 = 19$. From Table 4.4 at 95% probability, $t_{19.95}$ is 2.093. Then, the interval of values in which 95% of the measurements of x should lie is given by equation 4.15:

$$x_i = \bar{x} \pm (2.09 \times 0.16) = 1.02 \pm 0.33 \quad (95\%)$$

Accordingly, if a 21st data point were to be taken, there is a 95% probability that its value would lie within this interval.

The true mean value is estimated by the sample mean value. However, the precision interval for this estimate is $\pm t_{19.95} S_{\bar{x}}$, where

$$S_{\bar{x}} = \frac{S_x}{N^{1/2}} = \frac{0.16}{(20)^{1/2}} = 0.04$$

Then, from equation 4.17

$$x' = \bar{x} \pm t_{19.95} S_{\bar{x}} = 1.02 \pm 0.08 \quad (95\%)$$

Pooled Statistics

As discussed in Chapter 1, a good test plan uses replication, as well as repetition, in the measurement. Since replications are independent estimates of the same measured value, their data represent separate data samples that can be combined to provide a better statistical estimate of a measured variable than are obtained from a single sample. Samples that are grouped in a manner so as to determine a common set of statistics are said to be *pooled*.

Consider M replicates of a measurement of variable x, each of N repeated readings so as to yield the data set x_{ij}, where $i = 1, 2, \ldots, N$ and $j = 1, 2, \ldots, M$. It is assumed that bias error remains negligible in each replication. The *pooled sample mean* of x is defined by

$$\langle \bar{x} \rangle = \frac{1}{MN} \sum_{j=1}^{M} \sum_{i=1}^{N} x_{ij} = \frac{1}{M} \sum_{j=1}^{M} \left[\frac{1}{N_j} \sum_{i=1}^{N_j} x_{ij} \right] \tag{4.18}$$

The *pooled standard deviation* of x is defined by

$$\langle S_x \rangle = \sqrt{\frac{1}{M(N-1)} \sum_{j=1}^{M} \sum_{i=1}^{N} (x_{ij} - \bar{x}_j)^2} = \sqrt{\frac{1}{M} \sum_{j=1}^{M} S_{xj}^2} \tag{4.19a}$$

with degrees of freedom, $v = M(N - 1)$ or, if N is not constant in each test, by

$$\langle S_x \rangle = \sqrt{\frac{v_1 S_{x_1}^2 \pm v_2 S_{x_2}^2 + \cdots + v_M S_{x_M}^2}{v_1 + v_2 + \cdots + v_M}} \tag{4.19b}$$

with $\nu = \sum\limits_{j=1}^{M} (\nu_j) = \sum\limits_{j=1}^{M} (N_j - 1)$. The *pooled standard deviation of the means* of x is estimated by

$$\langle S_{\bar{x}} \rangle = \frac{\langle S_x \rangle}{(MN)^{1/2}} = \frac{\langle S_x \rangle}{\left(\sum\limits_{j=1}^{M} N_j \right)^{1/2}} \tag{4.20}$$

4.5 REGRESSION ANALYSIS

The discussion so far has dealt solely with the statistical determination of a measurand under fixed operating conditions. The measurand is usually a function of one or more independent variables which are controlled during the measurement. When the measurand is sampled these variables are controlled, to the extent possible, as are all the other operating conditions. Following the sample, one of these variables is changed and a sample is made under the new operating conditions. This is the procedure used to document the relationship between the measurand and an independent process variable. Regression analysis can often be used to establish a functional relationship between the dependent measurand and the independent variable which will hold on the average. It is used in those situations where the relationship anticipated is either polynomial in form or can be approximated by a Fourier series. This discussion pertains directly to polynomial fits. More information on Fourier series fits can be found in [3].

Regression analysis assumes that the variation found in the dependent variable, the measurand, follows a normal distribution about each fixed value of the independent variable. Such behavior is illustrated in Figure 4.7 by considering

---

FIGURE 4.7 Distribution of measurand y about each fixed value for independent variable x and a possible functional relation.

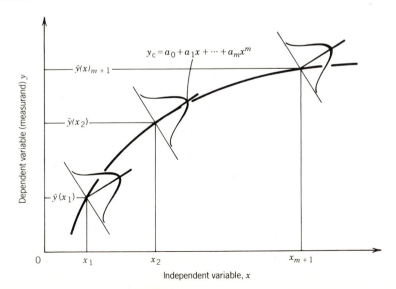

There is no rule that can be used to estimate which order fit will yield an acceptable value of S_{yx} without trial and error. This is the attractive feature of having a least-squares software package available. The choice of the actual order of fit used is always a compromise between the precision needed and the convenience of using a low-order polynomial.

EXAMPLE 4.5

The following data are suspected to follow a linear relationship. Find an appropriate equation of the first-order form.

x [cm]	y[V]
1.0	1.2
2.0	1.9
3.0	3.2
4.0	4.1
5.0	5.3

KNOWN

Independent variable, x
Dependent variable, y
$N = 5$

ASSUMPTIONS

Linear relation

FIND

$y_c = a_0 + a_1 x$

SOLUTION

We seek a polynomial of the form $y_c = a_0 + a_1 x$ which minimizes the term $D = \sum_{i=1}^{N} (y_i - y_{c_i})^2$:

$$\frac{\partial D}{\partial a_0} = 0 = -2 \sum_{i=1}^{N} [y_i - (a_0 + a_1 x_i)]$$

$$\frac{\partial D}{\partial a_1} = 0 = -2 \sum_{i=1}^{N} [y_i - (a_0 + a_1 x_i)] x_i$$

yielding

$$\sum_{i=1}^{N} [y_i - (a_0 + a_1 x_i)] = 0$$

$$\sum_{i=1}^{N} (y_i - [a_0 + a_1 x_i]) x_i = 0$$

Solving simultaneously for the coefficients a_0 and a_1 yields

$$a_0 = \frac{\Sigma\, x_i\, \Sigma\, x_i y_i - \Sigma\, x_i^2\, \Sigma\, y_i}{(\Sigma\, x_i)^2 - N\, \Sigma\, x_i^2}$$

$$a_1 = \frac{\Sigma\, x_i\, \Sigma\, y_i - N\, \Sigma\, x_i y_i}{(\Sigma\, x_i)^2 - N\, \Sigma\, x_i^2} \tag{4.26}$$

From the data set, one finds from (4.26) $a_0 = 0.02$ and $a_1 = 1.04$. Hence,

$$y_c = 0.02 + 1.04 x \text{ V}$$

COMMENT

Although the polynomial described by y_c is the linear curve fit for this data set, we still have no idea of how good this curve fits this data set or even if a first-order fit is appropriate. This is studied below.

Linear Polynomials

For linear polynomials a correlation coefficient, r, can be found by

$$r = \sqrt{1 - \frac{S_{yx}^2}{S_y^2}}$$

where

$$\tag{4.27}$$

$$S_y^2 = \frac{1}{N-1} \sum_{i=1}^{N} (y_i - \bar{y})^2$$

The correlation coefficient represents a quantitative measure of the linear association between x and y. It is bounded by ± 1 which represent perfect correlation; the sign indicates that y increases or decreases with x. For $\pm 0.9 < r \le \pm 1$, a linear regression can be considered as a reliable relation between y and x. Alternatively, the value r^2 is often reported, which is indicative of how well the variance in y is accounted for by the fit. It is a ratio of the variation assumed by the linear fit to the actual measured variations in the data. However, the correlation coefficient and the r^2 value are only indicators of the hypothesis that y and x are linearly related. They are not effective precision indicators of y_c. The S_{yx} value is used for that purpose.

The precision estimate of the slope of the fit can be estimated by

$$S_{a_1} = S_{yx} \sqrt{\frac{N}{N \sum\limits_{i=1}^{N} x_i^2 - \left(\sum\limits_{i=1}^{N} x_i \right)^2}} \tag{4.28}$$

For example, S_{a_1} would provide a measure of the static sensitivity error of a measurement system based on a linear fit of the calibration data. The precision estimate of the zero intercept can be estimated by

$$S_{a_0} = S_{yx} \sqrt{\frac{N \sum\limits_{i=1}^{N} x_i^2}{N\left[N \sum\limits_{i=1}^{N} x_i^2 - \left(\sum\limits_{i=1}^{N} x_i\right)^2\right]}} \qquad (4.29)$$

An error in a_0 would offset a calibration curve about its y intercept.

EXAMPLE 4.6

Compute the correlation coefficient and the standard error of the fit for the data in Example 4.5.

KNOWN

$y_c = 0.02 + 1.04 x$ V

ASSUMPTIONS

Normal distribution

FIND

r and S_{yx}

SOLUTION

Direct application of equation 4.27 with the data set yields for the correlation coefficient $r = 0.996$. An equivalent estimator is r^2. Here $r^2 = .99$, which indicates that 99% of the variance in y is accounted for by the fit, whereas only 1% is unaccountable. These values suggest that a linear fit is a reliable relation between x and y. The precision error between the data and this fit can be quantified through S_{yx}. Using equation 4.25, $S_{yx} = 0.09$ with degrees of freedom, $v = N - (m + 1) = 3$.

COMMENT

The t estimator, $t_{3,95} = 3.18$, establishes a precision or confidence interval about the fit of $\pm(t_{3,95}S_{yx}) = \pm0.29$. Accordingly, the polynomial fit can be stated at 95% confidence as

$$y_c = 1.04 x + 0.02 \pm 0.29 \text{ V} \quad (95\%)$$

This curve is plotted in Figure 4.8 with its 95% confidence interval. The regression polynomial with its precision interval is the only acceptable way to report a curve fit to a data set.

FIGURE 4.8   Results of regression analysis for
Example 4.6.

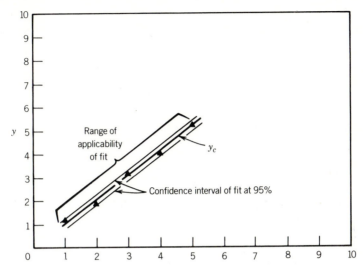

EXAMPLE 4.7

A velocity probe provides a voltage output that is related to velocity, U, by
the form $E = a + bU^m$. A calibration is run and the data ($n = 5$) are recorded
below. Estimate an appropriate curve fit.

U [ft/s]	E [V]
0.0	3.19
10.0	3.99
20.0	4.30
30.0	4.48
40.0	4.65

KNOWN

$n = 5$

ASSUMPTIONS

Data related by $E = a + bU^m$

FIND

a and b

SOLUTION

The equation can be transformed into

$$\log(E - a) = \log b + m \log U$$

which has the linear form

$$Y = B + mX$$

The value of a is evaluated at $U = 0$ ft/s to be 3.19 V. The values for Y and X are computed below with the corresponding deviations from the resulting fit:

U	Y	X	$E_i - E_{ci}$
0.0	—	—	0.0
10.0	−0.097	1.0	−0.01
20.0	0.045	1.30	0.02
30.0	0.111	1.48	0.0
40.0	0.164	1.60	−0.01

From equation 4.26, the values for Y and X are substituted with the result: $a_0 = B = -0.525$ and $a_1 = m = 0.43$. The standard error of the fit is found from equation 4.25 to be $S_{yx} = 0.01$. From Table 4.4, $t_{3.95} = 3.18$ so that the precision interval $\pm t_{v,P} S_{yx}$ is found to be ± 0.03 V. This implies that

$$Y = -0.525 + 0.43X \pm 0.03 \text{ V} \quad (95\%)$$

The polynomial is transformed back to the form, $E = a + bU^m$

$$E = 3.19 + 0.30U^{0.43} \text{ V}$$

The curve fit with its 95% precision interval is denoted on Figure 4.9.

FIGURE 4.9   A curve fit for Example 4.8.

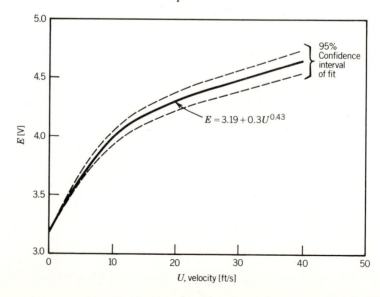

Regression analysis of multiple variables of the form $y = f(x_1, x_2, \ldots)$ is also possible. It will not be discussed here, but the concepts generated for the single-variable analysis are carried through for multiple-variable analysis. The interested reader is referred to references [2,4].

4.6 DATA OUTLIER DETECTION

It is not uncommon for a measured data set to contain one or more spurious data points that appear to not be related to the expected tendency of the data set. Normal variation in measured data is a result of measurement precision. However, data that lie outside the probability of normal variation can bias the sample mean value estimate and increase the precision estimates of the measured data set. Statistical techniques can be used to detect extraneous data points, which are known as *outliers* because their measured values fall outside the probable range of values for a measurand. Once detected, a decision as to whether to remove the data points from the data set can be made. One is strongly cautioned to be certain that these extraordinary deviations from central tendency are truly spurious, and not a unique physical tendency of the measurand, prior to their removal. Data set histograms are particularly effective for visual detection of suspected outliers. If outliers are removed from a data set, the data statistics should be recomputed using the remaining data.

Several methods exist for the detection of outliers. The specific methods presented here were chosen for their ease of use. The objective of the methods is to detect those data points that fall outside the normal range of variation expected in a data set based on the variance of that data set. This range is defined by some multiple of the standard deviation of the data set. Since a multiple of the standard deviation is related to the area under the probability density function, the basis for the outlier detection lies with the probability of occurrence for the value in question.

The simplest method for outlier detection would be to calculate the data set statistics and to label all data points that lie outside the range of 99.8% probability of occurrence, $\bar{x} \pm t_{v,99.8}S_x$, as outliers. This *three-sigma*[7] test works well with small data sets of 10 or more points and is easily programmed for use with computer-based data reduction schemes.

A more rigorous variation of the above test, particularly well suited to large data sets, is the *modified three-sigma* test for outliers. The sample mean and standard deviations are estimated and a modified z variable is computed for each data point by

$$z_0 = \left| \frac{x_i - \bar{x}}{S_x} \right| \qquad (4.30)$$

The probability that x lies outside the one-sided range defined by 0 and z_0 is $0.5 - P(z_0)$ where $P(z_0)$ is found from the one-sided z chart of Table 4.3. For N data points, if $N[0.5 - P(z_0)] \leq 0.1$, the data point can be considered as an

---

[7]So named since, as $v \to \infty$, $t \to 3$ for 99.8% probability.

outlier. Other methods of outlier detection can be found in statistics texts and measurement codes (e.g., [2,5]).

EXAMPLE 4.8

Consider the data given below for 10 measurements of tire pressure using an inexpensive hand-held gauge. Compute the statistics of the data set, then test for outliers using the modified three-sigma test.

n_i	x [psi]	n_i	x [psi]
1	28	6	24
2	31	7	29
3	27	8	28
4	28	9	18
5	29	10	27

KNOWN

$N = 10$

ASSUMPTIONS

Each measurement obtained under fixed conditions

FIND

$\bar{x}$ and S_x; apply outlier detection tests

SOLUTION

Based on the 10 data points, the sample mean and sample variance can be found from equations 4.14a and 4.14b to be $\bar{x} = 27$ psi, $S_x = 3.8$. But one would not expect tire pressure to vary much between two readings beyond the precision capabilities of the measurement system and technique, so data point 9 looks suspect as an outlier. Apply the modified three-sigma test to this data point.

For $x = 18$, $z_0 = 2.34$ and, from Table 4.3, $P(z_0) = 0.4904$, so we find $0.5 - P(z_0) = 0.0096$. Hence, $N[0.5 - P(z_0)] = 0.096$, a value that is less than 0.1. Data point 9 can be considered an outlier; that is, the magnitude of its deviation from the mean value is greater than that expected by normal variation. Since this data point clearly stands out as unusual in light of the measurement undertaken, it can be eliminated as a bad point. The remaining data variations reflect the precision of the measurement (variable, procedure, and instrument). The data set statistics become $N = 9$, $\bar{x} = 28$ psi, $S_x = 2.0$, $S_{\bar{x}} = 0.7$. When bias errors are neglected, the best estimate of the tire pressure can be stated as

$$x' = \bar{x} \pm t_{8,95}S_{\bar{x}} = 28 \pm 1.6 \text{ psi} \quad (95\%)$$

4.7 NUMBER OF MEASUREMENTS REQUIRED

Statistics can be used to assist in the planning and design of measurements. As an example, we want to determine the minimum number of degrees of freedom required to reduce the precision estimate in the true mean value of the measurand to within a certain constraint. A couple of common scenarios can be studied that assist in such estimations using the statistical understanding and methodology developed so far.

It is desired to estimate the number of measurements, N, required to give an acceptable precision in the mean value. The precision interval, CI, will be estimated by

$$\text{CI} = x' - \bar{x} = \pm t_{v,95} \frac{S_x}{N^{1/2}} \quad (95\%)$$

where S_x should be a conservative estimate based on prior experience, manufacturer's information, or a test standard. The precision interval is two-sided about $\bar{x}$, so we define $d = \text{CI}/2$. Then, it follows that

$$N \approx \left(\frac{t_{v,95}S_x}{d}\right)^2 \quad (95\%) \tag{4.31}$$

The accuracy of equation 4.31 will depend on how well S_x^2 approximates σ^2. The value for our constraint, CI, must be assigned based on the particular application needs. When the precision interval is considerably smaller than the variance, the value of N required will be somewhat large ($N > 60$) and the t estimator will have a value of approximately 2. However, in other situations, since the t estimator is a function of the degrees of freedom in the sample variance, a direct solution will not be possible. Instead, a trial and error iteration is used to estimate N.

The obvious deficiency in the above method is that an estimate for the sample variance is needed. One way around this is to make a preliminary small number of measurements, N_1, to obtain an estimate of the sample variance, S_1, to be expected. Then S_1 is used to estimate the number of measurements required. The total number of measurements, N_T, will be estimated by

$$N_T \approx \left(\frac{t_{N_1-1,95}S_1}{d}\right)^2 \quad (95\%) \tag{4.32}$$

This establishes that $N_T - N_1$ additional measurements will be required.

___

EXAMPLE 4.9

Determine the number of measurements required to reduce the precision interval of the mean value of a variable to within 1 unit if the variance of the variable is estimated to be about 64 units.

KNOWN

$CI = 1$ unit $P = 95\%$
$d = \frac{1}{2}$ $= 64$ units

ASSUMPTIONS

$\sigma^2 \approx S_x^2$

FIND

N required

SOLUTION

Begin this problem by guessing at some value for N. Then, using this guessed value, compute the t variable at the probability level desired. An updated value for N can then be found from the formulation

$$N = \left(\frac{t_{v,95}S_x}{d}\right)^2 \quad (95\%)$$

Then use trial and error iteration to converge on a value for N.
   Suppose we begin by guessing that $N = 500$. Then

$$v = 499 \qquad t_{499,95} = 1.96 \Rightarrow N = 983$$

So, now guess $N = 983$. Then

$$v = 982 \qquad t_{982,95} = 1.96 \Rightarrow N = 983$$

We have converged on $N = 983$. Thus, at least 983 measurements must be made to achieve the desired precision interval in the measured variable. An analysis of the results after 983 measurements should be made to ensure that the variance level used was representative of the actual data set.

COMMENT

Since the precision interval is reduced as $N^{1/2}$, the procedure of increasing N to decrease the precision interval becomes one of diminishing returns.

EXAMPLE 4.10

From 51 measurements of a variable, S_1 is found to be 160. For a 95% precision interval of 60 units in the mean value, estimate the total number of measurements required.

KNOWN

$S_1 = 160$ units $N_1 = 51$
$d = CI/2 = 30$ $t_{50,95} = 2.01$

ASSUMPTIONS

$\sigma \approx S_1$

FIND

N_T

SOLUTION

The total number of measurements required is estimated by

$$N_T \approx \left(\frac{t_{N_1-1,95}\, S_1}{d}\right)^2 \quad (95\%)$$

$$= \left(\frac{2.01 \times 160}{30}\right)^2 = 115$$

Thus, as a first guess a total of 115 measurements are estimated to be necessary. This means that an additional 64 measurements should be taken. The N_T data must then be reanalyzed to be certain that the constraint is met.

4.8 SUMMARY

The normal scatter of data about some central mean value is brought about through several contributing factors, including the process variable's own temporal unsteadiness and spatial distribution under nominally fixed operating conditions, as well as, precision errors in the measurement system and in the measurement procedure. Data scatter introduces an additional element, an uncertain vagueness, into the measurement scheme which requires statistical methods to sort out. Statistics, then, becomes a powerful tool used to interpret and present data. In this chapter, we developed the most basic methods used to understand and quantify finite data sets. Methods to estimate the true mean value based on a limited number of data points and the precision in such estimates were presented along with treatment of data curve fitting. A summary table of these statistical estimators is given as Table 4.5.

TABLE 4.5   Summary Table for a Sample of N Data Points

Sample mean	$\bar{x} = \dfrac{1}{N} \displaystyle\sum_{i=1}^{N} x_i$
Sample standard deviation	$S_x = \sqrt{\dfrac{1}{N-1} \displaystyle\sum_{i=1}^{N} (x_i - \bar{x})^2}$
Standard deviation of the means	$S_{\bar{x}} = \dfrac{S_x}{(N)^{1/2}}$
Precision interval for a single data point, x_i	$\pm t_{v,P} S_x \quad (P\%)$
Precision interval for a mean value, $\bar{x}$	$\pm t_{v,P} S_{\bar{x}} \quad (P\%)$
Precision interval for a curve fit between x and y	$\pm t_{v,P} S_{yx} \quad (P\%)$

REFERENCES

1. Kendal, M. G., and A. Stuart, *Advanced Theory of Statistics*, Vol. 2, Griffin, London, 1961. See also Bendat, J., and A. Piersol, *Random Data Analysis*, Wiley, New York, 1971.
2. Lipson, C., and N. J. Sheth, *Statistical Design and Analysis of Engineering Experiments*, McGraw-Hill, New York, 1973.
3. James, M. L., G. M. Smith, and J. C. Wolford, *Applied Numerical Methods For Digital Computation*, 2d ed., Harper & Row, New York, 1977.
4. Miller, I., and J. E. Freund, *Probability and Statistics for Engineers*, 3d ed., Prentice-Hall, Englewood Cliffs, NJ, 1985.
5. Measurement Uncertainty PTC 19.1-1985, *ASME Power Test Codes*, American Society of Mechanical Engineers, New York, 1985.

NOMENCLATURE

$a_0, a_1, \ldots, a_m$ polynomial regression coefficients

f_j frequency of occurrence

$p(x)$ probability density function of x

r correlation coefficient

t Student-t variable

x variable

x' true mean value of the measurand x

$\bar{x}$ sample mean value of x

$\langle \bar{x} \rangle$ pooled sample mean of x

y variable

z_1 z variable

N total number of measurements

$P(x)$ probability of measuring a value of x

$P\%$ percent probability

S_x sample standard deviation of x

$S_{\bar{x}}$ sample standard deviation of the means of x

S_x^2 sample variance of x

$\langle S_x \rangle$ pooled sample standard deviation of x

$\langle S_{\bar{x}} \rangle$ pooled standard deviation of the means of x

$\langle S_x^2 \rangle$ pooled sample variance of x

S_{yx} standard error of the fit between y and x

β normalized standard variate

σ true standard deviation of the measurand

σ^2 true variance of the measurand

ν degrees of freedom

PROBLEMS

4.1 For a large data set ($N > 1000$) found to have a mean value of 9.2 units and a standard deviation of 1.1 units, determine the range of values in which 50% of the data set should be found assuming normal behavior.

4.2 For a very large data set ($N > 10\ 000$), a mean value of 204 units with a standard deviation of 18 units has been determined. Determine the range of values in which 90% of the data fall.

4.3 At a fixed operating setting, the pressure in a line downstream of a reciprocating compressor is found to have a mean value of 121.6 psi with a standard deviation of 14 psi based on a very large data set obtained from continuous monitoring. What is the probability that the line pressure will exceed 150 psi during any measurement?

4.4 Consider the toss of 3 pennies. There are 2^3 possible outcomes of a single toss. Develop a histogram of the number of heads that can appear on any toss. Does it look like a Gaussian distribution? Should this be expected? What is the probability that 3 heads will appear on any toss?

Problems 5 through 17 refer to the data in Table 4.6. Assume that the data have been recorded from a single process under a fixed operating condition.

TABLE 4.6 Sample Data for Exercise Problems

P [lb/in.^{2}]	T [°C]	P [lb/in.^{2}]
4.95	31.2	4.97
4.87	31.3	4.92
5.10	31.0	4.93
4.99	31.2	5.00
5.06	31.5	4.98
5.09	30.9	4.92
4.95	31.2	4.91
5.01	31.4	5.06
4.98	31.1	5.01
4.99	31.1	4.98
5.00	31.4	4.97
5.05	30.9	5.02
5.00	31.5	4.92
5.02	31.2	4.94
5.01	31.6	4.98
5.06	31.1	4.99
4.99	30.8	4.92
5.01	30.7	5.04
4.98	30.9	5.00

4.5 Develop a histogram for the pressure data listed in column 1.

4.6 Develop a frequency distribution plot for the temperature data given under column 2.

4.7 Develop a histogram for the pressure data in column 3.

4.8 For the data in column 1, determine the mean value and standard deviation. State the degrees of freedom in each.

4.9 For the data in column 3, determine the mean value and estimate the standard deviation of the means. State the degrees of freedom in each.

4.10 For the data in column 1, estimate the range of values for pressure in which you would expect 95% of all possible measured values for this operating condition to lie.

4.11 For the data in column 1, determine the best estimate of the mean value at a 95% confidence level.

4.12 For the data in column 2, estimate the range of values for temperature in which you would expect 50% of all possible values measured under this operating condition to lie.

4.13 For the data in column 2, determine the best estimate for the mean value at a 50% confidence level.

4.14 For the data in column 3, estimate the range of values for pressure in which you would expect 90% of all possible measured values for this operating condition to fall.

4.15 For the data in column 1, determine the best estimate of the mean value at a 90% confidence level.

4.16 If the pressure data in columns 1 and 3 are two replications, compute a pooled mean value for process pressure. State the range for the best estimate in mean pressure at 95% confidence based on these two data sets.

4.17 For the data in Column 3, if one additional measurement were made, estimate the interval of values in which the value of this measurement would fall with a 95% probability.

4.18 Consider the measurand having a known true mean of 20.0 N with variance of 4.0 N. Estimate the probability of a measurement indicating a value between 23.5 and 26.0 N; between 16.0 and 18.5 N.

4.19 A sample of 61 data points of force indicates a sample mean value of 44.2 N with sample variance of 4.0 N^2. Estimate the probability that an additional measurement would indicate a value between 45.6 and 48.2 N.

4.20 Suppose three sets of data are collected of some variable during a similar process operating condition. The statistics are found to be

$$
\begin{array}{lll}
N_1 = 16 & \overline{X}_1 = 32 & S_1 = 3 \text{ units} \\
N_2 = 21 & \overline{X}_2 = 30 & S_2 = 2 \text{ units} \\
N_3 = 9 & \overline{X}_3 = 34 & S_3 = 6 \text{ units}
\end{array}
$$

Determine the degrees of freedom in the pooled data. Compute an estimate of the mean value of this variable and the range in which the true mean should lie with 95% confidence.

4.21 Eleven core samples of fresh concrete are taken by a county engineer from the loads of 11 concrete trucks used in pouring a structural footing. After curing, the engineer tests to find a mean compression strength of 3027 lb/in.^{2} with a standard deviation of 53 lb/in.^{2}. State codes require a minimum strength of 3000 lb/in.^{2}. Should the footing be repoured based on a 95% confidence interval of the test data?

4.22 The following data were collected during the repeated measurement of the force load acting on a small area of a beam under "fixed" conditions:

Reading Number	Output [N]	Reading Number	Output [N]
1	923	6	916
2	932	7	927
3	908	8	931
4	932	9	926
5	919	10	923

Determine if any of these data points can be considered as outliers. If so reject the data point. Estimate the true mean value from this data set.

For Problems 4.23 through 4.25, it would be helpful either to use a least-squares regression analysis software package for personal computer or calculator or to implement the algorithm given in Appendix A.

4.23 Determine the static sensitivity for the following data by using a least-squares regression analysis. Plot the data and fit including the 95% precision interval.

Y	X
2.7	0.4
3.6	1.1
4.4	1.9
5.2	3.0
9.2	5.0

4.24 Using the following data, determine a suitable least-squares fit of the data. Which order polynomial best fits this data set? Plot the data and the fit with 95% confidence band on an appropriate graph.

Y	X
1.4	0.5
4.7	1.1
17.3	2.0
82.9	2.9
171.6	5.1
1227.1	10.0

4.25 The following data have the form $y = ax^b$. Plot the data and its best fit with 95% confidence band. Estimate the static sensitivity at each value of X.

Y	X
0.14	0.5
2.51	2.0
15.30	5.0
63.71	10.0

4.26   Estimate the number of specimens that should be tested to determine the mean failure strength of an alloy to within a precision interval of less than 5% of the mean using ASTM standard procedures. A sample run of 6 alloy specimens suggests a mean strength of 71 327 lb/in.^{2} with standard deviation of 8345 lb/in.^{2}. What sources contribute to this variation in ''measured'' strength?

4.27   Estimate the number of measurements of a time-dependent acceleration signal obtained from a vibrating vehicle that would lead to an acceptable precision interval about the mean of 0.1 g, if the standard deviation of the signal is expected to be 2 g.

4.28   The sample standard deviation of a time-dependent electrical signal, based on 60 measurements, is estimated to be 1.52 V. How many more measurements would be required to provide a precision interval in the mean of 0.28 V?

CHAPTER 5
UNCERTAINTY ANALYSIS

5.1   INTRODUCTION

Measurement is the act of assigning a value to some physical variable. In the ideal measurement, the value assigned by the measurement would be the actual value of the physical variable intended to be measured. However, measurement errors bring on an uncertainty in the correctness of the value resulting from the measurement. It should be recognized that measurement errors are, in part, responsible for the random, unbiased, scatter found in measured data. In addition, they contribute to the biased difference between a statistically arrived at mean value and the true value of a measured variable. But in real world measurements, errors are unavoidable. To give some measure of confidence to the measured value, measurement errors must be identified, and their probable effect on the result estimated. Uncertainty is simply an estimate of a possible value for the error in the reported results of a measurement. The process of systematically quantifying error estimates is known as *uncertainty analysis*.

In designing a measurement system, the engineer determines the variable to be measured, selects appropriate instruments and measuring techniques to quantify the variable, and performs the measurement. But as with any design process, there needs to be some tolerance in the result, some acceptable level of error in the measured value. Uncertainty analysis provides an engineer with a design tool for the evaluation of a measurement system and the measurement method. Both during and after a measurement, it becomes a tool to estimate how closely the measured value approximates the true value.

5.2   MEASUREMENT ERRORS

While no general discussion of errors can be complete in listing the elements contributing to error in a particular measurement, certain generalizations for error sources can be made to help in their identification. In the discussion that follows, errors will be grouped into two very general categories: bias error and precision (or random) error. We will not consider measurement blunders that result in obviously fallacious data. Such data should be discarded.

The relationship between a true value and its measured data set which contains both bias and precision errors is illustrated in Figure 5.1. It is shown that the total error in any single measurement is the sum of the bias and the precision errors in that measurement. The total error contained in a set of measurements obtained under seemingly fixed conditions can be described by an average bias error and a statistical estimate of the precision errors in the measurements. The bias errors will shift the sample mean away from the true mean of the measured variable by a fixed amount. And within a sample of many measurements, the precision errors bring about a normal distribution of measured values about the sample mean. The precision of a sample is measured by the variance of its distribution about its sample mean; good precision implies a small variance. Because accuracy deals specifically with the difference between the true value and the measured value, an accurate measurement is one having both small bias and precision errors.

The importance of measurement error is that it obscures our ability to ascertain the information that we desire: the true value of the variable measured. Because of errors, the accuracy of a measurement can never be certain. In Chapter 4, we stated that the best estimate of the true value sought in a measurement is provided by its sample mean value and the uncertainty in that value,

$$x' = \bar{x} \pm u_x \quad (P\%) \tag{4.1}$$

Uncertainty analysis is the method used to quantify the u_x term.

Certain assumptions are implicit in an uncertainty analysis:

1. The test objectives are known.
2. The measurement itself is a clearly defined process in which all known calibration corrections for bias error have already been applied.

FIGURE 5.1   Distribution of errors upon repeated measurements.

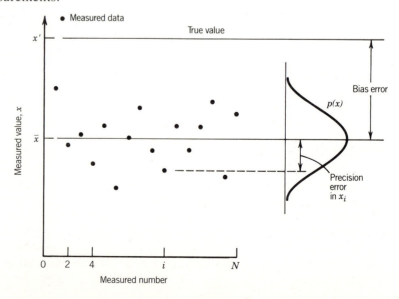

3. Data are obtained under fixed operating conditions.
4. Some system component experience is available.

Component experience is defined as an estimate of component bias and precision errors based on some evidence, such as personal experience through previous or simulated tests and calibrations, or someone else's experience, such as the manufacturer's performance literature, an NIST bulletin, a professional test code, or performance information discussed in the technical literature.

5.3 ERROR SOURCES

As a guide to looking for measurement errors, it is possible to consider the measurement process as consisting of three distinct steps: calibration, data acquisition, and data reduction. Errors that enter during each of these steps will be grouped under their respective error source heading:

1. Calibration errors
2. Data acquisition errors
3. Data reduction errors

Within each of these three error source groups, the objective should be to list the types of errors encountered. Such errors are the elemental errors of the measurement steps. These groupings are meant for convenience in establishing the errors encountered; the important thing is that those errors encountered be listed, regardless of grouping. If a step is not included in the actual measurement procedure (e.g., no calibration is explicitly performed), that source group can be excluded from consideration in the error analysis.

Calibration Errors

Calibration in itself does not eliminate measurement system errors; it merely reduces these errors to more acceptable values. *Calibration errors* include those elemental errors that enter the measuring system during the act of calibration. Calibration errors tend to enter through two principal sources: (1) the bias and precision errors in the standard used in the calibration and (2) the manner in which the standard is applied to the measuring system or system component. For example, the typical laboratory standard used for calibration is approximate. Accordingly, there may be a difference between the standard used and the primary standard that it represents. Hence, an uncertainty in the known input value on which the calibration is based will arise. In addition, there can be a difference between the value supplied by the standard and the calibration value actually sensed by the measuring system. Either of these effects will be built into the calibration data. In Table 5.1, a listing of common elemental errors contributing to this error source group is given.

Data Acquisition Errors

All errors that arise during the actual act of measurement are referred to as *data acquisition errors*. These errors include actual sensor and instrument error,

TABLE 5.1   Calibration Error Source Group

Element (j)	Error Source[a]
1	Primary to interlab standard
2	Interlab to transfer standard
3	Transfer to lab standard
4	Lab standard to measurement system
5	Calibration technique
Etc.	

[a]Bias and/or precision in each element.

changes or unknowns in measurement system operating conditions, such as power settings and environmental conditions that affect system performance, and sensor installation effects on the measurand. Also, the measured variable temporal and spatial variations contribute errors through the unknowns of finite statistics. A list of some elemental errors common to this source is given in Table 5.2.

Data Reduction Errors

The use of curve fits and correlations with their associated unknowns (Section 4.5) introduces *data reduction errors* into the reported test results. Also, the resolution of computational operations required to reduce the data into some desired result is another common contribution to this error source. A list of the elemental errors typical of this error source group are found in Table 5.3.

5.4   BIAS AND PRECISION ERRORS

Bias Error

A bias error remains constant during a given series of measurements under fixed operating conditions. Thus, in a series of repeated measurements, each measurement would contain the same amount of bias. Being a fixed value, the

TABLE 5.2   Data Acquisition Error Source Group

Element (j)	Error Source[a]
1	Measurement system operating conditions
2	Sensor–transducer stage (instrument error)
3	Signal conditioning stage (instrument error)
4	Output stage (instrument error)
5	Process operating conditions
6	Sensor installation effects
7	Environmental effects
8	Spatial variation error
9	Temporal variation error
Etc.	

[a]Bias and/or precision in each element.

Note: A total input-to-output measurement system calibration will combine elements 2, 3, 4, and possibly 1 within this error source group.

TABLE 5.3   Data Reduction Error
Source Group

Element (j)	Error Source[a]
1	Calibration curve fit
2	Truncation error
Etc.	

[a]Bias and/or precision in each element.

bias error cannot be directly discerned by statistical means. It can be difficult to estimate the value of bias or in many cases recognize the presence of a bias error. A bias error may cause either a high or a low estimate of the true value. Accordingly, an estimate of the bias error must be represented by a bias limit, noted as $\pm B$, a range within which the true value is expected to lie.

Bias can only be estimated by comparison. Various methodologies can be utilized:

1. Calibration
2. Concomitant methodology
3. Interlaboratory comparisons
4. Experience

When available, the most direct method is by calibration using a standard of known bias or, alternatively, using calibration methods of inherently negligible or small bias. Another procedure is the use of concomitant methodology, which is using different methods of estimating the same thing and comparing the results. Concomitant methods that depend on different physical measurement principles are preferable, as are methods that rely on calibrations that are independent of each other. In this regard, even accepted analytical methods could be used for comparison[1] or to estimate the bias due to influential sources such as environmental conditions, instrument response errors, and loading errors. Lastly, an elaborate but good approach is through interlaboratory comparisons of similar measurements, an excellent replication method. This approach introduces different instruments, facilities, and personnel into an otherwise similar measurement procedure in an effort to compare the bias errors between measuring facilities [1]. In lieu of the above, a value based on experience may have to be assigned.

Calibration against a standard provides the most direct method to estimate bias. But calibration cannot eliminate all bias; it can only reduce it. Consider the calibration of a temperature transducer against an NIST standard certified to be correct to within 0.01 °C. If the calibration data show that the transducer output has a bias error of 0.2 °C relative to the standard then we would just correct all the data obtained with this transducer by 0.2 °C. Simple enough! But the standard itself still has an intrinsic bias limit of 0.01 °C and this bias error must remain in the calibrated transducer.

The reader may be familiar with some bias errors in measurements. The measurement of the barefoot height of a person taken while the person is wearing

[1]Smith and Wenhofer [2] provide examples for determining jet engine thrust; several complementary measurements are used with an energy balance to estimate bias error.

shoes represents a simple example of one clear bias error in a measurement. In this case this bias error, a data acquisition error, is the height of the heels. But that is too simple! The real world is harsher.

Consider the home bathroom scale which *may* have a bias error. How might one estimate the magnitude of bias in the indicated weight? Perhaps you can calibrate the scale using calibrated masses, account for local gravitational acceleration, and correct the output, thereby estimating bias error of the measurement (direct calibration against a local standard). Or perhaps you can compare it to a measurement in a physician's office or one at the gym and can compare each reading (a sort of interlaboratory comparison). Or perhaps you can carefully measure your volume displacement in water and compare the results to estimate bias (concomitant methodology). Without any of the above what value would you assign? Would you even suspect bias error?

But let us think about this. An insidious aspect of bias error has been revealed. Why should one doubt a measurement indication and suspect a bias error? Figuratively speaking, there will be no shoe heels staring at you. Experience teaches us to think through each measurement carefully because bias error is always present at some magnitude. We see that it is difficult to estimate bias error without comparison, so a good design should include some means to estimate bias.

Precision Error

When repeated measurements are made under nominally fixed operating conditions, precision errors will be observed to manifest themselves as the scatter of the measured data. Precision error is affected by:

Measurement System
- Repeatability and resolution

Measurand
- Temporal and spatial variations

Process
- Variations in operating and environmental conditions

Measurement Procedure and Technique
- Repeatability

Because of the varying nature of precision errors, exact values cannot be given, but probable estimates can be made through statistical analyses. For example, the precision in any measured value of x can be estimated by $t_{v,P}S_x$. And the precision in the sample mean of x can be estimated by $t_{v,P}S_{\bar{x}}$.

In general, the classification of an error element as containing bias error, precision error, or both can be simplified by considering the methods used to quantify the error. *Treat an error as a precision error if it can be statistically estimated in some manner; otherwise treat the error as a bias error.* This will tend to simplify the classifications.

5.5 UNCERTAINTY ANALYSIS: ERROR PROPAGATION

We begin the design of a measurement system with an idea and some catalogs and end the project after data have been obtained and analyzed. As with any part of the design process, the uncertainty analysis will evolve as the design of the measurement process matures. We will study uncertainty analysis for the following measurement situations:

1. Design stage
2. Single measurement
3. Multiple measurement

Design-stage uncertainty analysis refers to an initial analysis performed prior to the measurement. It is useful for selecting instruments, selecting measurement techniques, and obtaining an approximate estimate of the uncertainty likely to exist in the measured data.

A not-so-unusual situation is the case where only one data point is obtained for a variable at any operating condition. Single-measurement uncertainty analysis is used to estimate the uncertainty in measurements in which a statistically viable data set has not or cannot be obtained. It is an advanced form of design-stage analysis that goes to the extent of conducting calibrations and surrogate measurements to estimate errors. It is useful for diagnosis of possible measurement errors both prior to and after a test is performed. The ANSI/ASME uncertainty code [3] refers to the methods used in such analysis as error replication analysis.

Multiple-measurement uncertainty analysis is performed after a statistically viable data set has been obtained. Errors are separated into bias and precision, with the precision estimates based on a statistical analysis and bias based on calibrations or other experience. It is assumed that all possible errors will be propagated within the data in a manner consistent with Figure 5.1.

Propagation of Error

Propagation of Elemental Errors

Each element of error present within a measurement will combine in some manner with other errors to increase the uncertainty of the measurement. Consider a measurement of measured variable x which is subject to, say, K elements of error, e_j, where $j = 1, 2, \ldots, K$. A realistic estimate of the uncertainty in the measurement, u_x, due to these elemental errors can be computed using the root-sum-squares method (RSS):

$$u_x = \pm \sqrt{e_1^2 + e_2^2 + \cdots + e_K^2}$$

$$= \pm \sqrt{\sum_{j=1}^{K} e_j^2} \quad (P\%) \tag{5.1}$$

It is imperative to maintain consistency in the units of each error. Ideally, each error should be estimated at the same probability level.

The RSS method of combining errors is based on the assumption that the possible variations in the values of an error encountered over repeated measurements will tend to follow a Gaussian distribution. As such, the RSS estimate of all contributing errors should provide a "probable" measure of the error in any measurement. While the simple arithmetic sum of the elemental errors would provide a larger estimate of possible measurement error, that method assumes that all errors can occur in their worst possible way for each and all measurements. As such, the method has no probable basis and would be appropriate in very few applications (such as catastrophic prediction).

A general, albeit somewhat arbitrary, rule is to use the 95% probability level ($P\% = 95\%$) throughout all uncertainty calculations. We will adhere to this rule here but point out that in some situations this may be neither possible nor desirable.

Propagation of Uncertainty to a Result

Measured variables are commonly used with a functional relationship to determine some resultant parameter. For example, the estimation of the normal stress acting perpendicular to a cross section of a beam would require measured information about the load force and the beam area, since $\sigma = f(F, A)$. But how would the uncertainties in the measured values of force and area contribute to the uncertainty in the stress? This general issue is explored below.

A general relationship between some dependent variable y and a measured variable x, $y = f(x)$, is illustrated in Figure 5.2. Now suppose we measure x a

---

FIGURE 5.2   Relationship between a measured variable and a resultant.

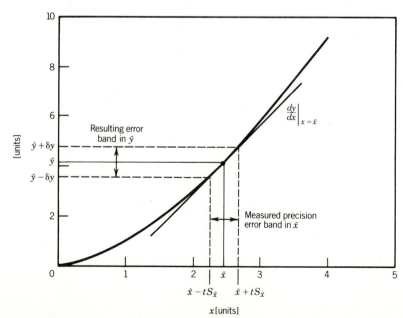

number of times at some operating condition so as to establish its sample mean value and the precision in this mean value, $tS_{\bar{x}}$. This implies that, neglecting other precision and bias errors, the true value for x lies somewhere within the interval $\bar{x} \pm tS_{\bar{x}}$. It is reasonable to assume that the true value of y, which is determined from the measured values of x, falls within the interval defined by

$$\bar{y} \pm \delta y = f(\bar{x} \pm tS_{\bar{x}})$$

Expanding this as a Taylor series yields

$$\bar{y} \pm \delta y = f(\bar{x}) \pm \left[\left(\frac{dy}{dx}\right)_{x=\bar{x}} tS_{\bar{x}} + \frac{1}{2}\left(\frac{d^2y}{dx^2}\right)_{x=\bar{x}} (tS_{\bar{x}})^2 + \cdots \right]$$

By inspection, the mean value for y must be $f(\bar{x})$ so that the term in brackets estimates $\pm\delta y$. A linear approximation for δy can be made, which is valid when $tS_{\bar{x}}$ is small, as

$$\delta y \approx \left(\frac{dy}{dx}\right)_{x=\bar{x}} tS_{\bar{x}}$$

The derivative term, $(dy/dx)_{x=\bar{x}}$, defines the slope of a line that passes through the point specified by $\bar{x}$. For small deviations from the value of $\bar{x}$, this slope predicts an acceptable, approximate relationship between $t_{v,P}S_{\bar{x}}$ and δy. It is a measure of the sensitivity of y relative to x. Since the slope of the curve can be different for different values of x, it is important to evaluate the slope using a representative value. The width of the interval defined by $\pm tS_{\bar{x}}$ defines a precision interval about y, that is, $\pm\delta y$, within which we should expect the true value of y to lie. Figure 5.2 illustrates the concept that errors in a measured variable are propagated through to a resultant variable in a predictable way. In general, errors that contribute to the uncertainty in x, u_x, will be related to the uncertainty in the estimate of resultant y by

$$u_y = \left(\frac{dy}{dx}\right)_{x=\bar{x}} u_x \tag{5.2}$$

This idea can be extended for multivariable relationships. Consider a result, R, which is determined through some functional relationship between the independent variables x_i, $x_1, x_2, \ldots, x_L$, defined by

$$R = f_1\{x_1, x_2, \ldots, x_L\} \tag{5.3}$$

where L is the number of independent variables involved. Each variable will contain some measure of uncertainty that will affect the result. The best estimate of the true mean value, R', would be stated as

$$R' = \bar{R} \pm u_R \quad (P\%) \tag{5.4}$$

where the sample mean of R is found from

$$\overline{R} = f_1\{\overline{x}_1, \overline{x}_2, \ldots, \overline{x}_L\} \tag{5.5}$$

and the uncertainty in $\overline{R}$ is found from

$$u_R = f_2\{u_{x_1}, u_{x_2}, \ldots, u_{x_L}\} \tag{5.6}$$

In equation 5.6, u_{x_i} represents the uncertainty associated with the best estimate of x_1 and so forth through x_L. The value of u_R reflects the individual contributions of the individual uncertainties as they are propagated through to the result. A general rule is to assign the uncertainties at the same probability level.

The most probable estimate of u_R is generally accepted as that value given by the Kline–McClintock second power law [5]. The second power law can be derived from the linearized approximation of the Taylor series expansion of the multivariable function defined by equation 5.3. A sensitivity index, θ_i, results from the Taylor series expansion and is given by

$$\theta_i = \left. \frac{\partial R}{\partial x_i} \right|_{x = \overline{x}_i} \qquad i = 1, 2, \ldots, L \tag{5.7}$$

This index is evaluated using either the mean values or, lacking these estimates, the expected nominal values of the variables. From a Taylor series expansion, the contribution of the uncertainty in x to the result, R, is estimated by the term $\theta_i u_{x_i}$, which is an extension of equation 5.2. The use of the partial derivative in (5.7) is necessary when R is a function of more than one variable. The propagation of uncertainty in the variables to the result will yield an uncertainty estimate given by

$$u_R = \pm \sqrt{\sum_{i=1}^{L} (\theta_i u_{x_i})^2} \quad (P\%) \tag{5.8}$$

5.6 DESIGN-STAGE UNCERTAINTY ANALYSIS

In every design there is that initial stage where the measurement system and its procedure is but a concept. At this point in the design, usually little is known about the instruments and in many cases they are still pictures in a catalog. Major facilities may need to be built and equipment ordered with considerable lead time. Uncertainty analysis should be used to assist in the selection of equipment and procedures based on their relative performance and even cost. In the design stage, distinguishing between bias and precision errors is too difficult to be of concern. Instead, consider only sources of uncertainty in general.

The initial step in design-stage analysis is to determine the minimum uncertainty in the measured value that would result from each of the proposed measurement methods. But a measurement system will usually consist of sensors

FIGURE 5.3   Design-stage uncertainty procedure.

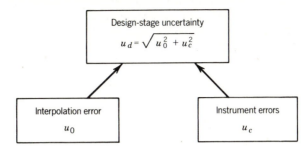

and instruments each with their respective contributions to system uncertainty. So let's first talk about individual contributions to uncertainty.

Even when all errors are otherwise zero, the value of the measurand must be affected by the ability to resolve the information provided by the instrument. We call this the *zero-order uncertainty* of the instrument, u_0. At the zero order, we assume that the variation expected in the measurand will be less than that due to instrument resolution and that all other aspects of the measurement are perfectly controlled.

As an arbitrary rule, assign a numerical value to u_0 of one-half of the instrument resolution[2] with a probability of 95%:

$$u_0 = \pm\tfrac{1}{2} \text{ resolution} \quad (95\%) \tag{5.9}$$

At 95% probability, we assume that only one measured value in twenty (20 to 1 odds) would have a value exceeding the interval defined by u_0.

The second piece of information that is usually available is the manufacturer's statement concerning instrument error. In a catalog, the value stated will be some typical value for that type of instrument under ideal conditions. We can assign this stated value as the uncertainty due to the instrument, u_c. Sometimes the instrument errors will be stated in parts, each part due to some contributing factor (e.g., Table 1.1). A probable estimate in u_c can be made by treating the propagation of errors to the instrument uncertainty using equation 5.1.

The design-stage uncertainty, u_d, for the instrument can be approximated by combining the instrument uncertainty with the zero-order uncertainty

$$u_d = \sqrt{u_0^2 + u_c^2} \quad (P\%) \tag{5.10}$$

This procedure for estimating the design-stage uncertainty is outlined in Figure 5.3. The design-stage uncertainty estimate is not meant to be a statement of the total uncertainty in a measurement. It is to be used only as a guide for selecting equipment and procedures.

A measurement system typically consists of a chain of sensors and instruments. The design-stage uncertainty for the measurement system is arrived at by com-

---

[2]In some situations, it may be possible to assign a value for u_0 that is smaller than 1/2 the scale resolution. Discretion should be used.

bining each of the design-stage uncertainties for each component in the system using the RSS method. When using the RSS method take care to use consistent units between each error or uncertainty! Similarly, the design-stage uncertainty of a resultant is found by inserting the design-stage uncertainty for each variable into equation 5.8.

EXAMPLE 5.1

A bulb thermometer listed in the catalog of an engineering supply house is described as follows:

Increment: 0.5 °F
Accuracy: within 0.5 °F (traceable to NIST standards)

Estimate the uncertainty in this instrument at the design stage.

KNOWN

Typical instrument specifications

ASSUMPTIONS

Information reliable at 95% confidence

FIND

u_d

SOLUTION

The design-stage uncertainty is given by equation 5.10:

$$u_d = \pm\sqrt{u_0^2 + u_c^2}$$

From the information provided,

$$u_0 = \pm0.25 \ °F \quad (95\%)$$

$$u_c = \pm0.5 \ °F \quad (95\% \text{ assumed})$$

Then,

$$u_d = \pm0.56 \ °F \approx \pm0.6 \ °F \quad (95\%)$$

The design-stage uncertainty in a measurement made using this instrument can be estimated to be ±0.6 °F (95%). We should expect that only one measured value out of every 20 will lie outside this 95% probability interval.

COMMENT

Keep in mind that catalog data provide for a representative value only. The calibration of a particular instrument may demonstrate performance superior or inferior to that represented by the specifications. Precision errors in the measurand and bias and precision errors related to the measurement procedure are not included in the above estimate.

EXAMPLE 5.2

Consider the force measuring instrument described by the catalog data below. Provide an estimate of the uncertainty attributable to this instrument and the instrument design stage uncertainty.

Resolution: 0.25 N
Range: 0 to 100 N
Linearity: within 0.20 N over range
Repeatability: within 0.30 N over range

KNOWN

Catalog specifications

ASSUMPTIONS

Values representative of instrument

FIND

u_c, u_d

SOLUTION

We will follow the procedure outlined in Figure 5.3. An estimate of the instrument uncertainty depends on each of the contributing elemental errors of linearity, e_1, and repeatability, e_2:

$$e_1 = 0.20 \text{ N} \qquad e_2 = 0.30 \text{ N}$$

Then using (5.1) with $K = 2$ yields

$$u_c = \pm \sqrt{(.20)^2 + (0.30)^2}$$

$$= \pm 0.36 \text{ N}$$

The instrument resolution is given as 0.25 N, from which $u_0 = \pm 0.125$ N. From (5.10), the design-stage uncertainty of this instrument would be

$$u_d = \sqrt{u_c^2 + u_0^2}$$

$$= \pm 0.38 \text{ N} \quad (95\%)$$

COMMENT

The value arrived at for the instrument is simply an estimate based on the "experience" at hand. Additional information, such as calibrations or prior use of the instrument, would provide justification for modifying this estimate.

EXAMPLE 5.3

A voltmeter is to be used to measure the output from a pressure transducer that outputs an electrical signal. The nominal pressure expected will be about 3 psi (3 lb/in.^{2}). Estimate the design-stage uncertainty in this combination. The following information is available:

Voltmeter
    Resolution: 10 μV
    Accuracy: within 0.001% of reading
Transducer
    Range: ±5 psi
    Sensitivity: 1 V/psi
    Input power: 10 VDC ± 1%
    Output: ±5 V
    Linearity: within 2.5 mV/psi over range
    Repeatability: within 2 mV/psi over range
    Resolution: negligible

KNOWN

Instrument specifications

ASSUMPTIONS

Values representative of instrument at 95% probability

FIND

u_c for each device and u_d for the measurement system

SOLUTION

The procedure in Figure 5.3 will be used for both instruments to estimate the design stage uncertainty in each. The resulting uncertainties will then be combined using the RSS approximation to estimate the system u_d.

The uncertainty in the voltmeter at the design stage is given by equation 5.10 as

$$(u_d)_E = \pm\sqrt{(u_0)_E^2 + (u_c)_E^2}$$

From the information available,

$$(u_0)_E = \pm 5\ \mu V \quad (95\%)$$

For a nominal pressure of 3 psi, we expect to measure an output of 3 V. Then,

$$(u_c)_E = \pm(3\ V \times 0.00001) = \pm 30\ \mu V \quad (95\%\ \text{assumed})$$

so that the design-stage uncertainty in the voltmeter is

$$(u_d)_E = \pm 30 \ \mu V \quad (95\%)$$

The uncertainty in the pressure transducer output at the design stage is also given by (5.10). Assuming that we operate within the input power range specified, the instrument output uncertainty can be estimated by considering each of the instrument elemental errors of linearity, e_1, and repeatability, e_2:

$$(u_c)_P = \sqrt{e_1^2 + e_2^2} \quad (95\% \text{ assumed})$$

$$= \pm\sqrt{(2.5 \text{ mV/psi} \times 3 \text{ psi})^2 + (2 \text{ mV/psi} \times 3 \text{ psi})^2}$$

$$= \pm 9.61 \text{ mV} \quad (95\%)$$

Since $(u_0) \approx 0$ V/psi, then the design-stage uncertainty in pressure in terms of indicated voltage is $(u_d)_P = \pm 9.61$ mV (95%). But since the sensitivity is 1 V/psi, this uncertainty can be stated as $(u_d)_P = \pm 0.0096$ psi (95%).

Finally, u_d for the combined system is found by use of the RSS method using the design-stage uncertainties of the two devices. The design-stage uncertainty in pressure as indicated by this measurement system is estimated to be

$$u_d = \sqrt{(u_d)_E^2 + (u_d)_P^2}$$

$$= \pm\sqrt{(0.030 \text{ mV})^2 + (9.61 \text{ mV})^2} \quad (95\%)$$

$$= \pm 10.06 \text{ mV} \quad (95\%) \quad \text{or} \quad \pm 0.010 \text{ psi} \quad (95\%)$$

5.7 MULTIPLE-MEASUREMENT UNCERTAINTY ANALYSIS

This section develops a method for the estimation of the uncertainty in the value assigned to a measured variable based on a set of measurements of that variable which are obtained under fixed operating conditions. The method parallels the uncertainty standards approved by professional societies and by NIST in the United States [3,4]. The procedures assume that the errors follow a normal probability function, although the procedures are actually quite insensitive to deviations away from such behavior [5]. Sufficient repetitions and replications must be present in the measured data, otherwise some estimate of the magnitude of expected variation must be provided in the analysis.

Propagation of Elemental Errors

The procedures for multiple-measurement uncertainty analysis consist of the following steps:

- Identify the elemental errors in each of the three source groups (calibration, data aquisition, and data reduction).

- Estimate the magnitude of bias and precision error in each of the elemental errors.

- Estimate any propagation of uncertainty through to the result.

Considerable guidance can be obtained from Tables 5.1–5.3 and from the design-stage analysis for identifying the elemental errors. In multiple-measurement analysis, it is possible to divide the estimates for elemental errors into precision and bias. Statistics are used to estimate the precision in each error.

Consider the measurement of variable x which is subject to elemental precision errors, P_{ij}, and bias, B_{ij}, in each of the three error source groups. Let subscript i, where $i = 1, 2, 3$, refer to the error source groups (calibration error, $i = 1$; data acquisition error, $i = 2$; data reduction error, $i = 3$), and subscript j, where $j = 1, 2, \ldots, K$, refer to each of any K elements of error, e_{ij}, within each group. Then, $e_{ij} = P_{ij} + B_{ij}$, as indicated in Figure 5.4. A method to estimate the uncertainty in x based on the elemental precision and bias errors is given below and outlined in Figure 5.5.

The propagation of precision errors within a source group is given by the precision index for the group called the *source precision index, P_i*. P_i is estimated by the RSS method of (5.1):

$$P_i = \sqrt{P_{i1}^2 + P_{i2}^2 + \cdots + P_{iK}^2} \qquad i = 1, 2, 3 \tag{5.11}$$

The *measurement precision index, P*, represents a basic measure of the elemental errors affecting the precision in the overall measurement of variable x. It is given by the RSS of each of the precision indices for the error groups by

$$P = \sqrt{P_1^2 + P_2^2 + P_3^2} \tag{5.12}$$

The propagation of elemental bias errors, B_{ij}, is treated in a similar manner. The *source bias limit, B_i*, for the error source groups is given by

$$B_i = \sqrt{B_{i1}^2 + B_{i2}^2 + \cdots + B_{iK}^2} \qquad i = 1, 2, 3 \tag{5.13}$$

The *measurement bias limit, B*, represents a basic measure of the elemental errors that affect the overall bias in the measurement of variable x. Let B be estimated by

$$B = \sqrt{B_1^2 + B_2^2 + B_3^2} \tag{5.14}$$

FIGURE 5.4   Multiple-measurement uncertainty method separates elemental errors into precision and bias.

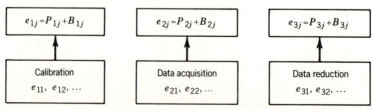

FIGURE 5.5   Multiple-measurement method for estimating measurement uncertainty.

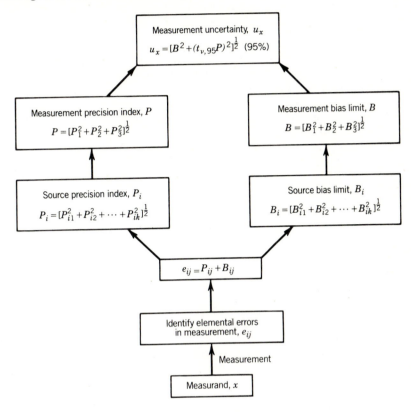

The measurement uncertainty in x, u_x, can be reported as a combination of the bias and precision uncertainty in x, as indicated in Figure 5.5:

$$u_x = \sqrt{(B)^2 + (t_{v.95}P)^2} \quad (95\%) \tag{5.15}$$

The product $t_{v.95}P$ is the estimate of the precision uncertainty in x at a 95% confidence.

Estimation of the degrees of freedom, v, in the precision index, P, requires some discussion since P is based on equations 5.11 and 5.12, each composed of elements that usually have different degrees of freedom. In this case, the degrees of freedom in the measurement precision index is estimated using the Welch-Satterthwaite formula [3]:

$$v = \frac{\left(\sum\limits_{i=1}^{3} \sum\limits_{j=1}^{K} P_{ij}^2 \right)^2}{\sum\limits_{i=1}^{3} \sum\limits_{j=1}^{K} (P_{ij}^4 / v_{ij})} \tag{5.16}$$

where i refers to the three source error groups and j to each elemental error within each source group with $\nu_{ij} = N_{ij} - 1$.

EXAMPLE 5.4

Ten repeated measurements of force, F, are made over time under fixed operating conditions. The data are listed below. Estimate the elemental error in estimating the mean value of the force that is introduced into the measured data through the data scatter.

n	F [lb]	n	F [lb]
1	123.2	6	119.8
2	115.6	7	117.5
3	117.1	8	120.6
4	125.7	9	118.8
5	121.1	10	121.9

KNOWN

Measured data set
$N = 10$

ASSUMPTIONS

Error due to data scatter (precision)

FIND

Estimate P_{ij} (equation 5.11)

SOLUTION

The mean value of the force based on this finite data set is computed from equation 4.14 as $\overline{F} = 120.1$ lb. A precision error is introduced in estimating the mean value because of data scatter. Since this error enters the measurement during data acquisition, it is listed under the data acquisition source errors ($i = 2$). Label this precision error as P_{2j} for the jth elemental error in source group 2. Then, P_{2j} can be computed through the standard deviation of the means, equation (4.16):

$$P_{2j} = \frac{S_F}{N^{1/2}}$$

$$= \frac{3.2}{10^{1/2}} = 1.01 \text{ lb}$$

COMMENT

It is common to call this elemental error, which arises due to data scatter, a temporal variation error as listed in Table 5.2.

EXAMPLE 5.5

Suppose that the force measuring device of Example 5.2 was used to generate the data set of Example 5.4. Estimate the bias in the measurement instrument.

KNOWN

$e_1 = 0.20$ N
$e_2 = 0.30$ N

ASSUMPTIONS

Manufacturer specifications reliable at 95% probability

FIND

B_{ij}

SOLUTION

Elemental errors due to instruments are considered as data acquisition source errors ($i = 2$). And since we have no information as to the statistics used to generate the numbers, e_1 and e_2, they must be considered as bias errors. Accordingly, the transducer bias error ($j = 2$, Table 5.2) in the measurement device, B_{22}, is estimated by

$$B_{22} = \sqrt{e_1^2 + e_2^2}$$

$$= 0.36 \text{ N}$$

COMMENT

Lacking specific calibration data, manufacturer specifications are always considered as bias errors contributing to the data acquisition error source. Note that the estimate for bias error due to the instrument is equal in value to an estimate that would be arrived at by design-stage analysis.

EXAMPLE 5.6

In Examples 4.5 and 4.6, a set of measured data was used to generate a polynomial curve fit equation between variables x and y. If during a test, variable x were to be measured and y computed from x through the polynomial, estimate the data reduction error involved. Assume that this is the only error in source group 3.

KNOWN

Data set of Example 4.5
Polynomial fit of Example 4.6
$i = 3$, $K = 1$ (equation 5.11)

ASSUMPTIONS

Data fits curve $y = 1.04x + 0.02$

FIND

P_3

SOLUTION

The curve fit was found to be given by $y = 1.04x + 0.02$ with a standard error of the fit, $S_{yx} = 0.09$. Accordingly, the precision error in this element, given by P_{31} (see Table 5.3), is

$$P_{31} = 0.09$$

Because this is the only elemental error in this source group, $P_3 = P_{31}$.

COMMENT

Likewise, the precision error in the power law equation fit of Example 4.7 can be computed. For example, at $E = 4.3$ V, $P_{31} = S_{yx} = 0.03$ V or 1.3 ft/s.

EXAMPLE 5.7

The measurement of x is found to contain three elemental precision errors under source group 2 (data acquisition errors). Each element is evaluated with the following conclusions:

$P_{21} = 0.60$ units, $\nu_{21} = 29$
$P_{22} = 0.80$ units, $\nu_{22} = 9$
$P_{23} = 1.10$ units, $\nu_{23} = 19$

Estimate the data acquisition error source precision index.

KNOWN

P_{ij} with $i = 2$ and $j = 1, 2, \ldots, K$
$K = 3$

FIND

P_2 (using equation 5.11)

SOLUTION

Using equation 5.11, the group 2 source precision index, P_2, is found by

$$P_2 = \sqrt{P_{21}^2 + P_{22}^2 + P_{23}^2}$$

$$= \sqrt{0.60^2 + 0.80^2 + 1.10^2}$$

$$= 1.49 \text{ units}$$

From equation 5.16 with $i = 2$, the degrees of freedom in P_2 is determined by

$$v_2 = \frac{\left(\sum\limits_{j=1}^{3} P_{2j}^2\right)^2}{\sum\limits_{j=1}^{3} (P_{2j}^4/v_{2j})} = \frac{(0.6^2 + 0.8^2 + 1.1^2)^2}{(0.6^4/29) + (0.8^4/9) + (1.1^4/19)}$$

or $v = 38.8 \approx 39$.

EXAMPLE 5.8

After an experiment to measure stress in a loaded beam, an uncertainty analysis reveals the following source errors in stress measurement whose magnitudes were computed from elemental errors using equations 5.11 and 13:

$$B_1 = 1.0 \text{ lb/in.}^2 \quad B_2 = 2.1 \text{ lb/in.}^2 \quad B_3 = 0 \text{ lb/in.}^2$$
$$P_1 = 4.6 \text{ lb/in.}^2 \quad P_2 = 10.3 \text{ lb/in.}^2 \quad P_3 = 1.2 \text{ lb/in.}^2$$
$$v_1 = 14 \quad\quad\quad v_2 = 37 \quad\quad\quad v_3 = 8$$

If the mean value of the stress in the measurement is $\bar{\sigma} = 223.4 \text{ lb/in.}^2$, determine the best estimate of the stress.

KNOWN

Experimental error source indices

ASSUMPTIONS

All elemental errors have been included

FIND

P, B and u_σ (using equations 5.12, 5.14–5.16)

SOLUTION

We seek values for the statement, $\sigma' = \bar{\sigma} \pm u_\sigma$ (95%), given that $\bar{\sigma} = 223.4$ lb/in.^{2}. The uncertainty estimate in the measurement is obtained from the source error statements through equations 5.12 and 5.14. The measurement precision index is given by (5.12) as

$$P = \sqrt{P_1^2 + P_2^2 + P_3^2} = 11.3 \text{ lb/in.}^2$$

The measurement bias limit is given by (5.14) as

$$B = \sqrt{B_1^2 + B_2^2 + B_3^2} = 2.3 \text{ lb/in.}^2$$

The degrees of freedom in P is found from (5.16) to be

$$\nu = \frac{\left(\sum\limits_{i=1}^{3} P_i^2\right)^2}{\sum\limits_{i=1}^{3} (P_i^4/\nu_i)} \approx 49$$

Therefore, the t estimator, $t_{49.95} \approx 2.0$. The uncertainty estimate is found using equation 5.15 to be

$$u_\sigma = \pm\sqrt{(2.3 \text{ lb/in.}^2)^2 + (22.6 \text{ lb/in.}^2)^2}$$

$$= \pm 22.7 \text{ lb/in.}^2 \quad (95\%)$$

The best estimate is given in the form of (4.1) as

$$\sigma' = 223.4 \pm 22.7 \text{ lb/in.}^2 \quad (95\%)$$

Propagation of Uncertainty to a Result

Consider the result, R, which is determined through the functional relationship between the measured independent variables x_i, x_1, x_2, . . . , x_L as defined by equation 5.3. Again, L is the number of independent variables involved and each x_i has an associated bias, B, given by the measurement bias limit determined for that variable by equations 5.13 and 5.14, and a measurement precision index, P, determined using equations 5.11 and 5.12. The best estimate of the true value, R', is given as

$$R' = \overline{R} \pm u_R \quad (P\%) \tag{5.17}$$

where

$$\overline{R} = f_1 \{\overline{x}_1, \overline{x}_2, \ldots, \overline{x}_L\} \tag{5.18}$$

but u_R is now given by

$$u_R = f_2 \{(B)_{x_1}, (B)_{x_2}, \ldots, (B)_{x_L}; (P)_{x_1} (P)_{x_2}, \ldots, (P)_{x_L}\}$$

where subscripts x_1 through x_L refer to the measurement bias limits and measurement precision indices in each of these variables.

The propagation of precision through the variables to the result will yield a *result precision index* given by

$$P_R = \pm\sqrt{\sum\limits_{i=1}^{L} [\theta_i(P)_{x_i}]^2} \tag{5.19}$$

where θ_i is the sensitivity index defined by equation 5.7. The propagation of bias through the variables to the result will yield a *result bias limit* given by

$$B_R = \pm \sqrt{\sum_{i=1}^{L} [\theta_i(B)_{x_i}]^2} \qquad (5.20)$$

The terms $\theta_i P_{x_i}$ and $\theta_i B_{x_i}$ represent the individual contributions of the *i*th term to the uncertainty in R. Comparisons between the magnitudes of each individual contribution will identify the uncertainty terms to which the result is most sensitive.

The measurement precision index and the bias limit are then combined to yield an estimate of the uncertainty in the result, u_R, by

$$u_R = \sqrt{B_R^2 + (t_{v,95}P_R)^2} \quad (95\%) \qquad (5.21)$$

Once again the t estimator is used to provide a reasonable weight to the interval defined by the precision index. However, the degrees of freedom in each of the variables, x_i, may not be the same. In such a case, the Welch-Satterthwaite formula is used to estimate the degrees of freedom in the result expressed as

$$v_R = \frac{\left\{ \sum_{i=1}^{L} [\theta_i(P)_{x_i}]^2 \right\}^2}{\sum_{i=1}^{L} \{[\theta_i(P)_{x_i}]^4/(v)_{x_i}\}} \qquad (5.22)$$

EXAMPLE 5.9

The density of a gas, ρ, which is believed to follow the ideal gas equation of state, $\rho = p/RT$, is to be estimated through separate measurements of pressure, p, and temperature, T. The gas is housed within a rigid impermeable vessel. The literature accompanying the pressure measurement system states an accuracy to within 1% of the reading and that accompanying the temperature measuring system suggests 0.6 °R. Twenty measurements of pressure, $N_p = 20$, and ten measurements of temperature, $N_T = 10$, are made with the following statistical outcome:

$$\bar{p} = 2253.91 \text{ psfa} \quad S_p = 167.21 \text{ psfa}$$
$$\bar{T} = 560.4 \text{ °R} \quad S_T = 3.0 \text{ °R}$$

where psfa refers to lb/ft² absolute. Determine a best estimate of the density. The gas constant is $R = 54.7$ ft-lb/lb$_m$-°R.

KNOWN

$\bar{p}, S_p, \bar{T}, S_T$
$\rho = p/RT \quad R = 54.7$ ft-lb/lb$_m$-°R

ASSUMPTIONS

Gas behaves as an ideal gas

FIND

$$\rho' = \bar{\rho} \pm u_\rho \quad (95\%)$$

SOLUTION

The measurement objective is to determine the density of an ideal gas through temperature and pressure measurements. The independent and dependent variables are related through the ideal gas law, $\rho = p/RT$. Equation 5.18 gives the mean value of density as

$$\bar{\rho} = \frac{\bar{p}}{R\bar{T}} = 0.074 \text{ lb}_m/\text{ft}^3$$

The next step must be to identify and estimate the errors and determine how they contribute to the uncertainty in the mean value of density. Since no calibrations are performed and the gas is considered to behave as an ideal gas in an exact manner, the measured values of pressure and temperature are subject only to elemental errors within the data acquisition error source group. That is, in equations 5.12 and 5.14,

$$(B_1)_T = (B_3)_T = (P_1)_T = (P_3)_T = 0$$
$$(B_1)_p = (B_3)_p = (P_1)_p = (P_3)_p = 0$$

The tabulated value of the gas constant is not without error. However, estimating the possible error in a tabulated value is sometimes difficult. According to [7] the uncertainty (bias) in the evaluation of the gas constant is on the order of ±(0.33 J/kg-K)/(gas molecular weight) or ±0.06 (ft-lb/lb$_m$-°R)/(molecular weight). Since this yields a small value for a reasonable gas molecular weight, here we will assume a zero bias limit.

Consider the pressure measurement. Within the data acquisition source group, this best estimate in the pressure is subject to two elemental errors: an instrument error, which will be based on the manufacturer's statement, and a temporal variation (data scatter) error, which will be based on the variation in the measured data obtained during presumably fixed operating conditions. The instrument error (Table 5.2) will be considered as a bias only (as discussed in Example 5.5):

$$(B_{22})_p = 22.5 \text{ psfa} \qquad (P_{22})_p = 0$$

where the first subscript 2 reflects a data acquisition error. The precision error in the mean value of pressure due to data scatter can be estimated directly from statistics:

$$(B_{29})_p = 0$$

$$(P_{29})_p = \frac{S_p}{20^{1/2}} = 37.4 \text{ psfa} \qquad (\nu)_p = 19$$

In a similar manner the data acquisition source error in temperature is computed. Elemental error due to instrument error is considered as bias only:

$$(B_{22})_T = 0.6 \text{ °R} \qquad (P_{22})_T = 0$$

The elemental error due to temporal variation is considered as precision only and estimated by

$$(B_{29})_T = 0 \qquad (P_{29})_T = \frac{S_T}{10^{1/2}} = 0.9 \text{ °R} \qquad (\nu)_T = 9$$

The data acquisition source errors in pressure and temperature are estimated from equations 5.13 and 5.11:

$$(B_2)_p = \sqrt{(22.5)^2 + (0)^2} = 22.5 \text{ psfa}$$

$$(B_2)_T = 0.6 \text{ °R}$$

and

$$(P_2)_p = \sqrt{(0)^2 + (37.4)^2} = 37.4 \text{ psfa}$$

$$(P_2)_T = 0.9 \text{ °R}$$

with degrees of freedom determined from equation 5.16 to be

$$(\nu)_p = N_p - 1 = 19$$

$$(\nu)_T = 9$$

Since only data acquisition errors are present in the problem, the measurement precision index and bias limit for pressure and temperature as found from equations 5.12 and 5.14 are

$$(B)_p = 22.5 \text{ psfa} \qquad (P)_p = 37.4 \text{ psfa} \qquad (\nu)_p = 19$$

$$(B)_T = 0.6 \text{ °R} \qquad (P)_T = 0.9 \text{ °R} \qquad (\nu)_T = 9$$

The propagation of bias and precision errors through to the result, the density, will be estimated using the second power law of equations 5.19 and 5.20:

$$P = \sqrt{\left[\frac{\partial \rho}{\partial T}(P)_T\right]^2 + \left[\frac{\partial \rho}{\partial p}(P)_p\right]^2}$$

$$= \sqrt{(1.2 \times 10^{-4})^2 + (1.2 \times 10^{-3})^2}$$

$$= 0.0012 \text{ lb}_m/\text{ft}^3$$

and

$$B = \sqrt{\left[\frac{\partial \rho}{\partial T}(B)_T\right]^2 + \left[\frac{\partial \rho}{\partial p}(B)_p\right]^2}$$

$$= \sqrt{(8 \times 10^{-5})^2 + (7 \times 10^{-4})^2}$$

$$= 0.0007 \text{ lb}_m/\text{ft}^3$$

The degrees of freedom in the density is determined from equation 5.22,

$$\nu = \frac{\{[(-\bar{p}/R\bar{T}^2)(P)_T]^2 + [(1/R\bar{T})(P)_p]^2\}^2}{[(\bar{p}/R\bar{T}^2)(P_T)]^4/(\nu)_T + [(1/R\bar{T})(P)_p]^4/(\nu)_p} = 23$$

From Table 4.4, $t_{23,95} = 2.06$.

The uncertainty in the mean value of density is estimated from equation 5.21:

$$u_\rho = \sqrt{(B)^2 + (t_{23.95}P)^2}$$

$$= 0.0025 \text{ lb}_m/\text{ft}^3 \quad (95\%)$$

The best estimate of the density is given in the form of equation 5.17:

$$\rho' = 0.074 \pm 0.0025 \text{ lb}_m/\text{ft}^3 \quad (95\%)$$

This measurement of density has an uncertainty of about 3.4%.

COMMENT

Examination of the magnitude of each group of terms in the equations for P and B shows that the pressure contributes more to the precision index and bias limit of density than does temperature. However, bias is essentially negligible compared to the precision of the measurements. The analysis reveals that the uncertainty in density could be reduced by improving the precision in the pressure measurement.

EXAMPLE 5.10

Consider the determination of the mean diameter of a shaft using a hand-held micrometer. The shaft was manufactured on a lathe presumably to a constant diameter. Identify possible elements of error that can contribute to the uncertainty in the estimated mean diameter.

SOLUTION

In the machining of the shaft, possible run-out during shaft rotation can bring about an eccentricity in the shaft cross-sectional diameter. Further, as the shaft is machined to size along its length possible run-out along the shaft axis can bring about a difference in the machined diameter. To account for such deviations, the usual measurement procedure is to repeatedly measure the diameter at one location of the shaft, rotate the shaft, and repeatedly measure again. Then the micrometer is moved to a different location along the axis of the shaft and the above procedure repeated.

It is unusual to calibrate a micrometer in a working machine shop, although an occasional bias check against accurate gauge blocks is normal procedure. Let us assume that the micrometer is used as is without calibration. Data acquisition errors are introduced from at least several elements:

1. Since the micrometer is not calibrated, the reading during any measurement can be affected by a possible bias in the micrometer markings. This can be labeled as an instrument error, B_{22}. Experience shows that this value will be on the order of the resolution of the micrometer. Ideally, the effects of using different micrometers and operators could be tested by replication so as to provide a precision estimate instead. However, this is rarely done.
2. The precision upon repeated readings will be affected by the resolution of the readout, eccentricity of the shaft, and the exact placement of the micrometer on any cross section along the shaft. It is not possible to separate these errors, so they are grouped as variation errors, a precision error, P_{25}. This value, P_{25}, can be discerned from the statistics of the measurements made at any cross section (replication).
3. Spatial variations in the diameter along its length will introduce scatter between the statistical values for each cross section. Since this affects the spatial precision of the overall mean diameter, its effect must be accounted for. Label this error as a spatial error, P_{28}. It can be estimated by the pooled statistical variance of the mean values at each measurement location (replication).

Since a diameter is the direct outcome from any reading, there will be no data reduction errors in such a measurement.

COMMENT

If we were interested in the uncertainty associated with the production of these shafts, note that machinist, machine, micrometer, and possibly day of the week would be extraneous variables in need of a randomized test plan.

FIGURE 5.6   Measurement locations for Example 5.11.

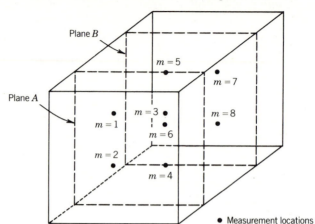

EXAMPLE 5.11

The mean temperature in an oven is to be estimated by using the information obtained from a temperature probe. The manufacturer offers a statement suggesting an error within 0.6 °C with this probe. Determine the oven temperature.

The measurement process is defined such that the probe is to be positioned at several strategic locations within the oven. The measurement strategy is to make four measurements within each of two equally spaced cross-sectional planes of the oven. The positions are chosen so that each measurement location corresponds to the centroid of an area of value equal for all locations so that spatial variations in temperature throughout the oven are accounted. The measurement locations within the square oven are shown in Figure 5.6. In this example, 20 measurements are made at each position, with the results given in Table 5.4. Assume that a temperature readout device with a resolution of 0.1 °C is used.

KNOWN

Data of Table 5.4

FIND

$$T' = \bar{T} \pm u_T \quad (95\%)$$

TABLE 5.4   Example 5.7: Oven Temperature Data, $N = 10$

Location	$\bar{T}_m$	S_{T_m}	Location	$\bar{T}_m$	S_{T_m}
1	342.1	1.1	5	345.2	0.9
2	344.2	0.8	6	344.8	1.2
3	343.5	1.3	7	345.6	1.2
4	343.7	1.0	8	345.9	1.1

NOTE:

$$\bar{T}_m = \frac{1}{N} \sum_{n=1}^{N} T_{m,n}; \; S_{T_m} = \left[\frac{1}{N-1} \sum_{n=1}^{N} (T_{m,n} - \bar{T}_m)^2 \right]^{1/2}$$

SOLUTION

The mean values at each of the locations will be averaged to yield a mean oven temperature using pooled averaging:

$$<\overline{T}> = -\sum_{m=1}^{8} \overline{T}_m = 344.4 \text{ °C}$$

Of the three error source groups only data acquisition errors remain, with $B_1 = B_3 = P_1 = P_3 = 0$. Elemental errors in the data acquisition group are due to (1) the temperature probe system (instrument error), (2) spatial variation errors, and (3) temporal variation errors. Consider first the elemental error in the probe system. A statement of 0.6 °C was provided by the manufacturer. Because no statistical information is provided, the total uncertainty due to the probe system will be considered to be due to bias. Thus, with $j = 2$ (Table 5.2)

$$B_{22} = 0.6 \text{ °C} \qquad P_{22} = 0$$

Consider next the spatial error contribution to the estimate in mean temperature, $\overline{T}$. This error arises from the use of discrete measurement locations within the oven coupled with the nonuniformity in the oven temperatures. An estimate of spatial temperature distribution within the oven can be made by examining the mean temperatures at each measured location. These temperatures suggest that the oven is not uniform in temperature. The mean temperatures within the oven display a standard deviation of

$$S_T = \sqrt{\frac{\sum_{m=1}^{8} (\overline{T}_m - <\overline{T}>)^2}{7}} = 1.26 \text{ °C}$$

Thus, the precision index of the oven mean temperature is,

$$P_{28} = \frac{S_T}{8^{1/2}} = 0.45 \text{ °C}$$

with degrees of freedom, $\nu = 7$. We assume no bias in this element:

$$B_{28} = 0$$

During each of the 10 measurements of temperature at each location, time variations in the probe output cause data scatter, as evidenced by the respective S_{T_m} values. Such time variations are caused by random local temperature variations as measured by the probe, probe system resolution, and oven temperature control variations during fixed operating conditions. It is rarely possible to separate these, so they are considered as one error. A precision index can be found.

Using the pooled standard deviation, equation 4.19,

$$<S_T> = \sqrt{\frac{\sum\limits_{m=1}^{8}\sum\limits_{n=1}^{10}(T_{mn} - <\overline{T}>)^2}{M(N - 1)}} = 1.08\ °C$$

the precision index is estimated by

$$P_{29} = \frac{<S_T>}{80^{1/2}} = 0.12$$

with degrees of freedom, $v = 72$ and $B_{29} = 0$.

The data acquisition source bias limit can now be estimated from (5.13); therefore,

$$B_2 = \sqrt{B_{22}^2 + B_{28}^2 + B_{29}^2} = 0.6\ °C$$

The data acquisition source precision index is estimated from (5.11) to be

$$P_2 = \sqrt{P_{22}^2 + P_{28}^2 + P_{29}^2} = 0.47\ °C$$

with degrees of freedom from (5.16) of $v = 8$.

Because error enters through only one source, the measurement bias limit and measurement precision index are, from equations 5.14 and 5.12, simply

$$B = B_2 = 0.6\ °C \qquad P = P_2 = 0.47\ °C \qquad \text{with } v = 8$$

The uncertainty in the mean oven temperature is estimated from (5.15):

$$u_T = \sqrt{B^2 + (t_{8.95}P)^2} \quad (95\%)$$

with $t_{8.95} = 2.306$.

The best estimate of the mean oven temperature is given by (4.1) as

$$T' = \overline{T} \pm u_T = 344.4 \pm 1.24\ °C \quad (95\%)$$

5.8 SINGLE-MEASUREMENT UNCERTAINTY ANALYSIS

In designing a measurement system a pertinent question is, how would it affect the result if this particular aspect of the technique or equipment were changed? It is an assumption in multiple-measurement analysis that, if enough measurements and replications are made, each of the elemental errors will manifest themselves in the measured data often enough to be accounted for in the bias and precision error estimates. But what if a large set of data are unavailable?

In design-stage uncertainty analysis, only errors due to a measurement system's resolution and estimated intrinsic errors are considered. Single-measurement uncertainty analysis permits taking that treatment further by considering other errors that affect the measurement. We will now consider a method for a thorough uncertainty analysis when a large data set is not available. Such an analysis, known as *single measurement uncertainty analysis* [6],[3] is to be used:

- In the advanced design stage of a test to estimate the expected uncertainty, beyond the initial design stage estimate.
- To report the results of a test program that involved measurements over a range of one or more parameters but with no or relatively few replications at each test condition.

In this section, the goal will be to estimate the uncertainty in some measured value, x, or in some general result, R, through an estimation of the uncertainty in each of the factors which may affect x or R. If all factors that influence a measurement were held constant, we might expect the measured value to remain constant upon repeated measurements. But measurement errors will replicate; that is, they will affect the measured value somewhat differently with each measurement. The overall uncertainty in any measurement is affected by this replication of errors and is an estimate of the possible value of these errors. A technique that uses a step by step approach for identifying and estimating errors is presented. We seek the combined value of the estimates at each step.

Zero-Order Uncertainty

At zero-order uncertainty all variables and parameters that affect the outcome of the measurement, including time, are assumed to be fixed except for the physical act of observation itself. Under such circumstances, any data scatter introduced upon repeated observations of the output value will be the result of instrument resolution alone. The value u_0 estimates the extent of variation expected in the measurand when all influencing effects are controlled. An estimate of u_0 is found using equation 5.9.

Because, at this level, the uncertainty value calculated would be the minimum possible, the use of zero-order analysis provides an estimate of the effect of instrument resolution on the measurement. Obviously, a zero-order uncertainty analysis is inadequate for the reporting of test results.

Higher-Order Uncertainty

Higher-order uncertainty estimates consider the controllability of the test operating conditions. At the first-order level, the effect of time as an extraneous variable in the measurement is considered. That is, what would happen if we started the test, set the operating conditions, and sat back and watched. If a variation in the measured value is observed, time is a factor in the test, presumably due to some extraneous variable.

---

[3]It is called *replication level analysis* in [3,6].

In practice, the uncertainty at this level would be evaluated for a particular measurand by operating the test facility at some single operating condition that would be within the range of conditions to be used during the actual tests. A set of data (say, $N \geq 30$) would be obtained under some set operating condition. The first-order uncertainty of that measurand could be estimated as

$$u_1 = \pm t_{v,95} S_x \qquad (5.23)$$

Only when $u_1 = u_0$ is time not a factor in the test. In itself, the first-order uncertainty is inadequate for reporting of test results.

At each successive order, each factor identified as affecting the measured value is introduced into the analysis, thus giving a higher but more realistic estimate of the uncertainty. For example, at the second level it might be appropriate to assess the limits of the ability to duplicate the exact operating conditions and the consequences on the test outcome.

Nth-Order Uncertainty

As the final estimate, the Nth-order uncertainty, instrument calibration characteristics are entered into the scheme through the instrument uncertainty, u_c. A practical estimate of the Nth-order uncertainty, u_N, is given by

$$u_N = \sqrt{(u_c)^2 + \sum_{i=1}^{N-1} u_i^2} \quad (95\%) \qquad (5.24)$$

Uncertainty estimates at the Nth order allow for the direct comparison between results of similar tests obtained either using different instruments or at different test facilities. The procedure for a single measurement analysis is outlined in Figure 5.7.

Ideally, the value for u_c would be based on the error estimates found during a series of calibrations for instruments representative of a pertinent type. Uncertainty estimates at this level are complicated by the fact that the u_c on hand may be based on a single model of instrument only. But so long as u_c is not the dominant term in (5.24), this approximation can be considered as reasonable.

Note that design-stage analysis includes only the effects found as a result of u_0 and u_c. It is those in-between levels that allow measurement procedure and control effects to be considered in the uncertainty analysis scheme. Only when $u_N = u_d$ are the test conditions under complete control. As such, *the Nth-order uncertainty estimate provides the uncertainty value sought in advanced design stage or in single measurement analyses.* It is the only appropriate value to be used in the reporting of results from single measurement tests.

EXAMPLE 5.12

As an exercise, obtain and examine a dial oven thermometer. How would you assess the zero- and first-order uncertainty in the measurement of the temperature of a kitchen oven using the device?

FIGURE 5.7   Single-measurement uncertainty procedure.

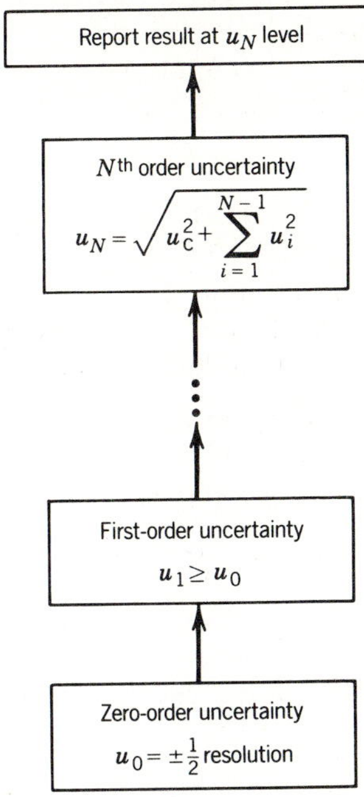

GIVEN

Dial thermometer

ASSUMPTIONS

Negligible bias in the instrument

FIND

Estimate u_0 and u_1 in oven temperature

SOLUTION

The zero-order uncertainty would be that contributed by the interpolation error of the measurement system only. For example, most gauges of this type have a resolution of 10 °C. So at the zero order from equation 5.9, we would estimate

$$u_0 = \pm 5\ ^\circ\text{C}\quad (95\%)$$

At the first order, the uncertainty would be affected by any time variation in the measurand and variations in the operating conditions. If we placed the thermometer in the center of an oven, set the oven to some relevant temperature, allowed the oven to preheat to a steady condition, and then proceeded to record the temperature indicated by the thermometer at random intervals, we could estimate our ability to control the oven temperature. For N measurements, the first-order uncertainty of the measurement in this oven's temperature using this technique and this instrument would be from (5.23)

$$u_1 = \pm t_{N-1,95} S_T \quad (95\%)$$

COMMENT

We could estimate the next order of uncertainty, u_2, by checking our ability to repeatedly set the oven to a desired mean temperature, (time-averaged temperature). This could be done by changing the oven setting and then resetting the thermostat back to the original operating setting. If we were to replicate the above sequence M times (reset thermostat, measure a set of data, reset thermostat, etc.), we could compute u_2 by

$$u_2 = \pm t_{M(N-1),95} {<}S_{\bar{T}}{>} \quad (95\%)$$

We could also check for spatial variation effects, u_3, such as those found in Example 5.11. The effects of instrument bias and precision would enter at the Nth order through u_c. The uncertainty in oven temperature at some setting would be well approximated by the estimate from (5.24)

$$u_N = \pm\sqrt{u_1^2 + u_2^2 + u_3^2 + u_c^2} \quad (95\%)$$

By inspection of the values u_0, u_c, and u_i, where $i = 1, 2, \ldots, N - 1$, single-measurement analysis provides the capability to pinpoint those aspects of a test that contribute most to the overall uncertainty in the measurement (compare these errors with those in Example 5.11).

EXAMPLE 5.13

A stopwatch is to be used to estimate the time between the start of an event and the end of this event. Event duration might range from several seconds to 10 min. Estimate the probable uncertainty in a time estimate using a hand-operated stopwatch that claims an accuracy of 1 min per month and a resolution of 0.01 s.

KNOWN

$u_0 = \pm 0.005$ s   (95%)
$u_c = \pm 60$ s/month   (95% assumed)

FIND

u_d, u_N

SOLUTION

The design-stage uncertainty will give an estimate of the suitability of an instrument for a measurement. At 60 s/month, the instrument accuracy works out to about 0.01 s per 10 min of operation. This would suggest a design stage uncertainty of

$$u_d = (u_0^2 + u_c^2)^2 = 0.01 \text{ s} \quad (95\%)$$

for an event lasting 10 min versus ± 0.005 s (95%) for an event lasting 10 s. Accuracy controls the longer duration measurement, whereas resolution controls the short duration measurement.

But do instrument resolution and accuracy control the uncertainty in this measurement? The design stage analysis does not include the data acquisition error involved in the act of physically turning the watch on and off. But a first-order analysis might be run to estimate the uncertainty that enters through the procedure of using the watch. Suppose a typical trial run of 20 tries of simply turning a watch on and off suggests that the uncertainty in determining the duration of an occurrence is estimated by

$$u_1 = 0.15 \text{ s} \quad (95\%)$$

The uncertainty in measuring the duration of an event would then be better estimated by equation 5.24:

$$u_N \approx \sqrt{u_0^2 + u_1^2 + u_c^2} = 0.15 \text{ s} \quad (95\%)$$

This estimate will hold for periods of up to about 2 h. Now a better decision as to whether the watch and procedure are suitable for the intended test can be made.

EXAMPLE 5.14

A flow meter can be calibrated by providing a known flow rate to the meter and measuring the meter output. One method of calibration with liquid systems is the use of a catch and time technique whereby a volume of liquid, after passing through the meter, is diverted to a tank for a measured period of time from which the flow rate, volume/time, is computed. There are two procedures that can be used to determine the known flow rate, Q, in, say, ft^3/min:

1. The volume of liquid, V, caught in known time, t, can be measured.

Suppose we arbitrarily set $t = 6$ s and assume that our available facilities can determine volume (at Nth order) to ± 0.001 ft^3. *Note:* The chosen time value depends on how much liquid we can accommodate in the tank.

2. The time, t, required to collect 1 ft^3 of liquid can be measured.

Based on Example 5.14, assume an Nth order uncertainty in time of ± 0.15 s. This uncertainty includes our ability to control the on/off time and watch accuracy, but assumes that the diversion is instantaneous. In either case the same instruments are used. Determine which method is better to minimize uncertainty over a range of flow rates based on these preliminary estimates.

KNOWN

$u_V = \pm 0.001 \text{ ft}^3$
$u_t = \pm 0.15 \text{ s}$
$Q = f(V,t) = V/t$

ASSUMPTIONS

Diversion instantaneous

FIND

Preferred method

SOLUTION

From the available information, the propagation of probable uncertainty to the result, Q, is estimated from equation 5.8:

$$u_Q = \pm \sqrt{\left(\frac{\partial Q}{\partial V}u_V\right)^2 + \left(\frac{\partial Q}{\partial t}u_t\right)^2}$$

or dividing through by Q, the percent uncertainty in flow rate, u_Q/Q, is

$$\frac{u_Q}{Q} = \pm \sqrt{\left(\frac{u_V}{V}\right)^2 + \left(\frac{u_t}{t}\right)^2}$$

Representative values of Q are needed to solve this relation. Consider a range of flow rates, say, 1, 10, and 100 ft³/min; the results are listed in the table for both methods.

Q [cfm]	t [s]	V [ft³]	$\frac{u_V}{V}$	$\frac{u_t}{t}$	$\frac{u_Q}{Q}$
			Method 1		
1	6	0.1	0.01	0.025	0.027
10	6	1.0	0.001	0.025	0.025
100	6	10.0	0.0001	0.025	0.025
			Method 2		
1	60.0	1.0	0.001	0.0025	0.0025
10	6.0	1.0	0.001	0.025	0.025
100	0.6	1.0	0.001	0.25	0.25

In method 1, it is clear the uncertainty in time contributes the most to the relative uncertainty in Q, provided that the flow diversion is instantaneous. But

FIGURE 5.8   Uncertainty plot for Example 5.15.

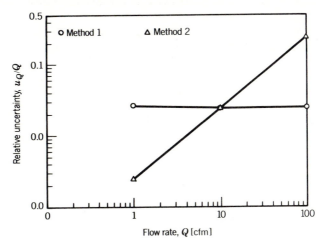

in method 2, uncertainty in either time or volume can contribute more to the uncertainty in Q, depending on the time sample length. The results for both methods are compared in Figure 5.8. It is apparent that for the conditions selected and preliminary uncertainty values used that method 2 would be a better procedure for flow rates up to 10 cfm. Beyond 10 cfm, method 1 would be better. However, the minimum uncertainty in method 1 is limited to 2.5%. The engineer may be able to reduce this uncertainty by trial and error selection of the operating parameters of t (method 1) and V (method 2) or improvements in the time procedure.

COMMENT

These results are without consideration of some other elemental errors that are present in the experimental procedure. For example, the diversion of the flow is assumed to occur instantaneously. A first-order uncertainty estimate could be used to estimate the added uncertainty intrinsic to the time required to divert the liquid to and from the catch tank. Operator influence should be randomized in actual tests. Accordingly, the above calculations may be used as a guide in procedure selection, but should not be used as the final estimate in the overall uncertainty in any results obtained from the actual calibration.

5.9 SUMMARY

In this chapter the manner in which various errors can enter into a measurement have been discussed. The procedures of uncertainty analysis have been introduced as a means of estimating error propagation within a measurement and the propagation of error among independent measurands that are used to determine a result through some functional relationship. The role and use of uncertainty analysis in both the design of measurement systems, through selection of equipment and procedures, as well as in the interpretation of measured data is discussed. The procedures will provide reasonable estimates of the uncertainty to be expected in measurements.

The reader is advised that measurement uncertainty by its very nature evades exact estimates. The important thing is that the uncertainty analysis be performed during the design and development stages of a measurement, as well as at its completion.

REFERENCES

1. International Standards Organization, ISO 5168, Measurement of fluid flow-estimation of uncertainty of a flow rate measurement, New York, 1976.
2. Smith, R. E., and S. Wenhofer, From measurement uncertainty to measurement communications, credibility and cost control in propulsion ground test facilities, *Journal of Fluids Engineering* 107: 165–172, 1985.
3. ANSI/ASME Power Test Codes-PTC 19.1, Measurement uncertainty, American Society of Mechanical Engineers, New York, 1983.
4. Abernathy, R. B., and J. W. Thompson, *Measurement Uncertainty Handbook*, Instrument Society of America, Research Triangle Park, NC, 1980.
5. Kline, S. J., and F. A. McClintock, Describing uncertainties in single sample experiments, *Mechanical Engineering* 75, 1953.
6. Moffat, R. J., Uncertainty analysis in the planning of an experiment, *Journal of Fluids Engineering* 107: 173–181, 1985.
7. Kestin, J., *A Course in Thermodynamics*, revised printing, Hemisphere, New York, 1979.

Suggested Reading

Sigurdson, S., and D. E. Kimball, Practical method for estimating number of test readings required, *Journal of Engineering Power*, 386–392, July 1976.
Taylor, J. R., *An Introduction to Error Analysis*, University Science Books, Mill Valley, CA, 1982.
Wang, T., and T. W. Simon, Development of a special purpose test surface: Introduction of a new uncertainty analysis step, AIAA paper 88-0169, New York, 1988.
Wyler, J. S., Estimating the uncertainty of spatial and time average measurements, *Journal of Engineering Power*, 473–480, October 1975.

NOMENCLATURE

e_j	jth elemental error	u_x	uncertainty of a measured variable
$f(\ )$	general functional relation	u_d	design-stage uncertainty
p	pressure [$m\text{-}l^{-1}\text{-}t^{-2}$]	u_c	instrument calibration uncertainty
$t_{v.95}$	t variable at 95% probability	u_0	zero-order uncertainty
t	time [t]		

u_1 first-order uncertainty

u_N Nth-order uncertainty

x measured variable

x' true value of x

$\bar{x}$ sample mean value of x

B measurement bias limit

B_{ij} jth elemental bias error in ith source group

B_1, B_2, B_3 source bias limit

F force $[m\text{-}l\text{-}t^{-2}]$

N number of measured values

R resultant value

S_x sample standard deviation of x

$S_{\bar{x}}$ sample standard deviation of the means

P measurement precision index

P_{ij} jth elemental precision error in ith source group

P_1, P_2, P_3 source precision index

$(P\%)$ percent probability

Q flow rate $[l^3\text{-}t^{-1}]$

T temperature $[°]$

V volume $[l^{-3}]$

θ_i sensitivity index

ρ gas density $[m\text{-}l^{-3}]$

ν degrees of freedom

$<\ >$ pooled statistic

PROBLEMS

5.1 How does bias error differ from precision error?

5.2 How can bias error be estimated for a measurement system? How is precision error estimated?

5.3 Explain what is meant by the terms true value, best estimate, mean value, uncertainty, and confidence interval.

5.4 Consider a common tire pressure gauge. How would you estimate the uncertainty in a measured pressure at the design stage and at the Nth order. Should the estimates differ? Explain.

5.5 A micrometer has graduations scribed at 0.001-in. (0.025-mm) intervals. Estimate the uncertainty due to resolution at a 95% probability.

5.6 A tachometer has an analog display dial graduated in 5 revolutions per minute (rpm) increments. The user manual states an accuracy of 1% of reading. Estimate the uncertainty in the reading at 10, 500, and 5000 rpm.

5.7 An automobile speedometer is graduated in 5-mph (8-kph) increments and has an accuracy rated to be within ±4%. Estimate the uncertainty in indicated speed at 60 mph (90 kph).

5.8 A temperature readout device has a claimed accuracy to within 0.6 °C of indicated reading with resolution of 0.1 °C. It is to be used with a sensor that has an off-the-shelf accuracy of better than 0.5 °C. Estimate a design-stage uncertainty in the temperature indicated by this combination.

5.9 Two resistors are to be combined to form an equivalent resistance of 1000 Ω. Readily available are two common resistors rated at 500 ± 50 Ω and two common resistors rated at 2000 Ω ± 5%. What combination of resistors (series or parallel) would provide the smaller uncertainty in an equivalent 1000-Ω resistance?

5.10   Heat transfer from a rod of diameter D immersed in a fluid can be described by the Nusselt number, Nu $= hD/k$, where h is the heat-transfer coefficient and k is the thermal conductivity of the fluid. If h can be measured to within $\pm 7\%$ (95%) estimate the uncertainty in Nu for the nominal value of $h = 150$ W/m²-K. Let $D = 20 \pm 0.5$ mm and $k = 0.6 \pm 2\%$ W/m-K.

5.11   Estimate the design-stage uncertainty in determining the voltage drop across a heater. The device has a nominal resistance of 30 Ω and power rating of 500 W. Available is an ohmmeter (accuracy: within 0.5%; resolution: 1 Ω) and ammeter (accuracy: within 0.1%; resolution: 100 mA). Recall $E = IR$.

5.12   A displacement transducer has the following specifications:

Linearity:	$\pm 0.25\%$ reading
Drift:	$\pm 0.05\%/°C$ reading
Repeatability:	$\pm 0.25\%$ reading
Excitation:	10 to 25 V dc
Output:	0 to 5 V dc
Range:	0 to 5 cm

The transducer output is to be indicated on a voltmeter having a stated accuracy of $\pm 0.1\%$ reading with a resolution of 10 μV. The system is to be used at room temperature, which can vary by ± 10 °C. Estimate an uncertainty in a nominal displacement of 2 cm at the design stage.

5.13   The displacement transducer of Problem 5.12 is used in measuring the displacement of a body impacted by a mass. Twenty measurements are made which yield

$$\bar{x} = 172 \text{ mm} \qquad S_x = 17 \text{ mm}$$

Determine a best estimate for the mass displacement at 95% probability.

5.14   A pressure transducer outputs a voltage to a readout device that converts the signal back to pressure. The device specifications are:

Resolution:	0.1 psi
Repeatability:	0.1 psi
Linearity:	within 0.1% of reading
Drift:	less than 0.1 psi/6 months (32–90 °F)

The transducer has a claimed accuracy of within 0.5% of reading. For a nominal pressure of 100 psi at 70 °F, estimate the uncertainty in a measured pressure.

5.15   For a thin-walled pressure vessel of diameter, D, and wall thickness, t, subjected to an internal pressure, p, the tangential stress is given by $\sigma_t = pD/2t$. During one test, 10 measurements of pressure yielded a mean of 8610 lb/ft² with a standard deviation of 273.1. Cylinder dimensions are to be based on a set of 10 measurements which yielded: $D =$

6.2 in., S_b = 0.18 and $\bar{t}$ = 0.22 in., S_t = 0.04. Determine the best estimate of stress.

5.16 Suppose the pressure measurand p contains three elemental precision errors in its calibration:

$$P_{11} = 0.90 \text{ psi} \quad P_{12} = 1.10 \text{ psi} \quad P_{13} = 0.09 \text{ psi}$$

$$\nu_{11} = 21 \qquad\qquad \nu_{12} = 10 \qquad\qquad \nu_{13} = 15$$

Estimate the source precision index for calibration errors.

5.17 An experiment to determine force acting at a point on a member contains the following source errors:

$$B_1 = 2 \text{ N} \quad B_2 = 4.5 \text{ N} \quad B_3 = 3.6 \text{ N}$$

$$P_1 = 0 \qquad P_2 = 6.1 \text{ N} \quad P_3 = 4.2 \text{ N}$$

$$\nu_2 = 17 \qquad \nu_3 = 19$$

If the mean value for force is estimated to be 200 N, determine the uncertainty in the mean value at 95% confidence.

5.18 The area of a flat, rectangular parcel of land is computed from the measurement of the length of two adjacent sides, X and Y. Measurements are made using a scaled chain accurate to within 0.5% over its indicated length. The two sides are measured several times with the following results:

$$\overline{X} = 556 \text{ ft} \qquad \overline{Y} = 222 \text{ ft}$$

$$S_x = 5.3 \text{ ft} \qquad S_y = 2.1 \text{ ft}$$

$$\nu = 8 \qquad\qquad \nu = 7$$

Estimate the area of the land and state the confidence interval of that measurement at 95%.

5.19 Estimate the precision in the measured value of stress for a data set of 23 measurements, if σ = 1061 psi and S_σ = 22 psi.

5.20 The density of a metal composite is to be determined from the mass of a cylindrical ingot. The volume of the ingot is determined from diameter and length measurements. It is estimated that mass, m, can be determined to within 0.1 lbm using an available balance scale; length, L, can be determined to within 0.05 in. and diameter, D, to within 0.0005 in. Estimate the zero-order uncertainty in the determination of the density. Which measurement contributes most to the uncertainty in the density? Which measurement method should be improved first if the uncertainty in density is unacceptable? Use the nominal values of m = 4.5 lbm, L = 6 in., D = 4 in.

5.21 For the ingot of Problem 5.20, the diameter of the ingot is measured 10 independent times at each of 3 cross sections. For the results below give a best estimate in the diameter with its uncertainty:

$$\overline{d}_1 = 3.9920 \text{ in.} \quad \overline{d}_2 = 3.9892 \text{ in.} \quad \overline{d}_3 = 3.9961 \text{ in.}$$

$$S_{d_1} = 0.005 \text{ in.} \quad S_{d_2} = 0.001 \text{ in.} \quad S_{d_3} = 0.0009 \text{ in.}$$

5.22 For the ingot of Problem 5.20, the following measurements are obtained from the ingot:

$$\overline{m} = 4.4 \text{ lb}_m \qquad \overline{L} = 5.85 \text{ in.}$$

$$S_m = 0.1 \text{ lb}_m \qquad S_L = 0.1 \text{ in.}$$

$$N = 21 \qquad N = 11$$

Based on the results of Problem 5.21, estimate the density and the uncertainty in this value. Compare your answer with that of 5.10. Explain any difference.

5.23 A temperature measurement system is calibrated against a standard system certified to an uncertainty of ± 0.05 °C at 95%. The system sensor is immersed alongside the standard within a temperature bath so that the two are separated by about 10 mm. The temperature uniformity of the bath is estimated at about 5 °C/m. The temperature system sensor is connected to a readout that indicates the temperature in terms of voltage. The following are calibration results between the temperature indicated by the standard and the indicated voltage:

T [°C]	E [mV]	T [°C]	E [mV]
0.1	0.004	40.5	1.624
10.2	0.399	51.2	2.147
19.5	0.771	99.6	4.121

a. Compute the calibration curve fit.
b. Estimate the uncertainty in using the output from the temperature measurement system for temperature measurements.

5.24 The power usage of a strip heater is to be determined by measuring heater resistance and heater voltage drop simultaneously. The resistance is to be measured using an ohmmeter having a resolution of 1 Ω and an error stated to be within 1% of its reading, and voltage is to be measured using a voltmeter having a resolution of 1 V and a stated error of 1% of reading. It is expected that the heater will have a resistance of 100 Ω and use 100 W of power. Determine the uncertainty in power determination to be expected with this equipment at the zeroth-order level. Recompute at the design stage.

5.25 The power usage of a dc strip heater can be determined in either of two ways: (1) heater resistance and voltage drop can be measured simultaneously and power computed, or (2) heater voltage drop and current measured simultaneously and power computed. Instrument manufacturer specifications are listed:

Instrument	Resolution	Error (% reading)
Ohmmeter	1 Ω	0.5%
Ammeter	0.5 A	1%
Voltmeter	1 V	0.5%

For loads of 10 W, 1 kW, and 10 kW, determine the best method based on an appropriate uncertainty analysis. Assume nominal values as necessary.

5.26 A set of tests is to be done to study how the cure temperature of a composite material will affect its strength. Tests will be done over a range of temperatures (20 to 60 °C). Temperature is to be measured at each of four quadrants within the cure chamber using temperature equipment for which the accuracy can be estimated. The test specimens are expensive, so only one test at any four temperatures can be performed. Suppose prior to testing any composite, the chamber thermostat is set to, say, 30 °C, and 20 measurements over time are taken. This exercise is then replicated 5 times so that the statistics can be computed. Compare this method to one in which a composite is placed in the chamber and cured while 100 measurements of chamber temperature are made (25 at each of the four locations). Would the uncertainty estimates for the cure temperature based on the two methods include the same effects? Explain.

5.27 Time variations in a signal require that the signal be measured often in time to determine its mean value. A preliminary sample is obtained by measuring the signal 50 times. The statistics from this sample show a mean value of 2.112 V with a standard deviation of 0.387 V. If it is desired to state the precision of the mean value to within 0.100 V at 95% confidence, how many measurements of the signal will be required based on the data variations.

5.28 The diameter of a shaft is estimated in a manner consistent with Example 5.10. A hand-held micrometer (resolution: 0.001 in.; accuracy: <.001 in.), is used to make measurements about four selected cross-section locations. Ten measurements of diameter are made around each location with the results noted below. Provide a reasonable statement as to the diameter of the shaft and its tolerance (the uncertainty in that estimate).

Location:	1	2	3	4
$\overline{D}$ (in.):	4.494	4.499	4.511	4.522
S_D:	0.006	0.009	0.010	0.003

5.29 The pressure in a large vessel is to be maintained at some set pressure for a series of tests. A compressor is to supply air through a regulating valve that is set to open and close at the set pressure. A dial gauge (resolution: 1 psi; accuracy: 0.5 psi) is used to monitor pressure in the vessel. Thirty trial replications of pressurizing the vessel to a set pressure of 50 psi are attempted to estimate pressure controllability and a standard deviation in set pressure of 2 psi is found. Estimate the uncertainty to be expected in the vessel set pressure.

CHAPTER 6
ELECTRICAL DEVICES, SIGNAL PROCESSING, AND DATA ACQUISITION

6.1 INTRODUCTION

The output at any stage of a measurement system is often an electrical signal. This signal typically originates from the measurement of a physical variable using a fundamental electromagnetic or electrical phenomenon and then propagates from stage to stage of the measurement system. Most transducers will output an electrical signal, and data acquisition, transmission, and processing are mostly accomplished using electronic devices.

This chapter provides for an introduction into

- Basic electrical measuring devices
- Sampling and signal analysis concepts
- Data acquisition methods

So many measurement and control system applications utilize electrical signals for data transmission and processing that we introduce the concepts of sampling, signal analysis, and data acquisition within that context. However, one should recognize that these concepts have a much broader range of application, including mechanical and optical signals. The basic electrical quantities of interest here are current, voltage, and resistance. Standards and definitions for these fundamental quantities are discussed in Chapter 1.

The integration of discrete (and/or digital) devices with analog electrical devices is common in engineering systems. As such, concepts related to the measurement and interpretation of both analog and discrete signals are included. The concepts and limitations of discrete measurement sampling as related to subsequent data interpretation are presented.

For $t_f = \pi$, the rms value is $\sqrt{50}$. Evaluation of the integral for the rms value with $t_f = 2\pi$ also yields $\sqrt{50}$.

COMMENT

Although the average value over the period 2π is zero, the power dissipated in the resistor must be the same over both the positive and negative half cycles of the sine function. Thus, the rms value of current is the same for the time period of π and 2π and is indicative of the power dissipated.

Effects of Signal-Averaging Period

For a complex waveform, averaging can serve to eliminate high-frequency contributions, to smooth fluctuations in data, or to extract long-term trends in fluctuating signals. Recall from Chapter 3 that the internal damping and inertia of a measurement system can affect its ability to follow the true time-dependent behavior of the input to the system. At input signal frequencies significantly higher than a system's frequency bandwidth the system output magnitude ratio will tend toward zero. The inertia and damping in such a system effectively serve to average the dynamic portion of an input signal. With an input signal of the general form

$$F(t) = A_0 + \sum_{i=1}^{\infty} A_i \cos \omega_i t + B_i \sin \omega_i t \qquad (6.7)$$

the resulting output from a measurement system will have the general form

$$y(t) = KA_0 + K \sum_{i=1}^{\infty} A_i' \cos[\omega_i t + (\phi_1)_i] + B_i' \sin[\omega_i t + (\phi_2)_i] \qquad (6.8)$$

where A_i' and B_i' are related to A_i and B_i through $M(\omega_i)$. The output signal will have an average value equal to KA_0.

When the ac component of the signal is of primary interest, the dc component can be removed. This procedure often allows a change of scale in the display or plotting of a signal such that fluctuations that were a small percentage of the dc signal can be more clearly observed without the superposition of the large dc component. The enhancement of fluctuations through the subtraction of the average value is illustrated in Figure 6.3. Figure 6.3a contains the entire complex waveform consisting of both static and dynamic parts. When the dc component is removed, the dynamic portion can be amplified for enhancement, as in Figure 6.3b.

6.3 ANALOG DEVICES: CURRENT MEASUREMENTS

This section deals specifically with the analog measurement of time-continuous static (dc) or dynamic (ac) current signals. Although a number of different

FIGURE 6.3   Effect of subtracting dc offset for a
dynamic signal.

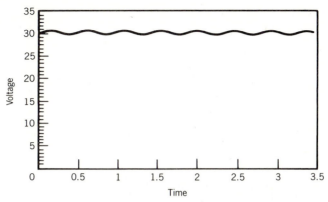

(*a*)   Signal prior to subtracting dc offset

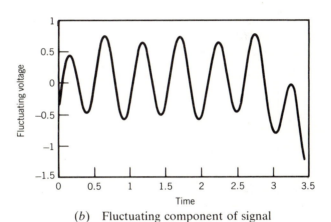

(*b*)   Fluctuating component of signal

methods are available for the measurement of current, the most common methods use a galvanometer with associated circuitry. In fact, most analog meters that may be calibrated to indicate some other variable actually sense and respond to current flow through a galvanometer.

Direct Current

The most basic analog measurements of electrical current are accomplished as a result of the force exerted on a current-carrying conductor in a magnetic field. If a positive electric charge in motion experiences a force that tends to deflect it from a straight line path, a magnetic field must be present. Because an electric current is a series of moving charges, a magnetic field will exert a force on a current-carrying conductor. The force exerted by a magnetic field can be utilized to measure the flow of current in a conductor, and to move a pointer on a display. Consider a straight length of a conductor through which

FIGURE 6.4   Current-carrying conductor in a magnetic field.

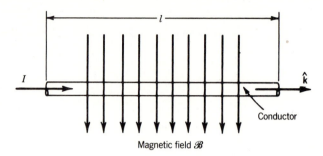

a current, I, flows, as shown in Figure 6.4. The force on this conductor due to a magnetic field strength $\mathscr{B}$ is

$$F = Il\mathscr{B} \tag{6.9}$$

This relation is valid only when the direction of the current flow and the magnetic field are at right angles. In the general case

$$\mathbf{F} = Il\hat{k} \times \mathscr{B} \tag{6.10}$$

where $\hat{k}$ is a unit vector along the direction of the current flow, and the force $\mathbf{F}$ and the magnetic field $\mathscr{B}$ are also vector quantities. Equation 6.10 provides the magnitude of the developed force $\mathbf{F}$, and, by the right-hand rule, also the direction in which the force on the conductor acts.

Similarly, a current loop in a magnetic field will experience a torque if the loop is not aligned with the magnetic field, as illustrated in Figure 6.5. The torque[2] on a loop composed of N turns is given by

$$T_\mu = NIA\,\mathscr{B} \sin \alpha \tag{6.11}$$

where

A = cross-sectional area defined by the perimeter of the current loop
$\mathscr{B}$ = magnetic field strength
I = current
α = angle between the normal the cross-sectional area of the current loop and the magnetic field

---

[2]In vector form, the torque on a current loop can be written

$$\mathbf{T}_\mu = \boldsymbol{\mu} \times \mathscr{B}$$

where $\boldsymbol{\mu}$ is the magnetic dipole moment.

FIGURE 6.5 Forces and resulting torque on a current loop in a magnetic field.

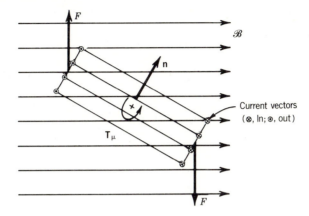

A galvanometer is a device that is sensitive to current flow through the torque exerted on a current loop in a magnetic field; the basic design of a galvanometer is shown in Figure 6.6. In this arrangement, the uniform radial magnetic field and torsional spring result in a steady angular deflection of the coil which corresponds to the existing current in the coil. The coil and fixed permanent magnet are called a D'Arsonval movement. In this arrangement, the normal direction to the current loop is such that $\alpha = 90°$. For increased sensitivity, the pointer can be replaced by a mirror and light beam arrangement. The highest sensitivity of commercial galvanometers ranges from approximately 10^{-5} μA/div for light beam galvanometers, to approximately 0.1 μA/div for a galvanometer with a pointer.

FIGURE 6.6 Basic D'Arsonval meter movement.

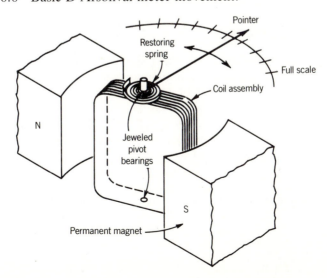

FIGURE 6.7   Simple multirange ammeter.

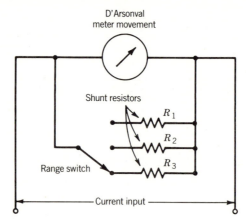

The sensitivity of the D'Arsonval movement to the flow of direct current makes it a suitable method for transforming an electrical signal into a visually discernible signal, the movement of a pointer on a dial. A typical circuit for an analog current measuring device, an ammeter, is shown in Figure 6.7. The range of current that can be measured is determined by selection of the combination of shunt resistor and the internal resistance of the galvanometer. The shunt resistor provides a bypass for current flow, reducing the current that flows through the galvanometer.

Errors inherent to a mechanical D'Arsonval movement include hysteresis and repeatability errors due to mechanical friction in the pointer-bearing movement, and linearity errors in the spring that provides the restoring force for equilibrium. In developing a torque, the D'Arsonval movement must extract energy from the current flowing through it. This draining of energy from the signal being measured changes the measured signal. Such an effect is a loading error. A quantitative analysis of loading errors will be provided later.

Alternating Current

The measurement of alternating current (ac) is accomplished in several ways in analog instruments. Common ac analog meter arrangements employ electromagnets whose magnetic field strength increases with increasing current flow. Either an iron plunger similar to the design of a solenoid switch, or an electrodynamometer may be used to transform the electromagnetic force into movement of a pointer. An electrodynamometer is basically a D'Arsonval movement modified for use with ac current by replacing the permanent magnet with an electromagnet in series with the current coil. These ac meters have upper limits on the frequency of the alternating current that they can effectively measure; many basic instruments are intended for use with standard line frequency.

EXAMPLE 6.2

A galvanometer consists of N turns of a conductor wound about a core of length l and radius r that is situated perpendicular to a magnetic field of uniform

FIGURE 6.8   Figure for Example 6.2.

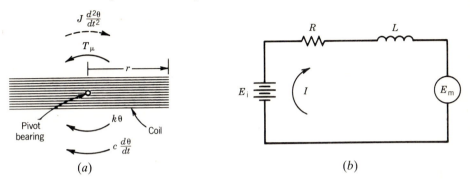

(a) (b)

flux density $\mathcal{B}$. A dc current passes through the conductor due to an applied potential, $E_i(t)$. The output of the device is the rotation of the core and pointer, θ, as shown in Figure 6.8. Develop a model relating pointer rotation and current.

FIND

A dynamic model relating θ and I.

SOLUTION

The galvanometer is a rotational system consisting of a torsional spring, a pointer, and core which are free to deflect; the rotation is subject to frictional damping. The mechanical portion of the device consists of a coil having moment of inertia, J, bearings that provide damping from friction, with a damping coefficient, c, and a torsional spring of stiffness, k. The electrical portion of the device consists of a coil with a total inductance, L_g, and a resistance, R_g.

Application of Newton's second law can be accomplished by examining the mechanical free-body diagram shown in Figure 6.8a. As a current passes through the coil, a force is developed that produces a torque on the coil

$$T_\mu = (2N\mathcal{B}lr)I \tag{6.12}$$

This torque, which tends to rotate the coil, is opposed by an electromotive force, E_m, due to the Hall effect

$$E_m = \left(\mathbf{r}\frac{d\theta}{dt} \times \mathcal{B}\right) \cdot l\hat{k} = (2N\mathcal{B}r)\frac{d\theta}{dt} \tag{6.13}$$

which produces a current in the direction opposite to that produced by E_i. The electrical free body is shown in Figure 6.8b. Application of Newton's law to the free body shown in Figure 6.8a yields

$$J\frac{d^2\theta}{dt^2} + c\frac{d\theta}{dt} + k\theta = T_\mu \tag{6.14}$$

Ohmmeter Circuits

The determination of resistance can be accomplished by imposing a voltage across an unknown resistance, and measuring the resulting current flow using a galvanometer. Clearly, from Ohm's law the value of resistance can be determined in a circuit such as in Figure 6.15 which forms the basis of an analog ohmmeter. Such analog ohmmeters would employ circuits similar to those shown in Figure 6.16, that use shunt resistors and a D'Arsonval mechanism for accurately measuring a wide range of resistance while limiting the flow of current through the ammeter.

The lower limit on resistance, R_1, which can be measured is determined by the current flow through R_1, I_1. Practical limitations on the maximum current flow through a resistance are imposed by the ability of the particular resistance to dissipate the power generated by the flow of current, (I^2R heating). For example, the ability of a thin metallic conductor to dissipate heat is the principle on which fuses are based. At too large a value of I_1, they melt. The same is true for many delicate, small resistance sensors!

Bridge Circuits

A variety of bridge circuits have been devised for the measurement of capacitance, inductance, and most often for the measurement of resistance. A purely resistive bridge, called a *Wheatstone bridge*, provides a means for accurately measuring resistance, and for detecting small changes in resistance. Figure 6.17 shows the basic arrangement for a bridge circuit, where R_1 may be a transducer that experiences a change in resistance with a change in the measured variable. A dc voltage is input across nodes A to D, and the bridge forms a parallel circuit arrangement between these two nodes. The currents flowing through the resistors R_1 to R_4 are I_1 to I_4, respectively. Under the condition that the current flow through the galvanometer, I_g, is zero, the bridge is said to exist

FIGURE 6.15   Basic analog ohmmeter. (Voltage is adjusted to yield full-scale deflection with terminals shorted.)

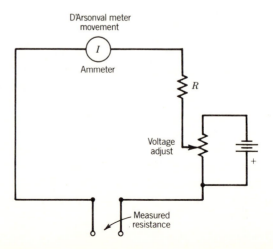

FIGURE 6.16   Multirange ohmmeter circuits.

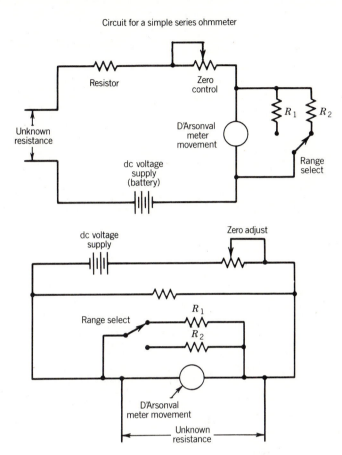

Circuit for a simple series ohmmeter

in a balanced condition. A specific relationship exists between the resistances that form the bridge at balanced conditions. To find this relationship, a circuit analysis is performed with $I_g = 0$. Under this balanced condition, there is no voltage drop from B to C and

$$I_1 R_1 - I_3 R_3 = 0 \qquad\qquad (6.17)$$
$$I_2 R_2 - I_4 R_4 = 0$$

Under balanced conditions, the current through the galvanometer is zero, and the currents through the arms of the bridge are equal:

$$I_1 = I_2 \quad \text{and} \quad I_3 = I_4 \qquad\qquad (6.18)$$

Solving equations 6.17 simultaneously, with the condition stated in equation 6.18, yields the necessary relationship among the resistances for a balanced bridge:

$$\frac{R_2}{R_1} = \frac{R_4}{R_3} \qquad\qquad (6.19)$$

FIGURE 6.17   Basic Wheatstone bridge circuit. (G, Galvanometer).

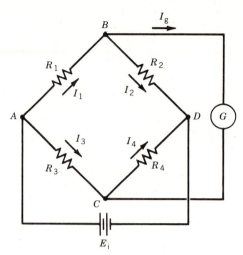

Measurement of resistance using this bridge circuit can be performed in two basic ways. If the resistor R_1 varies with changes in the measured physical variable, one of the other arms of the bridge can be adjusted to null the circuit. Another method for measurement using the Wheatstone bridge is to replace the galvanometer with a voltage measuring device, and to sense the unbalance in the bridge as an indication of the change in resistance. Both of these methods will be analyzed further.

Null Method

Consider the circuit shown in Figure 6.17, where R_2 is a variable resistance. If the resistance R_1 changes due to a change in the measured variable, the resistance R_2 can be adjusted to compensate so that the bridge is once again balanced. In this null method of operation, the resistance R_2 must be a calibrated variable resistor, such that adjustments to R_2 directly indicate the value of R_1. The balancing operation may be accomplished manually or controlled automatically, but a significant time is necessary to achieve balanced conditions making the null method most suitable for steady input signals. An advantage of the null method is that the input voltage need not be known, and changes in the input voltage do not affect the accuracy of the measurement. In addition, the galvanometer need only detect if there is a flow of current, not measure its value.

Earlier in this section, we assumed that the galvanometer current was exactly zero when the bridge was balanced. In fact, the resolution of the galvanometer is limited so that the current cannot be set exactly to zero. Consider a bridge that has been balanced within the sensitivity of the galvanometer, such that I_g is smaller than the smallest current detectable. This current flow will create an uncertainty in the measured resistance, R_1, due to the galvanometer resolution error, u_R. A basic analysis of the circuit with a current flow through the galvanometer yields

$$\frac{u_R}{R_1} = \frac{I_g(R_1 + R_g)}{E_i}$$

(6.20)

where R_g is the internal resistance of the galvanometer. This equation can serve to guide the choice of a galvanometer and a battery voltage for a particular application. Clearly, the error is reduced by increased input voltages. However, the input voltage is limited by the power dissipating capability of the resistance device, R_1. The power that must be dissipated by this resistance is $I_1^2 R_1$.

Deflection Method

In an unbalanced condition, the magnitude of the current or voltage drop for the meter or galvanometer portion of a bridge circuit is a direct indication of the change in resistance of one or more of the arms of the bridge. Consider first the case where the voltage drop from node B to node C in the basic bridge is measured by a meter having an infinite internal impedance, so that there is no current flow through the meter, as shown in Figure 6.18. Knowing the conditions for a balanced bridge given in equation 6.18, the voltage drop from B to C can be determined, since the current I_1 must equal the current I_2, as

$$E_o = I_1 R_1 - I_3 R_3$$

(6.21)

Under these conditions, substituting (6.17–6.19) into equation 6.21 yields

$$E_o = E_i \left(\frac{R_1}{R_1 + R_2} - \frac{R_3}{R_3 + R_4} \right)$$

(6.22)

FIGURE 6.18   Voltage-sensitive Wheatstone bridge.

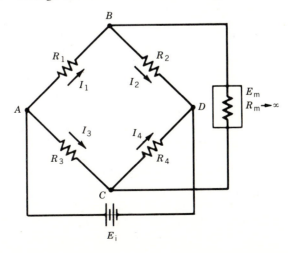

For many transducers, the bridge may be balanced at a reference condition. The transducer resistance changes, as a result of changes in the measured variable, would then cause a deflection in the bridge voltage away from the balanced condition. Assume that from an initially balanced condition where $E_o = 0$, a change in R_1 occurs to some new value, $R_1' = R_1 + \delta R$. The output from the bridge becomes

$$E_o = E_i \left(\frac{R_1'}{R_1' + R_2} - \frac{R_3}{R_3 + R_4} \right) = E_i \frac{R_1' R_4 - R_3 R_2}{(R_1' + R_2)(R_3 + R_4)} \quad (6.23)$$

In many designs the bridge resistances are initially equal to R, which allows equation 6.23 to be reduced to

$$\frac{E_o}{E_i} = \frac{\delta R/R}{4 + 2(\delta R/R)} \quad (6.24)$$

In contrast with the null method of operation of a Wheatstone bridge, the deflection bridge requires a meter capable of accurately indicating the output voltage, as well as a stable and known input voltage. Perhaps the biggest advantage of the deflection bridge is associated with its frequency response. The bridge output should follow any resistance changes over any frequency input, up to the frequency limit of the detection device! Such a bridge circuit may be connected to an oscilloscope to allow the observation and recording of signals having very high frequency content.

If the high-impedance voltage measuring device in Figure 6.18 is replaced with a relatively low-impedance current measuring device and the bridge is operated in an unbalanced condition, a current-sensitive bridge circuit results. Consider Kirchoff's laws applied to the Wheatstone bridge circuit for a galvanometer resistance R_g. The input voltage is equal to the voltage drop in each arm of the bridge,

$$E_i = I_1 R_1 + I_2 R_2$$

but

$$I_2 = I_1 - I_g$$

which implies

$$E_i = I_1(R_1 + R_2) - I_g R_2 \quad (6.25)$$

If we consider the voltage drops in the path through R_1, R_g, and R_3, the total voltage drop must be zero

$$I_1 R_1 + I_g R_g - I_3 R_3 = 0 \quad (6.26)$$

For the circuit formed by R_g, R_4, and R_2,

$$I_g R_g + I_4 R_4 - I_2 R_2 = 0$$

or with $I_2 = I_1 - I_g$ and $I_4 = I_3 + I_g$,

$$I_g R_g + (I_3 + I_g)R_4 - (I_1 - I_g)R_2 = 0 \tag{6.27}$$

Equations 6.25–6.27 form a set of three simultaneous equations in the three unknowns I_1, I_g, and I_3. Solving these three equations for I_g yields

$$I_g = \frac{E_i(R_3 R_2 - R_1 R_4)}{R_3(R_1 + R_2)(R_g + R_2 + R_4) + R_1 R_2 R_4 - R_3 R_2^2 + R_g R_4(R_1 + R_2)} \tag{6.28}$$

Then, the change in resistance of R_1 can be found in terms of the bridge deflection voltage, E_o, by

$$\frac{\delta R}{R_1} = \frac{(R_3/R_1)[E_o/E_i + R_2/(R_2 + R_4)]}{1 - E_o/E_i - R_2/(R_2 + R_4)} - 1 \tag{6.29}$$

Consider the case when all of the resistances in the bridge are initially equal to R, and subsequently R_1 changes by an amount δR. The current through the meter is given by

$$I_g = E_i \frac{\delta R/R}{4(R + R_g)} \tag{6.30}$$

and the output voltage is given by $E_o = I_g R_g$:

$$E_o = E_i \frac{\delta R/R}{4(1 + R/R_g)} \tag{6.31}$$

The bridge impedance can affect the output from a constant voltage source having an internal resistance R_s. The effective bridge resistance, based on a Thévenin equivalent circuit analysis, is given by

$$R_B = \frac{R_1 R_2}{R_1 + R_2} + \frac{R_3 R_4}{R_3 + R_4} \tag{6.32}$$

such that for a power supply of voltage E_s

$$E_i = \frac{E_s}{R_s + R_B} \tag{6.33}$$

In a similar manner, the bridge impedance can affect the voltage indicated by the voltage measuring device. For a voltage measuring device of internal impedance R_g, the actual bridge deflection voltage, relative to the indicated voltage, E_m, is

$$E_o = E_m\left(\frac{R_1}{R_1 + R_3} - \frac{R_2}{R_2 + R_4}\right) \tag{6.34}$$

EXAMPLE 6.3

A certain temperature sensor experiences a change in electrical resistance with temperature according to the equation

$$R = R_0[1 + \alpha(T - T_0)]$$

where

R = sensor resistance [Ω]
R_0 = sensor resistance at the reference temperature, T_0 [Ω]
T = temperature [°C]
T_0 = reference temperature [°C]
α = constant: 0.00395 °C^{-1}

This temperature sensor is connected in a Wheatstone bridge like the one shown in Figure 6.17, where the sensor occupies the R_1 location, and R_2 is a calibrated variable resistance. The bridge is operated using the null method. The fixed resistances R_3 and R_4 are each equal to 500 Ω. If the temperature sensor has a resistance of 100 Ω at 0 °C, determine the value of R_2 that would balance the bridge at 0 °C.

KNOWN

$R_1 = 100\ \Omega$

$R_3 = R_4 = 500\ \Omega$

FIND

R_2 for null balance condition

SOLUTION

From equation 6.19, a balanced condition for this bridge would be achieved when $R_2 = R_1$. Thus, a value of 100 Ω for R_2 would create a balanced condition. Notice that to be a useful circuit, R_2 must be adjustable and provide an indication of its resistance value at any setting.

EXAMPLE 6.4

Consider a deflection bridge, which initially has all arms of the bridge equal to 100 Ω, with the temperature sensor described in Example 6.3 again as R_1. The input or supply voltage to the bridge is 10 V. If the temperature of R_1 is changed such that the bridge output is 0.569 V, what is the temperature of the sensor? How much current flows through the sensor, and how much power must it dissipate?

KNOWN

$E_i = 10$ V

Initial state:

$$R_1 = R_2 = R_3 = R_4 = 100 \; \Omega$$

$$E_o = 0 \; V$$

Deflection state:

$$E_o = 0.569 \; V$$

ASSUMPTION

Voltmeter has infinite input impedance but source has negligible impedance ($E_s = E_i$).

FIND

$T_{R_1}, I_{R_1}, P_{R_1}$

SOLUTION

The change in the sensor resistance can be found from equation 6.24 as

$$\frac{E_o}{E_i} = \frac{\delta R}{4R + 2\delta R} \Rightarrow \frac{0.569}{10} = \frac{\delta R}{400 + 2\delta R}$$

$$\delta R = 25.67 \; \Omega$$

This results in a sensor resistance of 125.7 Ω, and implies a sensor temperature of 65 °C.

To determine the current flow through the sensor, consider first the balanced case where all the resistances are equal to 100 Ω. The equivalent bridge resistance, R_B, is simply 100 Ω, and the total current flow from the supply, $E_i/R_B = 100$ mA. Thus, at the initially balanced condition, through each arm of the bridge and through the sensor the current flow is 50 mA. If the sensor resistance changes to 125.67 Ω, the current will be reduced. If the output voltage is measured by a high-impedance device, such that the current flow to the meter is negligible, the current flow through the sensor is given by

$$I_1 = E_i \frac{1}{R_1 + R_2}$$

This current flow is then 44.3 mA.

The power that must be dissipated from the sensor is 0.25 W, which, depending upon the surface area and local heat transfer conditions, may cause a change in the temperature of the sensor.

COMMENT

The current flow through the sensor results in a sensor temperature higher than would occur with zero current flow, due to I^2R heating. This creates a loading error in the indicated temperature. There is a tradeoff between the increased sensitivity, dE_o/dR_1, and the correspondingly increased current associated with an increased E_1. The input voltage must be chosen appropriately for a given application.

6.6 LOADING ERRORS AND IMPEDANCE MATCHING

In an ideal sense, an instrument or measurement system should not in itself affect the variable being measured. Any such effect will alter the variable and be considered as a "loading" which the measurement system exerts on the measured variable. The resulting difference between the unaltered value of the measurand and the indicated value from the measurement is a *loading error*. Loading effects can be of any form: mechanical, electrical, or optical. When the insertion of a sensor into a process somehow changes the physical variable being measured, a *process loading error* occurs. Signal loading can occur along the signal path of a measurement system. If the output from one system stage is in any way affected by the subsequent stage, then the signal is affected by *interstage loading error*. A goal in measurement system design should be to minimize loading errors.

Consider the measurement of the temperature of a volume of a high-temperature liquid using a mercury-in-glass thermometer. Some finite quantity of energy must flow from the liquid to the thermometer to achieve thermal equilibrium between the thermometer and the liquid. As a result of this energy flow, the liquid temperature is reduced, and the measured value will not correspond to the initial liquid temperature. Because the goal of the measurement was to measure the temperature of the liquid in its initial state, the act of measurement has introduced a loading error.

Or consider the current flow that drives the galvanometer in the Wheatstone bridge of Figure 6.17. Under deflection conditions, some energy must be removed from the circuit to deflect the pointer. This reduces the current in the circuit, bringing about a loading error in the measured resistance. Under null balance conditions, there is no pointer deflection to within the resolution of the meter and a negligible amount of current is removed from the circuit. There is then negligible loading error.

In general, *null balance techniques will minimize the magnitude of loading error*, usually to negligible levels. However, deflection methods can derive considerable energy from the process being measured and therefore need careful consideration to keep loading to a minimum.

Loading Errors for Voltage Dividing Circuit

Consider the voltage dividing circuit shown in Figure 6.12, for the case where R_m is finite. Under these conditions, the circuit can be represented by the equivalent circuit shown in Figure 6.19. As the sliding contact at point A moves, it divides the full-scale deflection resistance, R_T, into R_1 and R_2, such that R_1 +

FIGURE 6.19   Instruments in parallel to signal path form
an equivalent voltage dividing circuit.

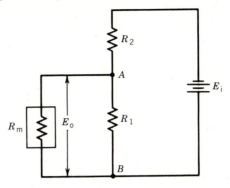

$R_2 = R_T$. The resistances R_m and R_1 form a parallel loop, yielding an equivalent
resistance R_L, given by

$$R_L = \frac{R_1 R_m}{R_1 + R_m} \tag{6.35}$$

The equivalent resistance for the entire circuit, as seen from the voltage source,
is R_{eq}:

$$R_{eq} = R_2 + R_L = R_2 + \frac{R_1 R_m}{R_1 + R_m} \tag{6.36}$$

The current flow from the voltage source is then

$$I = \frac{E_i}{R_{eq}} = \frac{E_i}{R_2 + R_1 R_m/(R_1 + R_m)}$$

and the output voltage is given by

$$E_o = E_i - IR_2$$

This result can be expressed as

$$\frac{E_o}{E_i} = \frac{1}{1 + (R_2/R_1)(R_1/R_m + 1)} \tag{6.37}$$

The limit of this expression as R_m tends to infinity is

$$\frac{E_o}{E_i} = \frac{R_1}{R_1 + R_2}$$

which is equation 6.16. Also, as R_2 approaches zero, the output voltage approaches the supply voltage, as expected. Expressing E_o/E_i found from (6.16) as $(E_o/E_i)'$, the loading error, e_1, may be given by

$$e_1 = E_i \left[\left(\frac{E_o}{E_i} \right)' - \frac{E_o}{E_i} \right]$$

$$= E_i \frac{R_1 - R_T + (R_T - R_1)[R_1/(R_m + 1)]}{R_T + (R_T^2/R_1 - R_T)(R_1/R_m - 1)} \qquad (6.38)$$

The loading error goes to zero as $R_m \rightarrow \infty$.

Interstage Loading Errors

Consider the common situation of Figure 6.20, in which the output voltage signal from one measurement system device provides the input to the following device. The open circuit potential, E_1, is present at the output terminal of device 1 with output impedance, Z_1. However, the output signal from device 1 provides the input to a second device, which, at its input terminals, has an input impedance, Z_m. As shown, the Thévenin equivalent circuit of device 1 consists of a voltage generator, of open circuit voltage E_1, with internal series impedance, Z_1. Connecting device 2 to the output terminals of device 1 is equivalent to placing Z_m across the Thévenin equivalent circuit. The finite impedance of the second device causes a current, I, to flow in the loop formed by the two terminals and Z_m acts as a load on the first device. The potential sensed by device 2 will be

$$E_m = IZ_m = E_1 \frac{1}{1 + Z_1/Z_m}$$

The original potential has been changed due to the interstage connection which has caused a loading error, $e_1 = E_1 - E_m$,

$$e_1 = E_1 \left(1 - \frac{1}{1 + Z_1/Z_m} \right) \qquad (6.39)$$

FIGURE 6.20   Equivalent circuit formed by interstage (parallel) connections.

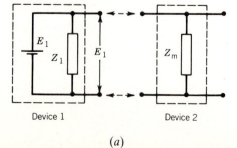

Device 1   Device 2

(a)

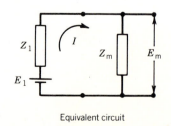

Equivalent circuit

(b)

FIGURE 6.21 Instruments in series with signal path.

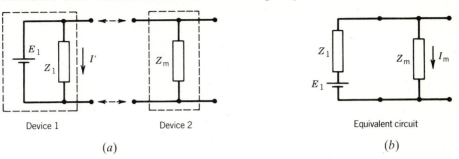

Device 1   Device 2   Equivalent circuit

(a) (b)

For maximum potential transfer between stages, it is required that $Z_m \gg Z_1$ such that $e_I \to 0$. As a practical matter, this can be difficult to achieve at reasonable cost and some design compromise is inevitable.

When the signal is current driven, the maximum current transfer between devices 1 and 2 is desirable. Consider the circuit shown in Figure 6.21 in which device 2 is current sensitive, such as with a current-sensing galvanometer. The current through the loop indicated by the device 2 is given by

$$I_m = \frac{E_1}{Z_1 + Z_m} \tag{6.40}$$

However, if the measurement device is removed from the circuit, the current is given by the short-circuit value

$$I' = \frac{E_1}{Z_1} \tag{6.41}$$

The loading error, $e_I = I' - I_m$, is

$$e_I = E_1 \frac{Z_m}{Z_1^2 + Z_1 Z_m} \tag{6.42}$$

From equation 6.42, it is clear that for maximum current transfer in which $I_m \to I'$, it is required that $Z_1 \gg Z_m$, such that $e_I \to 0$.

EXAMPLE 6.5

Consider the Wheatstone bridge shown in Figure 6.22. Find the open circuit output voltage (when $R_m \to \infty$) if the four resistances change by

$$\delta R_1 = +40 \ \Omega \qquad \delta R_2 = -40 \ \Omega \qquad \delta R_3 = +40 \ \Omega \qquad \delta R_4 = -40 \ \Omega$$

KNOWN

Measurement device resistance given by R_m

FIGURE 6.22   Figure for Example 6.5.

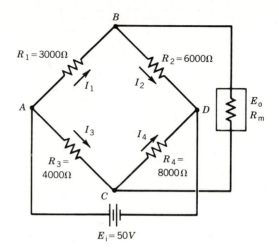

ASSUMPTIONS

$R_m \rightarrow \infty$

Negligible source impedance

FIND

E_o

SOLUTION

From equation 6.21,

$$E_o = E_i \left(\frac{R_1}{R_1 + R_2} - \frac{R_3}{R_3 + R_4} \right)$$

$$= E_i \left(\frac{5040 \ \Omega}{5040 + 4960 \ \Omega} - \frac{8040 \ \Omega}{8040 + 7960 \ \Omega} \right)$$

$$= 50 \ \text{V} \ (0.5040 - 0.5025) = +0.075 \ \text{V}$$

EXAMPLE 6.6

Consider the case where the output of the bridge in example 6.5 is measured by a meter with internal impedance, $R_m = 20\ 000 \ \Omega$. What is the output voltage with this particular meter in the circuit? Estimate the loading error.

KNOWN

$R_m = 20\ 000 \ \Omega$

FIND

E_o, e_1

SOLUTION

Because the output voltage, E_o, is equal to $I_m R_m$, equation 6.28 can be used:

$$I_m = \frac{E_i(R_3 R_2 - R_1 R_4)}{R_3(R_1 + R_2)(R_m + R_2 + R_4) + R_1 R_2 R_4 + R_m R_4(R_1 + R_2) - R_3 R_2^2}$$

For $R_m = 20\ 000\ \Omega$, this gives $E_o = -0.0566$ V.

The loading error, e_1, will be the difference between the output voltage assuming infinite impedance in the meter (Example 6.5) and the output voltage for the finite impedance value. This gives $e_1 = 18.4$ mV.

The bridge output may also be expressed as a ratio of the output voltage with $R_m \to \infty$, E_o', and the output voltage with a finite meter resistance, E_o

$$\frac{E_o}{E_o'} = \frac{1}{1 + R_e/R_m}$$

where

$$R_e = \frac{R_1 R_3}{R_1 + R_3} + \frac{R_2 R_4}{R_2 + R_4}$$

which yields for the present example

$$\frac{E_o}{E_o'} = \frac{1}{1 + 5143/20\ 000} = 0.795$$

The percent loading error, $100 \times [1 - (E_o/E_o')]$, is 20.4%.

6.7 ANALOG SIGNAL CONDITIONING: AMPLIFIERS

An amplifier is a device that scales the magnitude of an analog input signal according to the relation

$$E_o(t) = h\{E_i(t)\} \tag{6.43}$$

where $h\{E_i(t)\}$ defines a mathematical function. The simplest amplifier is the *linear scaling amplifier* in which

$$h\{E_i(t)\} = GE_i(t) \tag{6.44}$$

where the gain G is a constant that may be any positive or negative value. Many other types of operation are possible, including the "base x" *logarithmic amplifier* in which

$$h\{E_i(t)\} = G\log_x[E_i(t)] \tag{6.45}$$

Amplifiers have a finite frequency response and limited input voltage range.

The most widely used type of amplifier is the operational amplifier. This device is characterized by

- Very high input impedance ($Z_i > 10^7\ \Omega$)
- Low output impedance($Z_o < 100\ \Omega$)
- High internal gain ($A \approx 10^5$ to 10^6)

As shown in the general diagram of Figure 6.23*a*, an operational amplifier has two input ports, a noninverting and an inverting input, and one output port.

FIGURE 6.23   Operational amplifier.

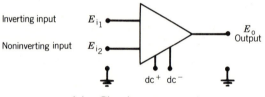

(*a*)   Circuit representation

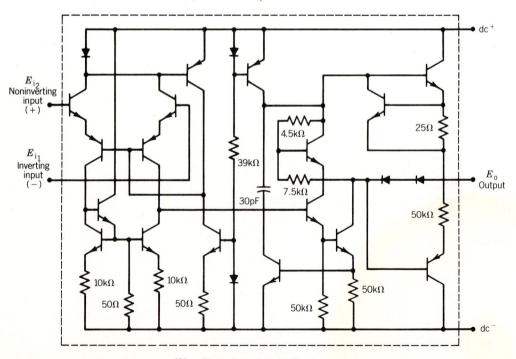

(*b*)   Type 741 circuit diagram

The signal at the output port will be in phase with a signal passed through the noninverting input port but will be 180° out of phase with a signal passed through the inverting input port. The amplifier requires dual polarity dc excitation power ranging from ± 5 V to ± 15 V. An internal schematic diagram of a common operational amplifier circuit, the type 741, is shown in Figure 6.23b. As shown, each input port is attached to the base of an npn transistor.

The high internal open loop gain, A, of an operational amplifier is given as

$$E_o(t) = A[E_{i_2}(t) - E_{i_1}(t)]$$

The magnitude of A, flat at low frequencies, falls off rapidly at high frequencies, but this intrinsic gain curve is overcome by using external input and feedback resistors which will actually set the circuit gain, G, and circuit response.

Possible amplifier configurations using an operational amplifier are shown in Figure 6.24. Because the amplifier has a very high internal gain and negligible current draw, resistors R_1 and R_2 are used to form a feedback loop and control the overall amplifier circuit gain, called the closed loop gain, G. The noninverting linear scaling amplifier circuit of Figure 6.24a has a closed loop gain of

$$G = \frac{E_o(t)}{E_i(t)} = \frac{R_1 + R_2}{R_2} \tag{6.46}$$

The inverting linear scaling amplifier circuit of Figure 6.24b provides a gain of

$$G = \frac{E_o(t)}{E_i(t)} = \frac{R_2}{R_1} \tag{6.47}$$

By utilizing both inputs, the arrangement forms a differential amplifier, Figure 6.24c, in which

$$E_o(t) = [E_{i_1}(t) - E_{i_2}(t)] \frac{R_2}{R_1} \tag{6.48}$$

The differential amplifier circuit is effective as a voltage comparator for many instrument applications.

A voltage follower circuit is commonly used to isolate an impedance load from other stages of a measurement system, such as might be needed with a high output impedance transducer. A schematic diagram of a voltage follower circuit is shown in Figure 6.24d. Note that the feedback signal is sent directly to the inverting port. For such a circuit, then,

$$E_o(t) = A[E_i(t) - E_o(t)]$$

or in terms of the circuit gain, $G = E_o(t)/E_i(t)$,

$$G = \frac{A}{1 + A} \approx 1 \tag{6.49}$$

FIGURE 6.24 Amplifier circuits using operational amplifiers.

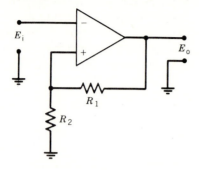

(a) Noninverting amplifier

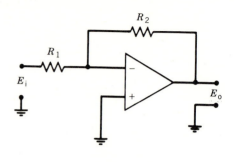

(b) Inverting amplifier

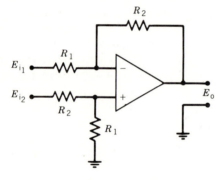

(c) Differential amplifier

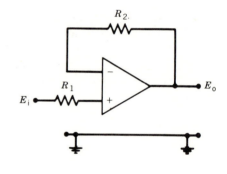

(d) Voltage follower

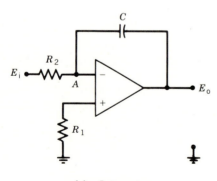

(e) Integrator

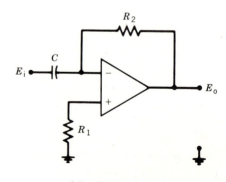

(f) Differentiator

Using Kirchhoff's law about the noninverting input loop yields,

$$I_i(t)R_1 + E_o(t) = E_i(t)$$

Then the circuit input resistance, R_i, is

$$R_i = \frac{E_i(t)}{I_i(t)} = \frac{E_i(t)R_1}{E_i(t) - E_o(t)}$$

$$= (1 + A)R_1 \tag{6.50}$$

Likewise the output resistance is found to be

$$R_o = \frac{R_2}{1 + A} \tag{6.51}$$

Because A is large, equations 6.49–6.51 show that the input impedance, developed as a resistance, of the voltage follower can be large, its output impedance can be small, and the circuit will have near unity gain. Acceptable values for R_1 and R_2 range from 10 kΩ to 100 kΩ to maintain stable operation.

Input signal integration and differentiation can be performed with the operational amplifier circuits. For the integration circuit in Figure 6.24e, the currents through R_2 and C are given by

$$I_{R_2}(t) = \frac{E_i(t) - E_A(t)}{R_2}$$

$$I_c(t) = C\frac{d}{dt}[E_o(t) - E_i(t)]$$

Summing currents at node A yields

$$E_o(t) = -R_2C \int E_i(t)\, dt \tag{6.52}$$

Following a similar analysis, the differentiator circuit shown in Figure 6.24f, will perform the operation,

$$E_o(t) = -R_2C\dot{E}_i(t) \tag{6.53}$$

6.8 ANALOG FILTERS

A *filter* is used to remove undesirable frequency information from a dynamic signal. Filters can be broadly classified as being low pass, high pass, bandpass, and notch. The ideal gain characteristics of filters can be described by the magnitude ratio plots shown in Figure 6.25. It is shown that a *low-pass* filter permits frequencies below a prescribed cut-off frequency to pass while blocking the passage of frequency information above the cutoff frequency, f_c. Similarly, a

FIGURE 6.25   Ideal filter characteristics.

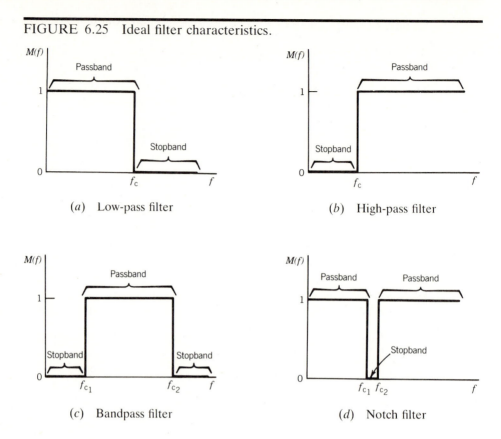

(a)   Low-pass filter

(b)   High-pass filter

(c)   Bandpass filter

(d)   Notch filter

high-pass filter permits only frequencies above the cutoff frequency to pass. A *bandpass* filter combines features of both the low- and high-pass filters. It is described by a low cutoff frequency, f_{c_1}, and a high cutoff frequency, f_{c_2}, to define a band of frequencies that are permitted to pass through the filter. A *notch* filter permits the passage of all frequencies except those within a narrow frequency band. An intensive treatment of filters for analog and digital signals can be found in many specialized texts (e.g., [3–5]).

Filters perform a well-defined mathematical operation, as specified by their transfer function, on the input signal. Passive analog filter circuits consist of combinations of resistors, capacitors, and inductors. Active filters incorporate operational amplifiers into the circuit [6].

The sharp cutoff of the ideal filter cannot be realized in a practical filter. As an example, a plot of the magnitude ratio for a real low-pass filter is shown in Figure 6.26. All filter response curves will contain a transition band over which the magnitude ratio decreases relative to the frequency. This rate of transition is known as the filter *roll-off*, usually specified in dB/decade. In addition, the filter will introduce a phase shift between its input and output signal. Depending on the transfer function used, filters can be designed to achieve certain desirable response features. For example, a relatively flat magnitude ratio over its pass-band with a moderately steep initial roll-off is a characteristic of a *Butterworth* filter response. A characteristic of a *Bessel* filter response is a linear phase shift

FIGURE 6.26   Magnitude ratio of low-pass RC
Butterworth filter.

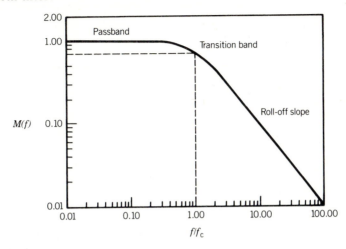

over its passband, but with a relatively nonflat magnitude ratio and a gradual
initial roll-off.

Butterworth Filter Design

A simple passive low-pass filter can be constructed using the resistor and
capacitor (RC) circuit of Figure 6.27. Application of Kirkoff's law about the
input loop gives the model relating the input voltage, E_i, to the output voltage,
E_o:

$$RC\dot{E}_o + E_o = E_i$$

This real filter model is a first-order system. The magnitude ratio for this filter
when subjected to a sinusoidal waveform input is given by equation 3.10 with
$\tau = RC$ and $\omega = 2\pi f$ and is described by Figure 6.26. The phase shift is given
in equation 3.9.

The cutoff frequency of a filter, f_c, is usually defined as the frequency at which
the magnitude ratio is reduced to 70.7%. In terms of the decibel (dB)

$$dB = 20 \log M(f) \tag{6.54}$$

FIGURE 6.27   Low-pass RC Butterworth filter.

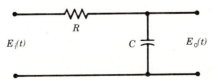

FIGURE 6.28   Multistage or cascading RC filters.

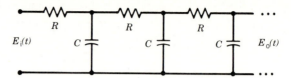

f_c occurs at -3 dB. For filter design purposes, this requires that

$$\tau = \frac{1}{2\pi f_c} \tag{6.55}$$

The roll-off characteristics of a filter can be improved by the placement of several filters in series, known as "cascading" or "multipole" filters. In such an arrangement, the respective transfer functions for each filter stage will multiply. For example, for k low-pass filters, each designed for the same value of f_c, and placed in series as in Figure 6.28, the transfer function will have the form

$$G(s) = \frac{1}{1 + (s/2\pi f_c)^k} \tag{6.56}$$

The transfer function of (6.56) defines a k-stage low-pass Butterworth filter. The magnitude ratio and phase shift become

$$M(f) = \frac{1}{[1 + (f/f_c)^{2k}]^{1/2}} \tag{6.57}$$

$$\Phi(f) = \sum_{i=1}^{k} \Phi_i(f)$$

Hence, the RC filter of Figure 6.27 falls into the Butterworth class of filters.

A k-stage high-pass RC Butterworth filter is shown in Figure 6.29. The transfer function for this filter is

$$G(s) = \left[\left(\frac{2\pi f_c}{s} \right)^k + 1 \right]^{-1} \tag{6.58}$$

with f_c defined from (6.55).

FIGURE 6.29   Multistage high-pass RC Butterworth filter.

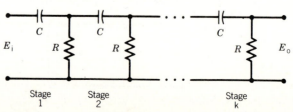

EXAMPLE 6.7

Design a one-stage Butterworth RC low-pass filter with a cutoff frequency of -3 dB set at 100 Hz. Calculate the attenuation at 60 and 200 Hz.

KNOWN

$f_c = 100$ Hz $k = 1$

$M(100$ Hz$) = 0.707$

FIND

R, C, and δ

SOLUTION

A single-stage Butterworth RC filter circuit is shown in Figure 6.27. Using $f = 2\pi\omega$, the magnitude ratio for this circuit is given by

$$M(f) = \frac{1}{[1 + (2\pi f \tau)^2]^{1/2}}$$

Setting $M(f) = 0.707 = -3$ dB with $f = f_c = 100$ Hz yields, from (6.55),

$$\tau = \frac{1}{2\pi f_c} = RC = 0.0016 \text{ s}$$

One possible combination might be $R = 800$ Ω and $C = 2$ μF.

The attenuation is found directly from the dynamic error, $\delta(f) = M(f) - 1$. The attenuation at 60 Hz would be

$$\delta(60) = M(60 \text{ Hz}) - 1 = 0.86 - 1$$

$$= -0.14$$

So the amplitude of the input signal at 60 Hz will be down 14% after passing through this filter. Ideally, there would be no attenuation for frequencies below the cutoff frequency. But practically this is not possible.

At 200 Hz,

$$\delta(200) = -0.55$$

or over 55% of the signal amplitude at 200 Hz has been eliminated.

EXAMPLE 6.8

For the low-pass filter design of Example 6.7, determine the filter response improvement obtained at 60 and at 200 Hz by using two stages.

KNOWN

$f_c = 100$ Hz $k = 2$

$M(100$ Hz$) = 0.707$ $\tau = 0.0016$ s

ASSUMPTIONS

Filters have identical design

FIND

$\delta(60)$ and $\delta(200)$

SOLUTION

For k low-pass RC filters in series, the magnitude ratio, $M(f)$, is given by equation 6.57 and the attenuation by $\delta(f) = M(f) - 1$. For the values given,

$$\delta(60) = -0.06$$

$$\delta(200) = -0.76$$

COMMENT

These numbers illustrate the improved filter characteristics available using cascaded filters. The use of two cascaded filters relative to one, as in Example 6.7, has decreased the undesirable signal attenuation at 60 Hz while providing an even better signal attenuation at 200 Hz. However, the use of cascaded Butterworth filters increases phase shift and phase distortion.

Bessel Filter Design

A Bessel filter sacrifices a flat gain over its passband in exchange for a linear phase shift. A k-stage low-pass Bessel filter has the transfer function

$$G(s) = \frac{a_0}{a_0 + a_1 s + \cdots + a_k s^k} \tag{6.59}$$

For design purposes, this can be rewritten as

$$G(s) = \frac{a_0}{D_k(s)}$$

where

$$D_k(s) = (2k - 1)B_{k-1}(s) + s^2 B_{k-2}(s) \tag{6.60a}$$

and

$$D_1(s) = s + 1 \tag{6.60b}$$

FIGURE 6.30   Multistage low-pass LC Bessel filter.

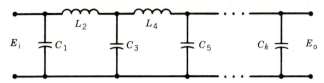

A k-stage LC low-pass Bessel filter is shown in Figure 6.30. Table 6.1 lists the required values for elements L and C for a 1- through 5-stage filter based on (6.60).

6.9 SAMPLING CONCEPTS

Consider the analog signal and its discrete time series representation in Figure 6.31. There would appear to be quite a difference in the information contained in the analog and discrete representations. However, the important analog signal information concerning amplitude and frequency can be well represented by a discrete series. Just how well represented will depend on:

1. The frequency content of the measured analog signal.
2. The size of the time increment taken between each discrete number.
3. The total sample period of the measurement.

It was discussed in Chapter 2 how a continuous dynamic signal could be represented by a Fourier series. The discrete Fourier transform (DFT) was also introduced as a method for the reconstruction of a dynamic signal from a discrete set of data. The DFT, through equation 2.32, conveys all the necessary information needed to reconstruct the Fourier series of a continuous dynamic signal from a representative discrete time series. Hence, Fourier analysis will provide certain guidelines for sampling continuous data. In this section, the concept of converting a continuous analog signal into an equivalent discrete time series is examined. Extended discussions on this subject of signal analysis can be found in many texts (e.g., [6,7,10]).

Sample Rate

Just how frequently must a time-dependent signal be measured to determine its frequency content? Consider Figure 6.32a in which the magnitude variation

TABLE 6.1   Values for Low-Pass LC Bessel Filter[a]

k	C_1	L_2	C_3	L_4	C_5
1	2.0				
2	1.577	0.423			
3	1.255	0.553	0.192		
4	1.060	0.512	0.318	0.110	
5	0.930	0.458	0.331	0.209	0.072

[a]Values for C are in farads and for L are in henries.

FIGURE 6.31   Analog and discrete representations of a time-varying signal.

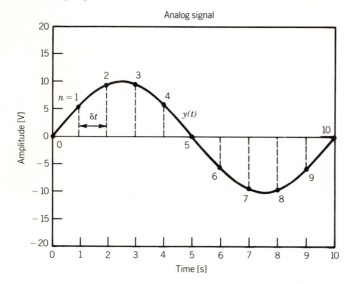

n	Discrete data
0	0
1	5.9
2	9.5
3	9.5
4	5.9
5	0
6	−5.9
7	−9.5
8	−9.5
9	−5.9
10	0

of a 10-Hz sine wave is plotted versus time over time period, t_f. Suppose this sine wave is measured repeatedly at successive sample time increments, δt. This corresponds to measuring the signal with a sample frequency or *sample rate* of $f_s = 1/\delta t$. For this discussion, we will assume that the signal measurement occurs at a constant sample rate. At each measurement, the sine wave is converted into a number. For comparison, in Figures 6.32*b–d* the resulting series versus time plots are given when the signal is measured using sample time increments (or sample rates) of (*b*) 0.010 s (f_s = 100 Hz), (*c*) 0.037 s (f_s = 27 Hz), and (*d*) 0.083 s (f_s = 12 Hz). It is apparent that the sample rate has a significant effect on our perception and reconstruction of the continuous analog signal in the time domain. As sample rate decreases, the amount of information per unit time describing the signal decreases. In Figures 6.32*b* and *c*, we can still discern the 10-Hz frequency content of the original signal. But as seen in Figure 6.32*d*, an interesting phenomenon occurs if the sample rate becomes too slow: The frequency of the original sine wave appears to be a lower frequency.

It can be concluded that the sample time increment or the corresponding sample rate plays a significant role in signal frequency representation. The *sampling theorem* states that to reconstruct the frequency content of a measured signal accurately, *the sample rate must be more than twice the highest frequency contained in the measured signal*. Denoting the maximum frequency in the analog signal as f_m, the sampling theorem requires

$$f_s > 2f_m \qquad (6.61)$$

or in terms of sample time increment,

$$\delta t < \frac{1}{2f_m} \qquad (6.62)$$

FIGURE 6.32   Effect of sample rate on signal frequency
interpretation.

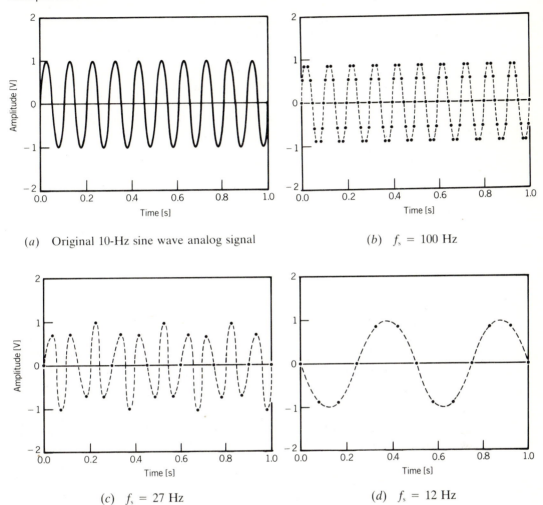

(*a*)   Original 10-Hz sine wave analog signal

(*b*) $f_s = 100$ Hz

(*c*) $f_s = 27$ Hz

(*d*) $f_s = 12$ Hz

When signal frequency content is important, equations 6.61 and 6.62 provide
a criterion for the minimum sample rate or maximum sample time increment,
respectively, to be used in converting data from a continuous to a discrete form.
The frequencies that will be defined by the DFT of the resulting discrete series
will provide an accurate representation of the original signal frequencies regardless of the sample rate used, provided that the requirements of the sampling
theorem are satisfied.

Alias Frequencies

When a signal is sampled at a sample rate that is less than $2f_m$, the higher
frequency content of the analog signal will take on the false identity of a lower
frequency in the resulting discrete series. This is seen to occur in Figure 6.32*d*,

where, because $f_s < 2f_m$, the 10-Hz analog signal is observed to take on the false identity of a 2-Hz signal. A misinterpretation of the frequency content of the original signal results! Such a false frequency is called an *alias frequency*.

The alias phenomenon is an inherent consequence of a discrete sampling process. To illustrate this, consider a simple periodic signal which can be described by the one-term Fourier series:

$$y(t) = C \sin[2\pi ft + \phi(f)]$$

Suppose $y(t)$ is measured with a sample time increment of δt, so that its discrete time signal is given by

$$y(r\,\delta t) = C \sin[2\pi fr\,\delta t + \phi(f)] \qquad r = 0, 1, 2, \ldots, (N-1)$$

Now using the identity, $\sin x = \sin(x + 2\pi q)$, where q is any integer, we rewrite $y(r\,\delta t)$ as

$$C \sin[2\pi fr\,\delta t + \phi(f)] = C \sin[2\pi fr\,\delta t + 2\pi q + \phi(f)]$$
$$= C \sin\left[2\pi\left(f + \frac{m}{\delta t}\right) r\,\delta t + \phi(f)\right]$$

where $m = 0, 1, 2, \ldots$ (and hence, $mr = q$ is an integer). This implies that for any value of δt, the frequencies of f and $f + m/\delta t$ must be indistinguishable. Hence, all frequencies given by $f + m/\delta t$ are the alias frequencies of f. However, by adherence to the sampling theorem criterion, all $m \geq 1$ will be eliminated from the sampled signal and, thus, this ambiguity between frequencies is avoided.

This same discussion must apply to complex periodic, aperiodic, and non-deterministic waveforms. This is immediately seen by considering the general Fourier series used to represent such signals. Such a discrete series,

$$y(r\,\delta t) = \sum_{n=1}^{\infty} C_n \sin[2\pi nfr\,\delta t + \phi_n(f)]$$

for $r = 0, 1, 2, \ldots, (N-1)$, can be rewritten as

$$y(r\,\delta t) = \sum_{n=1}^{\infty} C_n \sin\left[2\pi\left(nf + \frac{nm}{\delta t}\right) r\,\delta t + \phi_n(f)\right]$$

which displays the same aliasing phenomenon.

In general, the Nyquist frequency defined by

$$f_N = \frac{f_s}{2} = \frac{1}{2\delta t} \tag{6.63}$$

represents a folding point for the aliasing phenomenon. All actual frequency content in the analog signal that is at frequencies above f_N will appear as alias frequencies of less than f_N; that is, such frequencies will be folded back and perceived as lower frequencies.

An alias frequency, f_a, can be computed from the folding diagram of Figure 6.33 in which the original frequency axis is folded back over itself at the folding point of f_N and its harmonics, mf_N, where $m = 1, 2, \dot{.} \ldots$ Use of the folding diagram is illustrated in Example 6.9.

How does one avoid the alias phenomenon when sampling a signal of unknown frequency content? If the frequency range of interest is limited to a certain maximum frequency, the signal should be filtered at and above this frequency prior to the sampling process. Otherwise, the maximum frequency is to be limited by the maximum sample rate possible and the sampling theorem. In all cases, once a sample rate is chosen *the signal must be filtered at and above f_N to avoid the problem of alias frequencies.*

EXAMPLE 6.9

A 10-Hz sine wave is sampled at 12 Hz. Compute the maximum frequency that can be represented in the resulting discrete signal. Compute the alias frequency.

KNOWN

$$f = f_m = 10 \text{ Hz}$$

$$f_s = 12 \text{ Hz}$$

ASSUMPTIONS

Constant sample rate

FIGURE 6.33   Folding diagram.

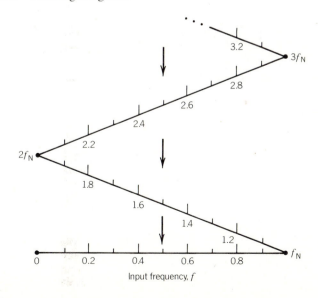

Input frequency, f

FIND

f_N and f_a

SOLUTION .

The Nyquist frequency, f_N, sets the maximum frequency that can be represented in a resulting data set. Using (6.63) with $f_s = 12$ Hz gives the Nyquist frequency, $f_N = 6$ Hz.

All frequency content in the analog signal above f_N will appear as an alias frequency, f_a, between 0 and f_N. Because $f = 10$ Hz and $f_N = 6$ Hz, $f/f_N \approx 1.67$, so that f must be folded back and appear as a frequency between 0 and f_N. From Figure 6.33, an $f = 1.67f_N$ will be folded back to $f_a = 0.33f_N = 2$ Hz. As a consequence, a 10-Hz sine wave sampled at 12 Hz will be completely indistinguishable from a 2-Hz sine wave sampled at 12 Hz in discrete time series.

EXAMPLE 6.10

A complex periodic signal has the form

$$y(t) = A_1 \sin 2\pi(25)t + A_2 \sin 2\pi(75)t + A_2 \sin 2\pi(125)t$$

If the signal is sampled at 100 Hz, determine the frequency content of the resulting discrete series.

KNOWN

$f_s = 100$ Hz $f_2 = 75$ Hz

$f_1 = 25$ Hz $f_3 = 125$ Hz

ASSUMPTIONS

Constant sample rate

FIND

f_{a_1}, f_{a_2}, and f_{a_3}, in $\{y(n\,\delta t)\}$

SOLUTION

From (6.63) for a sample rate of 100 Hz, the Nyquist frequency is 50 Hz. All frequency content in the analog signal above f_N will take on an alias frequency between 0 and f_N in the sampled data series. With $f_1 = 0.5f_N$, $f_2 = 1.5f_N$, and $f_3 = 2.5f_N$, we can use Figure 6.33 to determine the respective alias frequencies.

i	f_i	f_{a_i}
1	25 Hz	25 Hz
2	75 Hz	25 Hz
3	125 Hz	25 Hz

So in the resulting series, $f_{a_1} = f_{a_2} = f_{a_3} = 25$ Hz. Due to aliasing, the 75- and 125-Hz components would be completely indistinguishable from the 25-Hz component. The discrete series would be described by

$$y(n \, \delta t) = (A_1 + A_2 + A_3) \sin 2\pi(25)n \, \delta t \qquad n = 0, 1, 2, \ldots$$

Clearly, this misinterprets the original analog signal.

Amplitude Ambiguity

For simple and complex periodic waveforms, the discrete Fourier transform of the sampled discrete time signal will remain unaltered by a change in the total sample period, $N \, \delta t$, provided that

1. The total sample period remains an integer multiple of the fundamental period, T_1, of the measured continuous waveform.
2. The sample time increment meets the sampling theorem criterion.

As a consequence, the amplitudes associated with each frequency in the periodic signal, the spectral amplitudes, will be accurately represented by the DFT. This suggests that an original periodic waveform can be completely reconstructed from a discrete time series regardless of the sample time increment used. The total sample period will define the frequency resolution of the DFT,

$$\delta f = \frac{1}{N \, \delta t} = \frac{f_s}{N} \tag{6.64}$$

The frequency resolution plays a crucial role in the reconstruction of the signal amplitudes, as noted below.

An important difficulty appears when $N \, \delta t$ is not coincident with an integer multiple of the fundamental period of $y(t)$: The resulting DFT cannot exactly represent the spectral amplitudes of the sampled continuous waveform. The problem is brought on by the truncation of a complete cycle of the signal (see Figure 6.34), and from spectral resolution, because the associated fundamental frequency and its harmonics will not be coincident with a center frequency of the DFT. However, this error in the spectral amplitudes will decrease as the value of $N \, \delta t$ more closely approximates an exact integer multiple of T_1.

This situation is illustrated in Figure 6.34, which compares the amplitude spectrum resulting from sampling the signal, $y(t) = 10 \cos 628t$ over different sample periods. Two sample periods of 0.0256 and 0.1024 s were used with $\delta t = 0.1$ ms, and a third sample period of 0.08 s with $\delta t = 0.3125$ ms.[4] These sample periods provide for a DFT frequency resolution, δf, of about 39, 9.8, and 12.5 Hz, respectively.

The two spectra shown in Figures 6.34a and b display a spike near the correct 100 Hz with surrounding noise spikes, known as leakage, at adjacent frequencies.

---

[4]Most DFT algorithms require that $N = 2^M$, where M is an integer. This affects the selection of δt.

FIGURE 6.34   Amplitude spectra for $y(t)$.

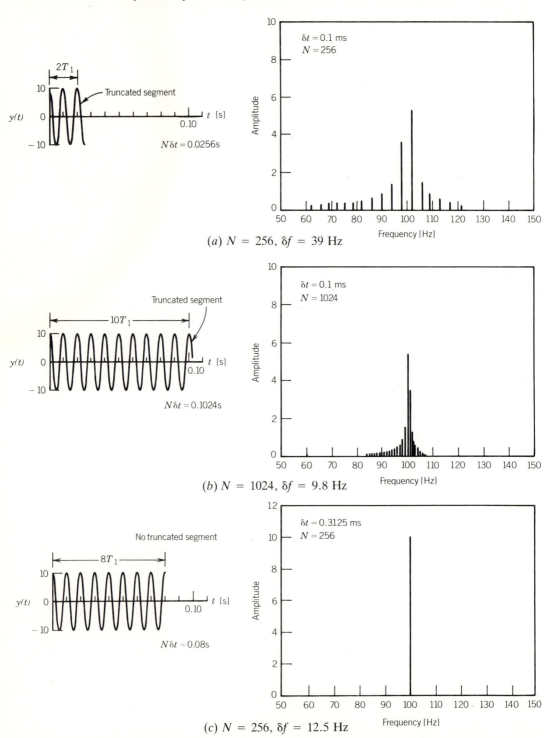

(a) $N = 256$, $\delta f = 39$ Hz

(b) $N = 1024$, $\delta f = 9.8$ Hz

(c) $N = 256$, $\delta f = 12.5$ Hz

From the standpoint of minimum leakage and maximum amplitude, Figure 6.34*c* provides the most accurate representation of the amplitude of $y(t)$. Its sample period corresponds to exactly 8 periods of $y(t)$ and the frequency of 100 Hz corresponds to the center frequency of the eighth frequency interval of the DFT. As seen in Figure 6.34, the loss of accuracy of a DFT representation occurs in the form of amplitude "leakage" to adjacent frequencies. To the DFT, the truncated segment of the sampled signal appears as an aperiodic signal. The DFT returns the correct spectral amplitudes for both the periodic and aperiodic signal portions. But as a result, the spectral amplitudes for any truncated segment will be superimposed onto portions of the spectrum adjacent to the correct frequency. The effects of leakage are also apparent in the time signal of Figure 6.32c. By varying the sample period or its equivalent, the DFT resolution, one can minimize leakage and control the accuracy of the spectral amplitudes.

If $y(t)$ is an aperiodic or nondeterministic waveform, there may not be a fundamental period. In such situations, one controls the accuracy of the spectral amplitudes by varying the DFT resolution, δf, to minimize leakage. An associated method uses a cut and try approach whereby δf is varied and the resulting signal statistics estimated and compared. The statistics will approach an acceptable level of precision when the DFT resolution is reduced to an adequate level.

In summary, *the reconstruction of a measured waveform from a discrete signal is controlled by the sampling rate and the DFT resolution.* By adherence to the sampling theorem, one controls the frequency content of both the measured signal and the resulting spectrum. By variation of δf, one can control the accuracy of the spectral amplitude representation.

6.10 DIGITAL DEVICES: BITS AND WORDS

Digital signals are discrete in both time and amplitude. Nearly all digital measuring systems will use some variation of a binary numbering system to represent and to transmit the signal information in digital form. A binary system is a number system using base 2.

Binary systems use the binary digit or *bit* as the smallest unit of information. A bit will consist of only a single digit, either a 1 or a 0. Bits are electrical switches used to convey both logical and numerical information. From a logic standpoint, the 1 or 0 are represented by the on or off switch settings. By appropriate action a bit can be reset to either on or off positions thereby possessing a value of 1 or 0. By combining bits it is possible to define integer numbers greater than a 1 or 0. Such a collection of bits are used to express numerical information in the entity called a *word*. Typically, word length may range from 4 bits for a small dedicated digital device up to 32 bits or more for a large microcomputer. A particular physical location of the collection of bits used to define a particular word is known as a *register*.

The number of bits that define a word determines the size of the maximum number that the word can represent. For example, since any bit can represent either a 1 or 0 then a combination of two bits can represent 2^2 or 4 possible combinations of bit arrangements: 00, 01, 10, or 11. We can alternatively reset this 2-bit word to produce these 4 different arrangements which represent the decimal, that is, base 10, integer numbers 0, 1, 2, or 3, respectively. Thus, the M bits in an M-bit word can be combined to represent any of 2^M integers. An 8-bit word can represent the numbers 0 through 255; a 16-bit word can represent 0 through 65 535.

The numerical value for a word is computed by moving by bit from right to left. From the right side, each successive bit will increase the value of the word by a unit, a two, a four, an eight, and so forth through 2^M, provided that the bit is in its on (value of 1) position; otherwise, the particular bit increases the value of the word by zero. A weighting scheme of an M-bit word is given below:

Bit M-1	. . .	Bit 3	Bit 2	Bit 1	Bit 0
2^{M-1}	. . .	2^3	2^2	2^1	2^0
2^{M-1}	. . .	8	4	2	1

Using this scheme, bit M-1 is known as the most significant bit (MSB) because its contribution to the numerical level of the word is the largest relative to the other bits. Bit 0 is known as the least significant bit (LSB). This type of binary code is known as the *straight binary code*.

An alternative form is known as *binary coded decimal* or BCD. In this form, each digit of a decimal number is represented separately by a binary number. For example, the base 10 number 532_{10} is represented by the BCD number 0101 0011 0010 (i.e., binary 5, binary 3, binary 2). This form is often that used to transfer data between instruments.

EXAMPLE 6.11

The contents of a 4-bit register is the binary word 0101. Convert this to its decimal equivalent assuming a straight binary code.

KNOWN

4-bit register

ASSUMPTIONS

Straight binary code

FIND

Convert to base 10

SOLUTION

The content of the 4-bit register is the binary word 0101. This will represent

$$0*2^3 + 1*2^2 + 0*2^1 + 1*2^0 = 5$$

The equivalent of 0101 in decimal is 5.

6.11 DIGITAL DEVICES: VOLTAGE MEASUREMENTS

Digital measurement devices will consist of several components that interface the digital device with the analog world. In particular, the digital-to-analog converter and the analog-to-digital converter are discussed below. These form the major components of a digital voltmeter.

Digital-to-Analog Converter

A digital-to-analog converter is an M-bit digital device that converts a digital binary word into an analog voltage [3,7]. One possible scheme uses an M-bit register with a weighted resistor network and operational amplifier, as depicted in Figure 6.35. The network consists of M binary weighted resistors having a common summing point. The resistor associated with the register MSB will have a value of R. At each successive bit the resistor value is doubled, so that the

FIGURE 6.35   Digital-to-analog converter.

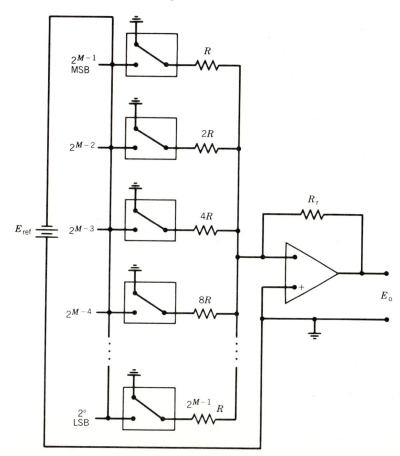

resistor associated with the LSB will have a value of $(2^{M-1}\,R)$. The network output is a current, I, given by

$$I = E_{ref} \sum_{m=1}^{M} \frac{c_m}{2_{m-1}R}$$

The values for c_m are either 0 or 1 depending on the associated mth bit value of the register that controls the switch setting. The output voltage from the amplifier is

$$E_o = IR_r \tag{6.65}$$

Note that this is equivalent to an operation in which an M-bit D/A converter would compare the magnitude of the actual input binary number, X, contained within its register to the largest possible number 2^M. This ratio determines E_o from

$$E_o = \frac{X}{2^M}$$

The D/A converter will have both digital and analog specifications, the latter expressed in terms of its full-scale analog voltage range output (E_{FSR}). Typical values for E_{FSR} are 0 to 10 V and ± 5 V and for M are 8, 12, 16, and 18 bits.

Analog-to-Digital Converter

An analog-to-digital converter is used to perform the actual conversion of an analog input value into a digital binary word. The conversion is a discrete process, taking place one number at a time. The A/D converter has an analog side and a digital side which require both analog and digital specifications. The analog side will have a full-scale analog voltage range which defines the minimum and the maximum input voltage magnitude that the device will convert. Typical E_{FSR} ranges include the unipolar 0 to 10 V and the bipolar ± 5 V. The digital side is specified in terms of the A/D converter bit number. An M-bit A/D converter can represent an analog magnitude in terms of up to 2^M integer binary numbers. This process of assigning a digital number to represent an analog signal magnitude is called *quantization*.

The resolution of an A/D converter is given in terms of volts per bit,

$$Q = \frac{E_{FSR}}{2^M} \tag{6.66}$$

Typical values for M are 8, 12, 14, and 16 bits. Resolution is sometimes specified in terms of a *dynamic range,* based on the unit of a decibel (dB), defined as

$$\text{dynamic range [dB]} = 20 \log Q \tag{6.67}$$

Many converters are bipolar and may use a variation of the straight binary code to enable them to distinguish a positive voltage from a negative voltage.

One such variation is called a *one complementary binary code*, which uses the MSB as a sign indicator, a 0 for positive numbers and a 1 for negative numbers. Three primary sources of error that are intrinsic to any A/D converter are

- Quantization error
- Saturation error
- Conversion error

Each will be discussed below.

Quantization Error

The relationship between the analog input and the digital output of an A/D converter is represented in Figure 6.36 for a 2-bit A/D converter having a full-scale range of 0 to 4 V and, therefore, $Q = 1$ V/bit. For such an A/D converter, an input voltage, of, say, 0.3 V would result in the same digital output of 0 as would an input of 0.0 or 0.8 V. So the limited resolution of the A/D converter can bring about an error in the analog-to-digital conversion. The difference, due to A/D converter resolution, between the digital value assigned by an A/D converter and the analog value represented is called a *quantization error*. The absolute quantization error, e_Q, is based on the analog resolution of the LSB and must be bounded by $\pm\frac{1}{2}Q$. The effect of quantization error becomes significant when measuring small voltages.

Saturation Error

An A/D converter is limited by both the minimum and maximum analog voltage that it can convert. If either limit is exceeded, the A/D converter output will not be able to represent the voltage in a correct digital form. Instead, the output from the A/D converter becomes saturated and will no longer change with an increase in input level. As noted in Figure 6.36, an input to the 0- to

---

FIGURE 6.36   Binary quantization and saturation.

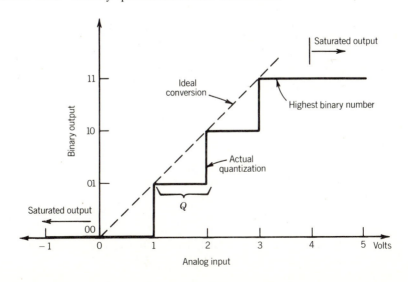

4-V, 2-bit A/D converter above 4 V results in an output of 11 and below 0 V of 00. A *saturation error* is defined by the difference between the input analog signal level and the equivalent digital value assigned by the A/D converter. If an analog signal level is unbounded, then the saturation error associated with it will be unbounded. Saturation error can be avoided by quantizing signals that remain within the limits of the A/D converter. This may require analog signal conditioning, particularly for highly dynamic signals.

Conversion Error

The ability of any measurement device to output a magnitude representative of its input value determines the device precision and bias. An A/D converter is not immune to elemental errors arising during the conversion process that can lead to a misrepresentation of the input value. As with any device, the A/D converter *conversion errors* can be delineated into hysteresis, linearity, sensitivity, zero, and repeatability errors. The extent of such errors depends on the particular method of analog-to-digital conversion. Factors that contribute to conversion error include A/D converter settling time, signal noise during the analog sampling, temperature effects, and device excitation power-setting effects [7].

EXAMPLE 6.12

Compute the relative effect of quantization error (e_Q/E_i) in the quantization of a 100-mV and a 1-V analog signal using an 8-bit and a 12-bit A/D converter, both having a full-scale range of 0 to 10 V.

KNOWN

E_i = 100 mV and 1 V

M = 8 and 12

E_{FSR} = 0 to 10 V

FIND

e_Q/E_i

SOLUTION

The resolutions for the 8-bit and 12-bit converters can be estimated from (6.66) as

$$Q_8 = \frac{E_{FSR}}{2^8} = \frac{10}{256} = 39 \text{ mV/bit}$$

$$Q_{12} = \frac{E_{FSR}}{2^{12}} = \frac{10}{4096} = 2.4 \text{ mV/bit}$$

The absolute quantization error is given by $\pm\frac{1}{2}Q \times$ LSB where the LSB $= 1$. The relative effect of the quantization error can be computed by the relative quantization error e_Q/E_i. The results are tabulated below:

E_i	M	e_Q	$100 \times e_Q/E_i$
100 mV	8	±19.5 mV	19.5%
100 mV	12	±1.2 mV	1.2%
1 V	8	±19.5 mV	1.95%
1 V	12	±1.2 mV	0.12%

From a relative point of view, the error in converting a voltage is more significant at lower voltage levels.

EXAMPLE 6.13

The A/D converter with the specifications listed below is to be used in an environment in which the A/D converter temperature may change by $\pm10\,°C$. Estimate the contributions of conversion and quantization errors to the uncertainty in the digital representation of an analog voltage by the converter.

Analog-to-Digital Converter

E_{FSR}	0–10 V
M	12 bits
Linearity	±3 bits/E_{FSR}
Temperature drift	1 bit/5 °C

KNOWN

See table

ASSUMPTIONS

Consider only the listed errors

FIND

$(u_c)_E$ measured

SOLUTION

We can estimate the instrument uncertainty as a combination of uncertainty due to quantization errors, e_Q, and due to conversion errors, e_c:

$$(u_c)_E = \sqrt{e_Q^2 + e_c^2}$$

The resolution of a 12-bit A/D converter with full-scale range of 0–10 V is given by

$$Q = \frac{E_{FSR}}{2^{12}} = \frac{10}{4096} = 2.4 \text{ mV/bit}$$

The quantization error is found to be

$$e_Q = \pm\tfrac{1}{2}Q = \pm1.2 \text{ mV}$$

Now the conversion error is affected by two elements:

$$\text{linearity error} = \pm3 \text{ bits} \times 2.4 \text{ mV/bit}$$
$$= \pm7.2 \text{ mV}$$

$$\text{temperature error} = e_2 = \frac{1 \text{ bit}}{5 °} \times 10 °C \times 2.4 \text{ mV/bit}$$
$$= \pm4.8 \text{ mV}$$

An estimate of the conversion error is found using the RSS method,

$$e_c = \pm\sqrt{e_1^2 + e_2^2}$$
$$= \pm\sqrt{(7.2 \text{ mV})^2 + (4.8 \text{ mV})^2} = \pm8.6 \text{ mV}$$

The combined uncertainty in the digital representation of the analog value due to these two errors is then

$$(u_c)_E = \pm\sqrt{(1.2 \text{ mV})^2 + (8.6 \text{ mV})^2}$$
$$= \pm8.7 \text{ mV} \quad (95\% \text{ assumed})$$

Note that the conversion errors dominate the uncertainty.

There are a number of different methods to perform the analog-to-digital conversion [3,7,8], but two common methods for converting voltage signals into binary words are discussed below.

Successive Approximation

The successive approximation technique for analog-to-digital conversion uses a trial and error approach for estimating the input level to the A/D converter. As depicted in Figure 6.37, this A/D converter uses an M-bit register, a D/A converter, and a voltage comparator. The A/D converter tests the input analog value against successive trial binary values to narrow in on the appropriate binary equivalent for the input value. At the start of the conversion sequence the A/D converter register is reset to zero and the MSB is set at 1. The D/A converter is used to convert the binary code of the register into an analog voltage. The comparator compares the D/A converter voltage to the input voltage signal.

FIGURE 6.37   Successive approximation A/D converter.

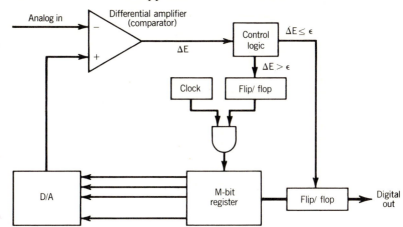

If the D/A converter value is found to be higher than the input signal, the register MSB is set to 0, otherwise it is kept at 1. The next lower significant bit is then set at 1, the register value converted to an analog value and compared. The process is continued through to the LSB. The final register value is used to represent the analog input voltage.

The successive approximation converter is typically used when conversion speed is important in A/D converter selection. It is the most popular design for fast conversion. In theory, the number of steps required to perform the conversion should be about the same as the number of bits of the A/D converter register. Thus, for a 12-bit A/D converter operating with a typical 1-MHz clock, an analog-to-digital conversion could require a time of only 12 μs. However, in practice, software interaction will increase this time.

In the successive approximation conversion technique the final register value should ideally be numerically equivalent to the analog value to within one LSB. This difference is the quantization error produced by digitizing the analog input. However, the absolute accuracy is also affected by the conversion error which will ultimately depend on the accuracies of the D/A converter and of the comparator. Also, because this method does its comparisons bit by bit any change in the input voltage level during the conversion process can cause the comparator to make an incorrect decision. This is particularly bad in the more significant bit tests. For this reason a sample and hold circuit (SHC) is used which measures the analog voltage just prior to conversion and holds that value constant until the SHC is reset. The output of the SHC then sends a constant analog voltage to the input of the A/D converter. The SHC minimizes noise during the conversion process and therefore conversion error. Following the conversion the SHC is reset.

Ramp Converters

Ramp converter methods of analog-to-digital conversion use the voltage level of a linear reference ramp signal to discern the voltage level of the analog input signal and convert it to its binary equivalent. Principal components, as shown

systems are limited as to the voltage range that they can convert. This range is usually mandated by the analog-to-digital converter used. Accordingly, amplification or attenuation of the input analog signals may be necessary.

Anti-alias analog filters are generally required to remove any frequency components of the input analog signal that will exceed the Nyquist frequency prior to quantization. A digital filter operates on a digital signal and consists of a software algorithm or dedicated preprogrammed hardware, in place of physical analog components. It performs the same mathematical transfer function on the signal as its analog counterpart. Because a digital filter operates on data that have already been quantized, these filters cannot be used to prevent aliasing, but are useful for removing noise resulting from sampling and quantization.

Converters

The data acquisition system will use an analog-to-digital converter to sample and to convert the magnitude of the analog signal into a binary number.

A separate digital-to-analog converter permits the DAS to convert digital numbers into analog voltages. These devices may be used internally by the DAS controller as comparators to check on the conversion accuracy of the A/D converter. They are also used to communicate information to external analog devices.

Clock

In a data acquisition system all operations must take place at an orderly rate. The clock provides the master timing for the data acquisition process by providing a precise stream of pulses to the various system components. In that way the multiplexer is set at port x when the A/D converter is ready to convert signal x and the storage device is ready to receive the converted signal x. A typical clock is driven by an accurate pulsing device such as a crystal controlled oscillator.

Master Controller

The data acquisition system requires a master controller to provide the start and stop sequences for data acquisition, to control the actual data flow into and out of the system, and to organize that data for storage.

A common master controller is a microprocessor system that is composed of an arithmetic logic unit, memory, control circuits, and an internal clock. The arithmetic unit performs operations by combining the contents of two registers to perform addition, subtraction, division, multiplication, comparison, logical operations, or data management. The memory is the storage area within which each word of digital information is assigned a unique physical location. Memory will contain the digital data, as well as the software-driven and fixed execution instructions. The control circuits provide routing information and set up the functions of the arithmetic unit. The controller is driven by an executive program or operating system which supervises all tasks, compilations, and input/output from the master controller. The clock provides timing pulses for execution of each step.

The controller may be a stand-alone device dedicated to performing only data acquisition or it may be part of a microcomputer capable of performing different tasks. An advantage to a dedicated controller is simplicity of the operating system and cost. A dedicated system can usually send its data to a compatible computer when desired. But a microcomputer-based system can be programmed to apply different arithmetic data reduction algorithms on the sampled data or to interface the data with available software packages.

Digital Input/Output Device

Some transducers and measuring devices output a digital signal directly. This enables bypassing the A/D converter of the DAS. Certain standards exist for the manner in which digital information is communicated between digital devices [9,11]. Two common communication standards are RS-232C and IEEE-488. The RS-232C defines a serial form of communication whereby digital numbers are transmitted in serialized pieces, usually bit by bit. Data transfer rates are given in bits per second or *baud*, with rates of 300, 1200, 2400, and 9600 baud being common. The IEEE-488 defines a parallel form of communication whereby digital numbers (i.e., groups of bits) are transmitted number by number. Most data acquisition systems can accommodate either of these two communication standards for the input of digital data, as well as the output of acquired data to other digital devices.

Input/Output Buffer

The digital output from the analog-to-digital converter is sent to the input/output buffer, a digital holding tank. The buffer is digital *random access memory* (RAM) where the data are stored until the buffer receives a command to send the data to some other storage device. Some DAS permit double buffers, which allow one buffer to absorb the data stream from the A/D converter while the other is transfering data to a storage device. Data buffers offer certain capabilities to the user such as a real time display of data and limited mathematical processing. An alternative to a buffer is a capability known as *direct memory access* (DMA). With DMA the sampled and quantized information is sent directly to the storage device. Real time display or processing is not possible. DMA provides the fastest means of data handling and is used when speed of data acquisition is of prime importance.

Output Devices

Output devices are permanent storage or display devices such as floppy disks, hard disks, magnetic tape, printer, plotter, and/or terminal screens. They provide a means for inspection and retrieval of the acquired data.

Buses

A common method for information exchange between the master controller, including internally between controller, memory and arithmetic unit, and all peripheral devices is the universal central bus. The central bus is a collective term for three information paths: the address bus, the data bus, and the control bus. It can be considered as a three-lane highway, one lane for each bus, connecting all devices that reside along branch roads off the main highway. When information is to be sent between devices, the data bus will contain the execution instructions or data information that are to be relayed. The address bus contains the address of the device and the location within the device to which the information is being sent. The control bus sends status information over the central bus. While information is being sent between any two devices connected to the central bus, a busy flag is sent out over the control bus to alert all other devices that the data bus is busy. This allows devices of different speeds to operate over the same lines.

EXAMPLE 6.14

The output from an analog device (nominal output impedance of 600 Ω) is input to the 12-bit A/D converter (nominal input impedance of 1 MΩ) of a data acquisition system. For a 2-V signal, will interstage loading be a problem?

KNOWN

$Z_1 = 600\ \Omega \quad E_1 = 2$ V

$Z_m = 1$ MΩ

FIND

e_1

SOLUTION

The loading error is given by $e_1 = E_1 - E_m$, where E_m is the measured voltage. From equation 6.39,

$$e_1 = E_1 \left(1 - \frac{1}{1 + Z_1/Z_m} \right)$$

$$= 1.2 \text{ mV}$$

Although not negligible, the interstage loading error at 2 V input actually will be less than the quantization error of the 12-bit device.

EXAMPLE 6.15

A data acquisition system is to be operated at a 26.6-kHz sample rate. A low-pass filter is needed to provide no more than a 3-dB attenuation of the analog input signal from 0 to 5 kHz and with an attenuation of at least 23 dB for all $f \geq 10$ kHz in the analog signal. Specify a suitable RC Butterworth filter.

KNOWN

$f_s = 26.6$ kHz

$M(10 \text{ kHz}) \geq -23$ dB

$M(0 \text{ to } 5 \text{ kHz}) \leq -3$ dB

RC filter

FIND

R, C, and k

SOLUTION

For a Butterworth filter, signal response is gradually attenuated over the passband to a maximum of -3 dB at f_c. Hence, we can set $f_c = 5$ kHz and therefore from (6.55), $\tau = 32$ μs. Accordingly, one possible combination is $R = 100\ \Omega$ and $C = 0.32$ μF.

The number of filter stages required is found from (6.57)

$$M(10\ \text{kHz}) = -23\ \text{dB} = 0.07 = [1 + (f/f_c)^{2k}]^{-1/2}$$

which yields $k = 3.9$. Rounding to an integer stage gives $k = 4$. Four stages will provide 24-dB attenuation, which meets our needs.

EXAMPLE 6.16

Data acquisition is to be performed on an analog signal using a sample rate of 200 Hz with a 12-bit A/D converter that has a 0- to 10-V input range. The signal contains a 200-Hz component, f_1, with an amplitude of 100 mV. Design a suitable RC filter that will attenuate the 200-Hz component down to the A/D converter quantization resolution level and will act as an anti-alias filter for quantization.

KNOWN

$f_s = 200$ Hz $\quad f_1 = 200$ Hz

$M = 12 \quad\quad A_1 = 100$ mV

$E_{\text{FSR}} = 10$ V

FIND

τ and k

SOLUTION

From (6.63) for $f_s = 200$ Hz, f_N is 100 Hz, so an appropriate design for an anti-alias filter would have the properties of $f_c = 100$ Hz and $M(100\ \text{Hz}) = -3$ dB. For such a filter, $\tau = RC$ and from (6.55), $\tau = 1.6$ ms. A suitable combination of R and C should be selected as in Example 6.15.

The A/D converter quantization resolution level from (6.66) is

$$Q = \frac{E_{\text{FSR}}}{2^M} = \frac{10\ \text{V}}{4096} = 2.5\ \text{mV}$$

For a 100-mV signal at 200 Hz, this requires an attenuation of

$$M(200 \text{ Hz}) = \frac{100 \text{ mV}}{2.5 \text{ mV}} = 0.025 = -32 \text{ dB}$$

$$= [1 + (f/f_c)^{2k}]^{-1/2}$$

Setting $f_c = 100$ Hz yields $k = 5.4 \approx 6$.

6.13 SUMMARY

This chapter has focused on basic analog and digital electrical measurement devices, measurement sampling, and signal analysis concepts, as well as the components involved in automated data acquisition systems. Methods to represent and quantify both analog, discrete time, and digital signals were presented.

The selection of devices for the measurement and quantitative output of current, resistance, and voltage signals depends on the magnitude and dynamic content of those signals. Common devices for the measurement of each type of signal, whether static or dynamic, have been presented. Amplifiers and filters, two common signal conditioning devices, were introduced.

A measurement system and the act of measurement should not itself alter the value of the variable being measured. Such an alteration is called a loading error. Interstage loading error can also occur between the stages of a measurement system. A discussion concerning proper impedance matching between sensor and measurement system and between stages within a measurement system so as to minimize loading error has been presented.

Considerations for the reconstruction of waveforms from discrete data sets based on frequency and amplitude content were presented in detail. The reconstruction was shown to be controlled by the sampling theorem, the sample time period, and the discrete Fourier transform. These sampling concepts apply equally to any discrete data set, whether acquired manually or by automated sampling systems.

REFERENCES

1. Prentiss, S. R., *Oscilloscopes*, Reston (a Prentice Hall company), Reston, VA, 1981.
2. Bleuler, E., and R. O. Haxby, *Methods of Experimental Physics*, 2d ed., Vol. 2, Academic, New York, 1975.
3. Money, S. A., *Microprocessors in Instrumentation and Control*, McGraw-Hill, New York, 1985.
4. Stanley, W. D., G. R. Dougherty, and R. Dougherty, *Digital Signal Processing*, 2d ed., Reston (a Prentice Hall company), Reston, VA, 1984.
5. DeFatta, D. J., J. Lucas, and W. Hodgkiss, *Digital Signal Processing*, Wiley, New York, 1988.
6. Clayton, G. B., *Linear Integrated Circuit Applications*, Macmillan, London, 1975.

7. Vandoren, A., *Data Acquisition Systems*, Reston (a Prentice Hall company), Reston, VA, 1982.

8. Krutz, R., *Interfacing Techniques in Digital Design: Emphasis on Microprocessors*, Wiley, New York, 1988.

9. Hnatek, E., *A User's Handbook of D/A and A/D Converters*, Wiley-Interscience, New York, 1976.

10. Bendat, J., and A. Piersol, *Random Data: Analysis and Measurement Procedures*, Wiley-Interscience, New York, 1971.

11. Cline, B. E., *An Introduction to Automated Data Acquisition Systems*, Petrocelli Books, New York, 1983.

Suggested Reading

Ahmed, H., and P. J. Spreadbury, *Analogue and Digital Electronics for Engineers*, 2d ed., Cambridge University Press, Cambridge, UK, 1984.

Beauchamp, K. G., and C. K. Yuen, Data Acquisition for Signal Analysis, Allen & Unwin, London, 1980.

Evans, Alvis J., J. D. Mullen, and D. H. Smith, *Basic Electronics Technology*, Texas Instruments Inc., Dallas, TX, 1985.

Seitzer, D., G. Pretzl, and N. Hamdy, *Electronic Analog-to-Digital Converters: Principles, Circuits, Devices, Testing*, Wiley, New York, 1983.

Wobschall, D., *Circuit Design for Electronic Instrumentation: Analog and Digital Devices from Sensor to Display*, McGraw Hill, New York, 1979.

NOMENCLATURE

e	error	u_R	uncertainty in
e_Q	quantization error		resistance R
f	frequency $[t^{-1}]$	$y(t)$	analog signal
f_a	alias frequency $[t^{-1}]$	$\bar{y}$	average or mean value
f_c	filter cutoff frequency		of $y(t)$
	$[t^{-1}]$	$y_i, y(n\,\delta t)$	discrete values of a
f_m	maximum analog signal		signal, $y(t)$
	frequency $[t^{-1}]$	$\{y(n\,\delta t)\}$	complete discrete time
f_N	Nyquist frequency $[t^{-1}]$		signal of $y(t)$
f_s	sample rate frequency	y_{rms}	root-mean-square value
	$[t^{-1}]$	A	cross-sectional area of a
$h\{E_i(t)\}$	function		current-carrying loop;
k	cascaded filter stage		operational amplifier
	number		open loop gain
$\hat{k}$	unit vector along the	A_0	constant term in
	current direction		Fourier series
l	length $[l]$	A_n, A_n'	Fourier coefficients
$\mathbf{n}$	unit vector normal to	$\mathbf{B}$	magnetic field strength
	current loop	B_n, B_n'	Fourier coefficients
q	charge [C]	E	voltage [V]
t	time		

E_1 open circuit potential [V]

E_i input voltage [V]

E_m indicated output voltage [V]

E_o output voltage [V]

E_{FSR} full-scale analog voltage range

F force $[m-l-t^{-2}]$

G amplifier gain

I electric current

I_e effective current

I_g galvanometer current

L inductance [H]

M digital component bit size

$M(f)$ magnitude ratio at frequency, f

N data set size; number of turns in a current-carrying loop

$N\,\delta t$ total digital sample period

P power

R resistance $[\Omega]$

R_B effective bridge resistance

R_{eq} equivalent resistance

R_m meter resistance

Q A/D converter resolution [V/bit]

$\mathbf{T_\mu}$ torque on a current loop in a magnetic field $[m-l^2-t^{-2}]$

α angle

δf frequency resolution of DFT $[t^{-1}]$

δR change in resistance $[\Omega]$

δt sample time increment $[t]$

τ time constant $[t]$

$\phi(f)$ phase shift at frequency, f

PROBLEMS

6.1 Determine the average and rms values for the function

$$y(t) = 30 + 2 \cos 6\pi t$$

over the time periods
a.   0 to 0.1 s
b.   0.4 to 0.5 s
c.   0 to $\frac{1}{3}$ s
d.   0 to 20 s

Comment on the nature and meaning of the results in terms of analysis of dynamic signals.

6.2 The following values are obtained by sampling two time-varying signals once every 0.4 s:

t	$y_1(t)$	$y_2(t)$	t	$y_1(t)$	$y_2(t)$
0.0	0	0			
0.4	11.76	15.29	2.4	−11.76	−15.29
0.8	19.02	24.73	2.8	−19.02	−24.73
1.2	19.02	24.73	3.2	−19.02	−24.73
1.6	11.76	15.29	3.6	−11.76	−15.29
2.0	0	0	4.0	0	0

Determine the mean and the rms values for this discrete data. Discuss the significance of the rms value in distinguishing these signals.

6.3 Consider a spring–mass–damper system having a mass of 1 kg, a damping coefficient of 10 kN-s/m, and a spring constant of 60 N/m. Above what input frequency would the amplitude effectively be averaged by this system if the input is a simple periodic waveform?

6.4 Determine the maximum torque on a current loop having 20 turns, a cross-sectional area of 1 in.^{2} and experiencing a current of 20 mA. The magnetic field strength is 0.4 Wb/m^2.

6.5 A 5000-Ω voltage-dividing circuit is used in a particular application using the arrangement of Figure 6.19. What is the minimum allowable meter internal resistance such that the loading error in measuring the open circuit potential, E_o, will be less than 12% of the full-scale value? Note: a family of solutions will result.

6.6 Determine the loading error as a percentage of the output for the voltage dividing circuit of Figure 6.19, if $R_T = R_1 + R_2$ and $R_1 = kR_T$. The parameters of the circuit are

$$R_T = 500 \ \Omega \qquad E_i = 10 \ \text{V} \qquad R_m = 10 \ 000 \ \Omega \qquad k = 0.5$$

What would the loading error be if expressed as a percentage of the full-scale output? Show that the two answers for the loading error are equal when expressed in volts.

6.7 Consider the Wheatstone bridge shown in Figure 6.17. Suppose

$$R_3 = R_4 = 200 \ \Omega$$

$$R_2 = \text{variable calibrated resistor}$$

$$R_1 = \text{transducer resistance} = 40x + 100$$

    a.   When $x = 0$, what is the value of R_2 required to balance the bridge?
    b.   If the bridge is operated in a balanced condition in order to measure x, determine the relationship between R_2 and x.

6.8 For the Wheatstone bridge shown in Figure 6.17, R_1 is a sensor whose resistance is related to a measured variable x by the equation $R_1 = 20x^2$. If $R_3 = R_4 = 100 \ \Omega$ and the bridge is balanced when $R_2 = 46 \ \Omega$, determine x.

6.9 A force sensor has as its output a change in resistance. The sensor forms one leg (R_1) of a basic Wheatstone bridge. The sensor resistance with no force load is 500 Ω, and its static sensitivity is 0.5 Ω/N. Each arm of the bridge is initially 500 Ω.
    a.   Determine the bridge output for applied loads of 100, 200, and 350 N. The bridge is operated as a deflection bridge, with an input voltage of 10 V.
    b.   Determine the current flow through the sensor.
    c.   Repeat parts (a) and (b) with $R_m = 10 \ \text{k}\Omega$ and $R_S = 600 \ \Omega$.

FIGURE 6.43   Problem 6.10.

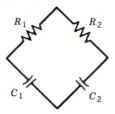

6.10   A reactance bridge arrangement replaces the resistor in a Wheatstone bridge with a capacitor or inductor. Such a reactance bridge is then excited by an ac voltage. Consider the bridge arrangement shown in Figure 6.43. Show that the balance equations for this bridge are given by

$$C_2 = C_1 \frac{R_1}{R_2}$$

6.11   Circuits containing inductance- and capacitance-type elements exhibit varying impedance depending upon the frequency of the input voltage. Consider the bridge circuit of Figure 6.44. For a capacitor and an inductor connected in a series arrangement, the impedance is a function of frequency, such that a minimum impedance occurs at the resonance frequency

$$f \doteq \frac{1}{2\pi(LC)^{1/2}}$$

where f is the frequency [Hz], L is the inductance [H], and C is the capacitance [F]. Design a bridge circuit that could be used to calibrate a frequency source at 500 Hz.

6.12   A Wheatstone bridge initially has resistances equal to $R_1 = 200\ \Omega$, $R_2 = 400\ \Omega$, $R_3 = 500\ \Omega$, and $R_4 = 600\ \Omega$. For an input voltage of 5 V, determine

FIGURE 6.44   Problem 6.11.

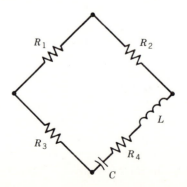

the output voltage at this condition. If R_1 changes to 250 Ω, what is the bridge output?

6.13 Construct a plot of the voltage output of a Wheatstone bridge having all resistances initially equal to 500 Ω, with a voltage input of 10 V for the following cases:

a. R_1 changes over the range 500 to 1000 Ω.

b. R_1 and R_2 change equally, but in opposite directions, over the range 500 to 600 Ω.

c. R_1 and R_3 change equally over the range 500 to 600 Ω.

Discuss the possible implications of these plots for using bridge circuits for measurements with single and multiple transducers connected as arms of the bridge.

6.14 Consider the simple potentiometer circuit shown in Figures 6.12 and 6.13. Perform a design-stage uncertainty analysis to determine the minimum uncertainty in measuring a voltage. The following information is available concerning the circuit (assume 95% confidence):

$$E_i = 10 \pm 0.1 \text{ V} \quad R_T = 100 \pm 1 \ \Omega \quad R_g = 100 \ \Omega \quad R_x = \text{reading} \pm 2\%$$

where R_g is the internal resistance of the galvanometer. The null condition of the galvanometer may be assumed to have negligible error. The uncertainty associated with R_x is associated with the reading obtained from the location of the sliding contact. Estimate the uncertainty in the measured value of voltage at nominal values of 2 and 8 V.

6.15 Convert the analog voltage, $E(t) = 5 \sin 2\pi t$ mV, into a discrete time signal. Specifically, using sample time increments of (a) 0.125 s, (b) 0.30 s, and (c) 0.75 s, plot each series as a function of time over at least one period. Discuss apparent differences between the discrete representations of the analog signal. Use of a personal computer or programmable calculator may be helpful.

6.16 Compute the DFT for each of the three discrete signals in Problem 6.15. Discuss apparent differences. Use a data set of 128 points.

6.17 Compute the DFT for the discrete time signal that results from sampling the analog signal, $T(t) = 2 \sin 4\pi t$ °C, at sample rates of 4 and 8 Hz. Use a data set of 128 points. Discuss and compare your results.

6.18 Determine the resulting alias frequency for f_1:

a. $f_1 = 50$ Hz; $f_s = 90$ Hz

b. $f_1 = 1.5$ kHz; $f_s = 2$ kHz

c. $f_1 = 8$ Hz; $f_s = 6$ Hz

d. $f_1 = 21$ Hz; $f_s = 8$ Hz

6.19 A particular data acquisition system is used to convert the analog signal, $E(t) = (\sin 2\pi t + 2 \sin 8\pi t)$ V, into a discrete time signal using a sample rate of 16 Hz. Use the DFT to reconstruct the Fourier series.

6.20 Consider the continuous signal found in Example 2.1. What would be an appropriate sample rate and sample period to use in sampling this signal if the resulting discrete series must have a size of 2^M where M is an integer and the signal is to be filtered at and above 2 Hz.

6.21 Convert the following binary numbers to positive integer base 10 numbers:
  a. 1010
  b. 11111
  c. 10111011
  d. 1100001

6.22 Convert the following bipolar (one complementary) binary numbers into integer base 10 numbers:
  a. 0111
  b. 1001
  c. 01111111
  d. 11111111

6.23 Convert the following decimal (base 10) numbers into bipolar binary numbers:
  a. 10
  b. -10
  c. -247
  d. 1013

6.24 List the possible sources of uncertainty in the dual-slope procedure for A/D conversion. Derive a relationship between the uncertainty in the digital result and the slope of the integration process.

6.25 Compute the resolution and dynamic range for an M-bit A/D converter having a full-scale range of ± 5 V, if M is 4, 8, 12 and 16.

6.26 A single stage RC filter with $f_c = 100$ Hz is used to filter an analog signal. Determine the attenuation of the filtered analog signal at 10, 50, 75, and 200 Hz.

6.27 A three-stage LC Bessel filter with $f_c = 100$ Hz is used to filter an analog signal. Determine the attenuation of the filtered analog signal at 10, 50, 75, and 200 Hz.

6.28 Design a cascading RC low-pass filter that has a magnitude ratio flat to within 3 dB from 0 to 5 kHz but with an attenuation of at least 30 dB for all frequencies at and above 10 kHz.

6.29 A complex periodic analog signal to be sampled at 500 Hz is passed through a low-pass RC filter rated at -3 dB at 250 Hz. What is the maximum frequency in the filtered signal for which the amplitude will be affected by a dynamic error of no more than 10%?

6.30 Choose an appropriate cascading low-pass filter to remove a 500-Hz component contained within an analog signal which is to be passed through an 8-bit A/D converter having 10-V range and 200-Hz sample rate. Attenuate the component to within the A/D converter quantization error.

6.31 The voltage output from a J-type thermocouple referenced to 0 °C is to be used to measure temperatures from 50 to 70 °C. The output voltages will vary linearly over this range from 2.585 to 3.649 mV.
  a.   If the thermocouple voltage is input to a 12-bit A/D converter having a ± 5-V range, estimate the percent quantization error in the digital value.

b. If the analog signal can be first passed through an amplifier circuit, compute the amplifier gain required to reduce the quantization error to 5% or less.

c. If the ratio of signal-to-noise level (SNR) in the analog sigal is 40 dB, compute the magnitude of the noise after amplification. Discuss the results of (b) in light of this.

6.32 Specify an appropriate ± 5-V M-bit A/D converter (8- or 12-bit), sample rate (up to 100 Hz) and signal conditioning to convert these analog signals into digital series. Estimate the quantization error and dynamic error resulting from the system specified:

a. $E(t) = 2 \sin 20\pi t$ V

b. $E(t) = 1.5 \sin \pi t + 20 \sin 32\pi t - 3 \sin (60\pi t + \pi/4)$ V

c. $P(t) = -10 \sin 4\pi t + 5 \sin 8\pi t$ psi; $K = 0.4$ V/psi

CHAPTER 7
TEMPERATURE MEASUREMENTS

7.1   INTRODUCTION

Temperature is one of the most commonly used and measured engineering variables. Much of our lives is affected by the diurnal and seasonal variations in ambient temperature, but the fundamental, scientific definition of temperature and a scale for the measurement of temperature are not commonly understood. Although temperature is one of the most familiar engineering variables, it is unfortunately not easy to define. This chapter explores the establishment of a practical temperature scale and common methods of temperature measurement. In addition, errors associated with the design and installation of a temperature sensor are discussed.

Historical Background

Early exploration of thermodynamic temperatures was accomplished by a French scientist Guillaume Amontons (1663–1705). His efforts examined the behavior of a constant volume of air which was subject to temperature changes. The modern liquid-in-glass bulb thermometer traces its origin to Galileo (1565–1642) who attempted to use the volumetric expansion of liquids in tubes as a relative measure of temperature. Unfortunately, this open tube device was actually sensitive to both barometric pressure and temperature changes, and thus could be called a "barothermoscope." A major advance in temperature measurement occurred in 1630 as a result of a seemingly unrelated event: the development of the technology to manufacture capillary glass tubes. These tubes were then used with water and alcohol in a thermometric device resembling the bulb thermometer, and eventually led to the development of a practical temperature measuring instrument.

A temperature scale proposed by Sir Isaac Newton (1642–1727) used the freezing point of water and the armpit temperature of a healthy man as extremes of temperature on a linear 0 to 12 scale. The scale proposed by Gabriel D. Fahrenheit, a German physicist (1686–1736) in 1715 attempted to incorporate body temperature as the median point on a scale having 180 divisions between the freezing point and the boiling point of water. Fahrenheit also successfully used mercury as the liquid in a bulb thermometer, making significant improvements over the attempts of Ismael Boulliau in 1659.

In 1742, the Swedish astronomer Anders Celsius[1] (1701–1744) described a temperature scale that divided the interval between the boiling and freezing points of water at 1 atm pressure into 100 equal parts. The boiling point of water was fixed as 0, and the freezing point of water as 100. Shortly after Celsius' death, Carolus Linnaeus (1707–1778) reversed the scale so that the 0 point corresponded to the freezing point of water at 1 atm. And even though this scale may not have been originated by Celsius [1], in 1948 the change from degrees centigrade to degrees Celsius was officially adopted. It is also interesting to note that despite the many practical applications for temperature measurement, a practical temperature scale and measuring devices were initially developed as measurement tools for research. As stated by H. A. Klein in *The Science of Measurement: A Historical Survey*[2]:

From the original thermoscopes of Galileo and some of his contemporaries, the measurement of temperature has pursued paths of increasing ingenuity, sophistication and complexity. Yet temperature remains in its innermost essence the average molecular or atomic energy of the least bits making up matter, in their endless dance. Matter without motion is unthinkable. Temperature is the most meaningful physical variable for dealing with the effects of those infinitesimal, incessant internal motions of matter.

We will begin our discussion of temperature by examining a method for measuring temperature, before attempting a precise definition.

7.2 TEMPERATURE STANDARDS AND DEFINITION

Temperature can be loosely described as the property of an object that describes its hotness or coldness, concepts that are clearly relative. Our experiences indicate that heat transfer tends to equalize temperature, or more precisely, systems that are in thermal communication will eventually have equal temperatures. The zeroth law of thermodynamics states that two systems in thermal equilibrium with a third system are in thermal equilibrium with each other. Thermal equilibrium implies that no heat transfer occurs between the systems, which also indicates equality of temperature. Although the zeroth law of thermodynamics essentially provides the definition of the equality of temperature, it provides no means for defining a temperature scale.

---

[1]It is interesting to note that in addition to his work in thermometry, Celsius published significant papers on the aurora borealis and the falling level of the Baltic Sea.
[2]Dover, Mineola, NY, 1988.

A *temperature scale* provides for three essential aspects of temperature measurement:

1. The definition of the size of the degree.
2. Fixed reference points for establishing known temperatures.
3. A means for interpolating between these fixed temperature points.

These provisions are consistent with the requirements for any standard, as described in Chapter 1. To construct a temperature scale, the three aspects listed above for a temperature scale must be established.

To begin, consider the definition of the triple point of water as having a value of 0.01 for our temperature scale, as is done for the Celsius scale (0.01 °C). This provides for an arbitrary starting point for a temperature scale; in fact, the number value assigned to this temperature could be anything. On the Fahrenheit temperature scale it has a value very close to 32. Consider another fixed point on our temperature scale. Fixed points are typically defined by phase-transition temperatures or the triple point of a pure substance. The point at which pure water boils at one standard atmosphere pressure is an easily reproducible fixed temperature. For our purposes let's assign this fixed point a numerical value of 100.

The next problem is to define the size of the degree. Since we have two fixed points on our temperature scale, we can see that the degree is 1/100th of the temperature difference between the ice point and the boiling point of water at atmospheric pressure. Conceptually, this defines a workable scale for the measurement of temperature; however, as yet we have made no provision for interpolating between the two fixed point temperatures.

The calibration of a temperature measurement device entails not only the establishment of fixed temperature points, but the indication of any temperature between fixed points. The operation of a mercury-in-glass thermometer is based on the thermal expansion of mercury contained in a glass capillary, where the level of the mercury is read as an indication of the temperature. Imagine that we submerged the thermometer in water at the ice point, made a mark on the glass at the height of the column of mercury, and labeled it 0 °C, as illustrated in Figure 7.1. Next we submerged the thermometer in boiling water, and again marked the level of the mercury, this time labeling it 100 °C. Clearly we want to be able to measure temperatures other than these two fixed points. How can we determine the appropriate place on the thermometer to mark say 50 °C?

The process of establishing 50 °C without a fixed point calibration is called *interpolation*. The simplest option would be to divide the distance on the thermometer between the marks representing 0 and 100 into equally spaced degree divisions. What assumption is implicit in this method of interpolation? It is obvious that we do not have enough information to appropriately divide the interval between 0 and 100 on the thermometer into degrees. Some theory of the behavior of the mercury in the thermometer, or many fixed points for calibration are necessary to resolve our dilemma. Even by the late eighteenth century, there was no standard for interpolating between fixed points on the temperature scale; the result was that different thermometers indicated different temperatures away from fixed points, sometimes with surprisingly large errors.

FIGURE 7.1   Calibration and interpolation for a liquid-
in-glass thermometer.

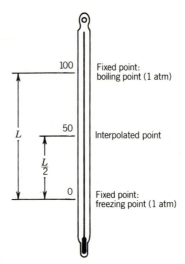

In practice, temperature cannot be measured directly. In all temperature
measurement devices a physical phenomenon or property is used to indicate
temperature. Ideally, this property is a known function of temperature, and
perhaps the relationship is theoretically predictable. The definite repeatable
relationship between electrical resistance of a metal and its temperature is a
phenomenon commonly used in temperature measuring devices. One goal of
this chapter is to describe the physical phenomena and properties that are most
successfully used to indicate temperature.

At this point, it is necessary to reconcile this arbitrary temperature scale with
the idea of thermodynamic and absolute temperature. Thermodynamics defines
a temperature scale that has an absolute reference, and defines an absolute zero
for temperature. For example, this absolute temperature governs the energy
behavior of an ideal gas, and is used in the ideal gas equation of state. The
behavior of real gases at very low pressure may be used as a temperature standard, to define a practical measure of temperature that approximates the thermodynamic temperature.

The modern engineering definition of the temperature scale is documented in
a published standard called the International Temperature Scale of 1990 (ITS–
90) [2]. This standard establishes fixed points for temperature, and provides
standard procedures and devices for interpolating between fixed points. Temperatures established according to ITS–90 do not deviate from the thermodynamic temperature scale by more than the uncertainty in the thermodynamic
temperature at the time of adoption of ITS–90. The primary fixed points from
ITS–90 are shown in Table 7.1. In addition to these fixed points, other fixed
points of secondary importance are available in ITS–90.

Along with the fixed temperature points established by ITS–90, a standard
for interpolation between these fixed points is necessary. Standards for acceptable thermometers and interpolating equations are provided in ITS–90. In the

TABLE 7.1   Temperature Fixed Points as Defined
by ITS–90

Defining State	Temperature	
	K	C
Triple point of hydrogen	13.8033	−259.3467
Liquid/vapor equilibrium for hydrogen at 25/76 atm	≈17	≈ −256.15
Liquid/vapor equilibrium for hydrogen at 1 atm	≈20.3	≈ −252.87
Triple point of neon	24.5561	−248.5939
Triple point of oxygen	54.3584	−218.7916
Triple point of argon	83.8058	−189.3442
Triple point of water	273.16	0.01
Solid/liquid equilibrium for gallium at 1 atm	302.9146	29.7646
Solid/liquid equilibrium for tin at 1 atm	505.078	231.928
Solid/liquid equilibrium for zinc at 1 atm	692.677	419.527
Solid/liquid equilibrium for silver at 1 atm	1234.93	961.78
Solid/liquid equilibrium for gold at 1 atm	1337.33	1064.18
Solid/liquid equilibrium for copper at 1 atm	1357.77	1084.62

range of interest for most engineering applications, the primary means of interpolating accurately between fixed points is provided by the variation of resistance with temperature for a platinum wire. For temperatures ranging from 13.8033 to 1234.93 K, ITS–90 establishes a platinum resistance thermometer as the standard interpolating instrument, and establishes interpolating equations that relate temperature to resistance. The practical application of these physical phenomena for temperature measurement is discussed in this chapter. Above 1234.93 K the temperature is defined in terms of blackbody radiation, without specifying an instrument for interpolation [2]. Prior to the adoption of ITS–90, the high-temperature interpolation standard was based on an optical pyrometer.

From 1968 through 1989 the International Practical Temperature Scale of 1968 (IPTS–68) established fixed temperature points and interpolation standards. The differences between IPTS–68 and ITS–90 are documented in [3], which provides the necessary information to assess the importance of changes for instruments calibrated under IPTS–68.

In summary, temperature measurement and the development of a practical temperature scale and standards for fixed points and interpolation have evolved over a period of about two centuries. Present standards for fixed-point temperatures and interpolation allow for practical and accurate measurements of temperature. In the United States, the National Institute of Standards and Technology (NIST, formerly the National Bureau of Standards) provides for a means to obtain accurately calibrated platinum wire thermometers for use as secondary standards in the calibration of a temperature measuring system to any practical level of uncertainty.

7.3 THERMOMETRY BASED ON THERMAL EXPANSION

Most materials exhibit a change in size with changes in temperature. Since this physical phenomenon is well-defined and repeatable, it is used for temperature measurement. The liquid-in-glass thermometer and the bimetallic thermometer are based on this phenomenon.

Liquid-in-Glass Thermometers

A liquid-in-glass thermometer measures temperature by virtue of the thermal expansion of a liquid. The construction of a liquid-in-glass thermometer is shown in Figure 7.2. The liquid is contained in a glass structure that consists of a bulb and a stem. The bulb serves as a reservoir, and provides sufficient fluid for the total volume change of the fluid to cause a detectable rise of the liquid in the stem of the thermometer. The stem contains a glass capillary tube, and the level of the liquid in the capillary is an indication of the temperature. The difference in thermal expansion between the liquid and the glass produces a practical change in the level of the liquid in the glass capillary.

During calibration, such a thermometer is subject to one of three measuring environments:

1. For a *complete immersion thermometer*, the entire thermometer is immersed in the calibrating temperature environment or fluid.

FIGURE 7.2   Liquid-in-glass thermometer.

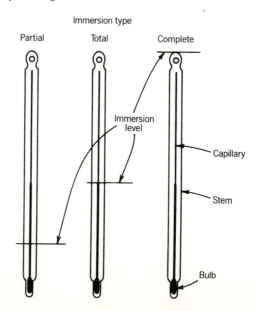

Immersion type

Partial   Total   Complete

Immersion level

Capillary

Stem

Bulb

2. For a *total immersion thermometer*, the thermometer is immersed in the calibrating temperature environment up to the liquid level in the capillary and,

3. For a *partial immersion thermometer*, the thermometer is immersed to a predetermined level in the calibrating environment.

For the most accurate temperature measurements, the thermometer should be immersed in the same manner in use as it was during calibration. In practice, it may not be possible to employ the thermometer in exactly the same way as when it was calibrated. In this case, stem corrections can be applied to the temperature reading [4].

Temperature measurements using liquid-in-glass thermometers can provide accuracies to ±0.01 °C under very carefully controlled conditions; however, such extraneous variables as pressure and changes in bulb volume over time can introduce significant errors in scale calibration. For example, pressure changes increase the indicated temperature by approximately 0.1 °C per atmosphere [5]. Practical measurements using liquid-in-glass thermometers typically result in total uncertainties that range from ±0.2 to 2 °C, depending upon the specific instrument.

Bimetallic Thermometers

The physical phenomenon employed in a bimetallic temperature sensor is the *differential* thermal expansion of two metals. Figure 7.3 shows the construction and response of a bimetallic sensor to an input signal. The sensor is constructed by bonding two strips of different metals, A and B. The resulting bimetallic strip may be in a variety of shapes, depending upon the particular application. Consider the simple linear construction shown in Figure 7.3. At the assembly tem-

FIGURE 7.3 Expansion thermometry: bimetallic strip.

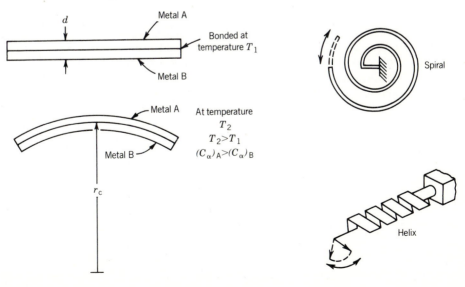

perature, T_1, the bimetallic strip will be straight; however, for temperatures other than T_1 the strip will have a curvature. The physical basis for the relationship between the radius of curvature and temperature is given as

$$r_c \propto \frac{d}{[(C_\alpha)_A - (C_\alpha)_B] (T_2 - T_1)} \tag{7.1}$$

where

r_c = radius of curvature
C_α = material thermal expansion coefficient
T = temperature
d = thickness

Bimetallic strips employ one metal having a high coefficient of thermal expansion with another having a low coefficient, providing increased sensitivity. Invar is often used as one of the metals, since for this material $C_\alpha = 1.7 \times 10^{-8}$ m/m-°C, as compared to typical values for other metals, such as steels, which range from approximately 2×10^{-5} to 20×10^{-5} m/m-°C.

The bimetallic sensor is used in many temperature control systems, and is the primary element in most dial thermometers. The geometries shown in Figure 7.3 serve to provide the desired deflection in the bimetallic strip for a given application. Dial thermometers using a bimetallic strip as their sensing element typically provide temperature measurements with uncertainties of ±1 °C.

7.4   ELECTRICAL RESISTANCE THERMOMETRY

As a result of the physical nature of the conduction of electricity, electrical resistance of a conductor or semiconductor varies with temperature. Using this behavior as the basis for temperature measurement is extremely simple in principle, and leads to two basic classes of resistance thermometers: resistance temperature detectors (conductors) and thermistors (semiconductors). Resistance temperature detectors (RTD) may be formed from a solid metal wire, which exhibits an increase in electrical resistance with temperature. The physical basis for the relationship between resistance and temperature is the temperature dependence of the resistivity of a material, ρ_e. The resistance of a conductor of length l and cross-sectional area A_c may be expressed in terms of the resistivity as

$$R = \frac{\rho_e l}{A_c} \tag{7.2}$$

Thermistors are semiconductor devices that display a very large decrease in resistance as temperature increases. Only recently have manufacturing techniques provided thermistors that are stable and accurate enough to function as temperature sensors.

Resistance Temperature Detectors

In the case of a resistance temperature detector, or RTD, the sensor is generally constructed by mounting a metal wire on an insulating support structure to eliminate mechanical strains, and encasing the wire to prevent changes in resistance due to influences from the sensor's environment, such as corrosion. Figure 7.4 shows such a typical RTD construction. Mechanical strains change a conductor's resistance and must be eliminated if accurate temperature measurements are to be made. This factor is essential since the resistance changes with mechanical strain are significant, as evidenced by the use of metal wire as sensors for the direct measurement of strain. Such mechanical stresses and resulting strains can be created by thermal expansion. Thus, provision for strain-free expansion of the conductor as its temperature changes is essential in the construction of an RTD. The support structure will expand as the temperature of the RTD increases, and the construction allows for strain-free differential expansion.

FIGURE 7.4   Construction of a platinum RTD. [From R. P. Benedict, *Fundamentals of Temperature, Pressure and Flow Measurements*, 3d ed., copyright © 1984 by John Wiley and Sons, New York.]

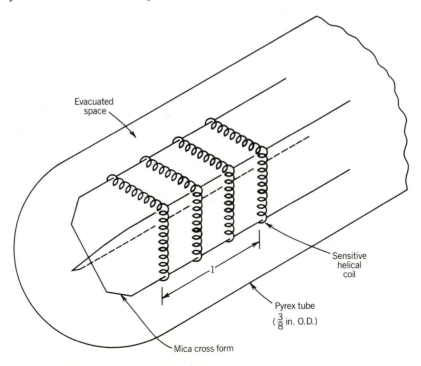

The relationship between the resistance of a metal conductor and its temperature can be expressed as a polynomial expansion in temperature:

$$R = R_0[1 + \alpha(T - T_0) + \beta(T - T_0)^2 + \cdots]$$

where R_0 is a reference resistance measured at temperature T_0. The coefficients $\alpha, \beta, \ldots$ are material constants. Figure 7.5 shows the relative relation between resistance and temperature for three common metals. This figure shows that the relationship between temperature and resistance over specific small temperature ranges can be expressed linearly as

$$R = R_0 [1 + \alpha(T - T_0)] \qquad (7.3)$$

where α is the temperature coefficient of resistivity. For example, for platinum conductors the linear approximation is accurate to within $\pm 0.3\%$ over the range 0 to 200 °C and $\pm 1.2\%$ over the range 200 to 800 °C. Table 7.2 lists a number of temperature coefficients of resistivity, α, for materials at 20 °C. With the assumed linear relationship between resistance and temperature, an appropriate value of α should be chosen for the temperature range of interest.

The most common material chosen for the construction of RTDs is platinum. The platinum RTD provides a means for the measurement of temperature, and historically provided the first interpolation standard for an internationally acceptable temperature scale. The RTD relies on the change in electrical resistance of a platinum wire to provide a precise measure of temperature, as expressed in equation 7.3. The principle of operation is quite simple: Platinum exhibits a predictable and reproducible change in electrical resistance with temperature,

FIGURE 7.5   Relative resistance of several pure metals (R_0 at 0 °C).

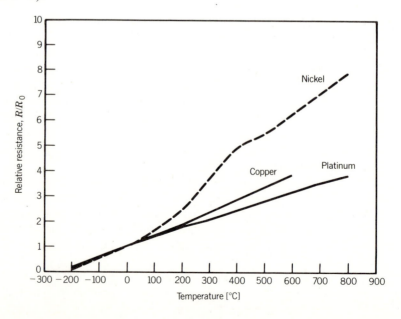

TABLE 7.2   Temperature Coefficients of
Resistivity for Selected Materials at 20 °C

Substance	$\alpha\ [\Omega/°C]$
Aluminum (Al)	0.00429
Carbon (C)	−0.0007
Copper (Cu)	0.0043
Gold (Au)	0.004
Iron (Fe)	0.00651
Lead (Pb)	0.0042
Nickel (Ni)	0.0067
Nichrome	0.00017
Platinum (Pt)	0.003927
Tungsten (W)	0.0048
Thermistors	−0.068 to +0.14

which can be calibrated and interpolated to a high degree of accuracy. The linear approximation for the relationship between temperature and resistance is valid over a wide temperature range, and platinum is highly stable. Considerations important to the accuracy of an RTD include ensuring that the wire is in a strain-free condition, preventing corrosion, and avoiding temperatures high enough to change the resistance characteristics of the platinum. To be suitable for use as a secondary temperature standard, a platinum resistance thermometer should have a value of α not less than 0.003925 °C^{-1}. This minimum value is an indication of the purity of the platinum. In general, RTDs may be used for the measurement of temperatures ranging from cryogenic to approximately 650 °C. By properly constructing an RTD, and correctly measuring its resistance, an uncertainty in temperature measurement of approximately ±0.005 °C is possible. Because of this potential for low uncertainties and the predictable and stable behavior of platinum, the platinum RTD is widely used as a local standard.

Temperature measurements using an RTD are accomplished by measuring the resistance of the platinum wire, and relating this resistance to temperature. For a National Institute of Standards and Technology (NIST) certified RTD, a table and interpolating equation would be available. The resistance of an RTD may be measured by a number of means, and the choice of an appropriate resistance measuring device must be made based on the required level of uncertainty in the final temperature measurement. Conventional ohmmeters cause a small current to flow during resistance measurements, creating self-heating in the RTD. An appreciable temperature change of the sensor may be caused by this current. This measuring procedure induced loading error is an important consideration for RTDs.

Bridge circuits are used to measure the resistance of RTDs, to minimize loading errors and provide low uncertainties in measured resistance values. Wheatstone bridge circuits are commonly used for these measurements. However, the basic Wheatstone bridge circuit does not compensate for the resistance of the leads in measuring resistance of an RTD, which are a major source of error in electrical resistance thermometers. When greater accuracies are required, three-wire and four-wire bridge circuits can be used.

Figure 7.6a shows a three-wire Callendar–Griffiths bridge circuit. The lead

FIGURE 7.6   Bridge circuits. Average of the two
readings (in *b* and *c*) eliminates the effect of lead wire
resistances.

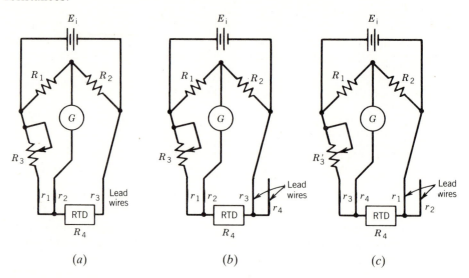

(*a*)   (*b*)   (*c*)

wires numbered 1, 2, and 3 have resistances r_1, r_2, and r_3, respectively. At
balanced conditions,

$$\frac{R_1}{R_2} = \frac{R_3}{R_4}$$

But with the lead wire resistances included in the circuit analysis,

$$\frac{R_1}{R_2} = \frac{R_3 + r_1}{R_4 + r_3} \tag{7.4}$$

and with $R_1 = R_2$, the resistance of the RTD, R_4, can be found as

$$R_4 = R_3 + r_1 - r_3 \tag{7.5}$$

If $r_1 = r_3$, the effect of these lead wires is eliminated from the determination of
the RTD resistance by this bridge circuit. Note that the resistance of lead wire
2 does not contribute to any error in the measurement at balanced conditions,
since no current flows through the galvanometer, G. It is good measurement
practice to twist the leads 1 and 3 to minimize interference.

The four-wire Mueller bridge, as shown in Figure 7.6*b*, provides increased
compensation for lead wire resistances compared to the Callendar–Griffiths
bridge, and is used with four-wire RTDs. The four-wire Mueller bridge is typically used when low uncertainties are desired, as in cases where the RTD is
used as a laboratory standard. A circuit analysis of the bridge circuit in the first
measurement configuration yields

$$R_4 + r_3 = R_3 + r_1 \tag{7.6}$$

and in the second measurement configuration

$$R_4 + r_1 = R'_3 + r_3 \tag{7.7}$$

where R_3 and R'_3 represent the indicated values of resistance in the first and second configurations, respectively. Adding equations 7.6 and 7.7 results in an expression for the resistance of the RTD in terms of the indicated values for the two measurements:

$$R_4 = \frac{R_3 + R'_3}{2} \tag{7.8}$$

With this approach, the effect of variations in lead wire resistances is minimized.

EXAMPLE 7.1

An RTD forms one arm of an equal-arm Wheatstone bridge, as shown in Figure 7.7. The fixed resistances, R_2 and R_3 are equal to 25 Ω. The RTD has a resistance of 25 Ω at a temperature of 0 °C and is used to measure a temperature that is steady in time.

The resistance of the RTD over a small temperature range may be expressed, as in equation 7.3:

$$R = R_0[1 + \alpha(T - T_0)]$$

Suppose the coefficient of resistance for this RTD is 0.003925 °C^{-1}. A temperature measurement is made by placing the RTD in the measuring environment and balancing the bridge by adjusting R_1. The value of R_1 required to balance the bridge is 37.36 Ω. Determine the temperature of the RTD.

FIGURE 7.7   RTD Wheatstone bridge arrangement.

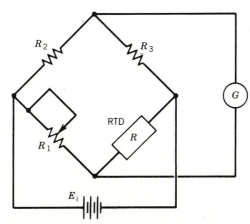

KNOWN

An RTD having a resistance of 25 Ω at 0 °C with α = 0.003925 °C^{-1} is used to measure a temperature. A value of R_1 = 37.36 Ω is required to balance the bridge circuit.

FIND

The temperature of the RTD.

SOLUTION

The resistance of the RTD is measured by balancing the bridge; recall that in a balanced condition,

$$R = R_1 \frac{R_3}{R_2}$$

The resistance of the RTD is found to be 37.36 Ω. With R_0 = 25 Ω at T = 0 °C, and α = 0.003925 °C^{-1}, equation 7.3 becomes

$$37.36 \ \Omega = 25(1 + \alpha T)\Omega$$

The temperature of the RTD is found to be 126 °C.

EXAMPLE 7.2

Consider the bridge circuit and RTD of Example 7.1. To select or design a bridge circuit for measuring the resistance of the RTD in this example, the required uncertainty in temperature would be specified. If the required uncertainty in the measured temperature is less than or equal to 0.5 °C, would a 1% total uncertainty in each of the resistors that make up the bridge be acceptable? Neglect the effects of lead wire resistances for this example.

KNOWN

A required uncertainty in temperature of ±0.5 °C, measured with the RTD and bridge circuit of Example 7.1.

FIND

The uncertainty level for a 1% total uncertainty in each of the resistors which make up the bridge circuit.

ASSUMPTIONS

All uncertainties are provided and evaluated at the 95% confidence level.

SOLUTION

Perform a design-stage uncertainty analysis. Assuming at the design stage that the total uncertainty in the resistances is 1%, then with initial values of the resistances in the bridge equal to 25 Ω,

$$(u_d)_{R_1} = (u_d)_{R_2} = (u_d)_{R_3} = (0.01)(25) = 0.25 \ \Omega$$

The second-power law is used to estimate the propagation of uncertainty in each resistor to the uncertainty in determining the RTD resistance by

$$u_{RTD} = \sqrt{\left(\frac{\partial R}{\partial R_1} (u_d)_{R_1}\right)^2 + \left(\frac{\partial R}{\partial R_2} (u_d)_{R_2}\right)^2 + \left(\frac{\partial R}{\partial R_3} (u_d)_{R_3}\right)^2}$$

where

$$R = R_{RTD} = \frac{R_1 R_3}{R_2}$$

Then, the design-stage uncertainty in the resistance of the RTD is

$$u_{RTD} = \sqrt{\left(\frac{R_3}{R_2} (u_d)_{R_1}\right)^2 + \left(\frac{-R_1 R_3}{R_2^2} (u_d)_{R_2}\right)^2 + \left(\frac{R_1}{R_2} (u_d)_{R_3}\right)^2}$$

$$u_{RTD} = \sqrt{(1 \times 0.25)^2 + (1 \times -0.25)^2 + (1 \times 0.25)^2}$$

$$= 0.433 \ \Omega$$

To determine the uncertainty in temperature, we know

$$R = R_{RTD} = R_0[1 + \alpha(T - T_0)]$$

and

$$u_T = \sqrt{\left(\frac{\partial T}{\partial R} u_{RTD}\right)^2}$$

Setting $T_0 = 0 \ °C$ with $R_0 = 25 \ \Omega$, and neglecting uncertainties in T_0, α, and R_0

$$\frac{\partial T}{\partial R} = \frac{1}{\alpha R_0}$$

$$\frac{1}{\alpha R_0} = \frac{1}{(0.003925 \ °C^{-1}) \ (25 \ \Omega)}$$

then the uncertainty in temperature is

$$u_T = u_{RTD} \left(\frac{\partial T}{\partial R} \right) = \frac{0.433 \ \Omega}{0.98 \ \Omega/°C} = 4.4 \ °C$$

The desired uncertainty in temperature is not achieved with the specified levels of uncertainty in the pertinent variables.

COMMENT

Uncertainty analysis, in this case, would have prevented performing a measurement that did not provide meaningful results.

EXAMPLE 7.3

Suppose the total uncertainty in the bridge resistances of Example 7.1 was reduced to 0.1%. Would the required level of uncertainty in temperature be achieved?

KNOWN

The uncertainty in each of the resistors in the bridge circuit for temperature measurement from Example 7.1 is ±0.1%.

FIND

The resulting uncertainty in temperature.

SOLUTION

The uncertainty analysis from the previous example may be directly applied, with the uncertainty values for the resistances appropriately reduced. The uncertainties for the resistances are reduced from 0.25 to 0.025, yielding

$$u_{RTD} = \pm\sqrt{(1 \times 0.025)^2 + (1 \times -0.025)^2 + (1 \times 0.025)^2}$$

$$=\cdot \pm 0.0433 \ \Omega$$

and the resulting uncertainty in temperature is ±0.44 °C, which satisfies the design constraint.

COMMENT

This result provides confidence that the effect of the resistors' uncertainties will not cause the uncertainty in temperature to exceed the target value. However, the uncertainty in temperature not only is a result of the uncertainty in the sensor, the RTD, but will depend on other aspects of the measurement system as well. The design-stage uncertainty analysis performed in this example may be viewed as ensuring that the factors considered do not produce a higher than acceptable uncertainty level. Additional sources of uncertainty will exist in any actual measurement employing an RTD.

Practical Considerations

The transient thermal response of typical commercial RTDs is generally quite slow compared to other temperature sensors, and for transient measurements bridge circuits must be operated in a deflection mode. For these reasons, RTDs are not generally chosen for transient temperature measurements. A notable exception is the use of very small platinum wires for temperature measurements in noncorrosive flowing gases. In this application, wires having diameters on the order of 0.01 cm can have frequency responses higher than any other temperature sensor, because of their extremely low thermal capacitance. Obviously, the smallest impact would destroy this sensor. Other resistance sensors in the form of thin metallic films provide fast transient response temperature measurements, often in conjunction with anemometry or heat flux measurements. Such platinum films are constructed by depositing a platinum film onto a substrate, and coating the film with a ceramic glass for mechanical protection [6]. Typical film thickness ranges from 1 to 2 μm, with a 10-μm protective coating. Continuous exposure at temperatures of 600 °C is possible with this construction. The range of applications for these thin-film sensors is increasing, especially for accuracies of ±0.5 to ±2 °C; practical uses include temperature control circuits for heating systems and cooking devices and surface temperature monitoring on electronic components subject to overheating.

Thermistors

Thermistors (from *therm*ally sensitive re*sistors*) are ceramic-like semiconductor devices. The resistance of a typical thermistor decreases rapidly with temperature, which is in contrast to the small increases of resistance with temperature for RTDs. The functional relationship between resistance and temperature for a thermistor is generally assumed to be of the form

$$R = R_0\, e^{\beta(1/T - 1/T_0)} \tag{7.9}$$

The parameter β ranges from 3500 to 4600 K, depending on the material, temperature, and individual construction for each sensor, and therefore must be determined for each thermistor. Figure 7.8 shows the variation of resistance with temperature for two common thermistor materials; the ordinate is the ratio of the resistance to the resistance at 25 °C. Thermistors exhibit large resistance changes with temperature in comparison to typical RTDs, as indicated by comparison of Figures 7.5 and 7.8. Equation 7.9 is not accurate over a wide range of temperature, unless β is taken to be a function of temperature; typically the value of β specified by a manufacturer for a sensor is assumed to be constant over a limited temperature range. A simple calibration is possible for determining β as a function of temperature, as illustrated in the circuits shown in Figure 7.9. Other circuits and a more complete discussion of measuring β may be found in the Electronic Industries Association standard Thermistor Definitions and Test Methods [7].

FIGURE 7.8   Representative thermistor resistance
variations with temperature.

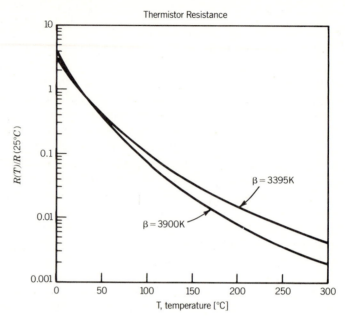

Thermistors are generally used when high sensitivity, ruggedness, or fast response time are required. Thermistors are often encapsulated in glass, and thus
can be used in corrosive or abrasive environments. The resistance characteristics
of the semiconductor material may change at elevated temperatures, and some
aging of a thermistor will occur at temperatures above 200 °C. The high resistance
of a thermistor, compared to that of an RTD eliminates the problems of lead
wire resistance compensation. But thermistors are not interchangeable, and

FIGURE 7.9   Circuits for determining β for thermistors.

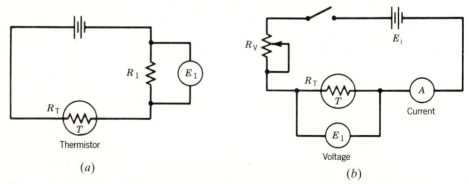

(a) Voltage divider method: $R_T = R_1 (E_i/E_1 - 1)$. *Note*: Both R_1 and E_i must be
known values. The value of R_1 may be varied to achieve appropriate values of
thermistor current.
(b) Volt ammeter method. *Note*: Both current and voltage are measured.

variations in room temperature resistances for ordinary thermistors having the same nominal characteristics may be as much as 20%.

The zero-power resistance of a thermistor is the resistance value of the thermistor with no flow of electric current. The dissipation constant for a thermistor is defined, at a given ambient temperature as

$$\delta = \frac{P}{T - T_\infty} \qquad (7.10)$$

where

δ = dissipation constant
P = power supplied to thermistor
T, T_∞ = thermistor and ambient temperatures

The zero power resistance should be measured such that a decrease in the current flow to the thermistor will result in not more than a 0.1% change in resistance.

EXAMPLE 7.4

The material constant β is to be determined for a particular thermistor using the circuit shown in Figure 7.9a. The thermistor has a resistance of 60 kΩ at 25 °C. The reference resistor in the circuit, R_1, has a resistance of 130.5 kΩ. The dissipation constant, δ, is 0.09 mW/°C. The voltage source used for the measurement is constant at 1.564 V. The thermistor is to be used at temperatures ranging from 100 to 150 °C. Determine the value of β.

KNOWN

The temperature range of interest is from 100 to 150 °C.

$$R_0 = 60\,000\ \Omega, \qquad T_0 = 25\ °C$$

$$E_i = 1.564\ V, \qquad \delta = 0.09\ mW/°C, \qquad R_1 = 130.5\ k\Omega$$

FIND

The value of β over the temperature range from 100 to 150°C

SOLUTION

The voltage drop across the fixed resistor is measured for three known values of thermistor temperature. The thermistor temperature is controlled and determined by placing the thermistor in a laboratory oven, and measuring the temperature of the oven. For each measured voltage across the reference resistor, the thermistor resistance, R_T, is determined from

$$R_T = R_1 \left(\frac{E_i}{E_1} - 1 \right)$$

The results of these measurements are:

Temperature [°C]	R_1 Voltage [V]	R_T [Ω]
100	1.501	5477.4
125	1.531	2812.9
150	1.545	1604.9

Equation 7.9 can be expressed in the form of a linear equation, $y = mx$, as

$$\ln \frac{R_T}{R_0} = \beta \left(\frac{1}{T} - \frac{1}{T_0} \right)$$

Applying this equation to the measured data, with $R_0 = 60\ 000$ Ω, the three data points above yield:

$\ln \dfrac{R_T}{R_0}$	$\dfrac{1}{T} - \dfrac{1}{T_0}$ [K^{-1}]	β [K]
-2.394	-6.75×10^{-4}	3546.7
-3.060	-8.43×10^{-4}	3629.9
-3.621	-9.92×10^{-4}	3650.2

COMMENT

These results are for constant β and are based on the behavior described by equation 7.9, over the temperature range from T_0 to the temperature T. The significance of the measured differences in β will be examined further.

The measured values of β in Example 7.4 are different at each value of temperature. If β were truly a temperature independent constant, and these measurements had negligible uncertainty, all three measurements would yield the same value for β. The variation in β may be due to a physical effect of temperature, or may be attributable to the uncertainty in the measured values.

Are the measured differences significant, and, if so, what value of β best represents the behavior of the thermistor over this temperature range? To perform the necessary uncertainty analysis, additional information must be provided concerning the instruments and procedures used in the measurement.

EXAMPLE 7.5

Perform an uncertainty analysis to determine the uncertainty in each measured value of β in Example 7.4, and evaluate a single best estimate of β for this temperature range. The measurement of β involved the measurement of voltages, temperatures, and resistances. For temperature there is a precision error associated with spatial and temporal variations in the oven temperature such that $S_{\bar{T}} = 0.19\ °C$ for 20 measurements. The interpolation error associated with instrument resolution ($\pm\frac{1}{2}$ least division) is $\pm0.3\ °C$. In addition, based on a manufacturer's specification, there is a known measurement bias limit for temperature of $\pm0.2\ °C$ in the thermocouple. All values are given at the 95% confidence level.

The bias errors in measuring resistance and voltage are negligible, and estimates of the instrument repeatability based on manufacturer's specifications in the measured values are $\pm1.5\%$ for resistance and $\pm0.002\ V$ for the voltage.

KNOWN

Standard deviation of the means for oven temperature, $S_{\bar{T}} = 0.19\ °C, N = 20$. The remaining values will be treated as bias limits, since no statistical analysis can be performed:

$$(B)_T = \pm\sqrt{0.2^2 + 0.3^2} = \pm0.36\ °C$$

$$(B)_R = \pm1.5\%$$

$$(B)_E = \pm0.002\ V$$

FIND

The uncertainty in β at each measured temperature, and a best estimate for β over the measured temperature range.

SOLUTION

Consider the problem of providing a single best estimate of β. One method of estimation might be to average the three measured values. This results in a value of 3609 K. However, since the relationship between $\ln(R_T/R_0)$ and $1/T - 1/T_0$ is expected to be linear, a least-squares fit can be performed on the three data points, and the point (0,0). The resulting value of β is 3638 K. Is this difference significant, and which value best represents the behavior of the thermistor? To answer these questions, an uncertainty analysis must be performed for β.

For each measured value

$$\beta = \frac{\ln(R_T/R_0)}{1/T - 1/T_0}$$

Errors in voltage, temperature, and resistance are propagated into the resulting value of β for each measurement.

Consider first the sensitivity indices, θ_i, for each of the variables R_T, R_0, T, and T_0. These may be tabulated by computing the appropriate partial derivatives of β, evaluated at each of the three temperatures, as

T [°C]	θ_{R_T} [K/Ω]	θ_{R_0} [K/Ω]	θ_T	θ_{T_0}
100	−0.270	0.0247	−37.77	59.17
125	−0.422	0.0198	−27.18	48.48
150	−0.628	0.0168	−20.57	41.45

The determination of the uncertainty in β, u_β, requires the uncertainty in the measured value of resistance for the thermistor, u_{R_T}. But R_T is determined from the expression

$$R_T = R_1 \left(\frac{E_i}{E_1} - 1 \right)$$

and thus requires an analysis of the uncertainty in the resulting value of R_T, from measured values of R_1, E_i, and E_1. All errors in R_T are treated as bias, yielding

$$(B)_{R_T} = \sqrt{\left[\frac{\partial R_T}{\partial R_1} (B)_{R_1} \right]^2 + \left[\frac{\partial R_T}{\partial E_i} (B)_{E_i} \right]^2 + \left[\frac{\partial R_T}{\partial E_1} (B)_{E_1} \right]^2}$$

To arrive at a representative value, we compute B_{R_T} at 125 °C. The uncertainty in R_1 is ±1.5% of 130.5 kΩ, or ±1.96 kΩ. The uncertainty in E_i and E_1 are each ±0.002 V. Using the appropriate values to compute the sensitivity indices, the value of $(B)_{R_T}$ is found as ±247 Ω. (See Problem 7.23.)

An uncertainty for β will be determined for each of the measured temperatures. The effect of the measurement bias limits for temperature and resistance is found by combining the individual contributions according to the second power law as

$$B_\beta = \sqrt{[\theta_T(B)_T]^2 + [\theta_{T_0}(B)_{T_0}]^2 + [\theta_{R_T}(B)_{R_T}]^2 + [\theta_{R_0}(B)_{R_0}]^2}$$

where

$$(B)_T = \sqrt{0.3^2 + 0.2^2} = \pm0.36 \qquad (B)_{R_T} = \pm247 \ \Omega$$

$$(B)_{T_0} = \pm0.36 \qquad\qquad\qquad (B)_{R_0} = \pm900 \ \Omega$$

The precision index for β contains contributions only from the statistically determined oven temperature characteristics and is found from

$$P_\beta = \sqrt{(\theta_T S_{\bar T})^2 + (\theta_{T_0} S_{\bar T_0})^2}$$

where both $S_{\bar T}$ and $S_{\bar T_0}$ are 0.19, as determined with $N = 20$.

The resulting values of uncertainty in β are found from

$$u_\beta = \sqrt{B_\beta^2 + (t_{19.95}\, P_\beta)^2}$$

where $t_{19.95}$ is 2.093. At each temperature the uncertainty in β is determined as shown in Table 7.3.

TABLE 7.3   Uncertainties in β

	Uncertainty (95%)		
	Precision	Bias	Total
T	P_β	B_β	u_β
[°C]	[K]	[K]	[K]
100	13.3	74.7	79.7
125	10.6	107.6	109.9
150	8.8	156.7	157.8

The effect of increases in the sensitivities, θ_i, on the total uncertainty is to cause increased uncertainty in β as the temperature increases.

The original results of the measured values of β must now be reexamined. The results, from Table 7.3, are plotted as a function of temperature in Figure 7.10, with uncertainty limits on each data point. Clearly, there is no justification for assuming that the measured values indicate a trend of changes with temperature, and it would be appropriate to use either the average value of β or the value determined from the linear least-squares curve fit.

FIGURE 7.10   Measured values of β and associated uncertainties for three temperatures.

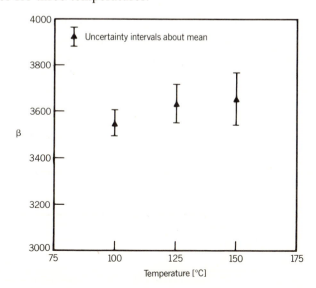

of a quantity of energy different than I^2R is required to maintain a constant temperature. The difference in I^2R and the amount of energy generated by the current flowing through the junction is due to the Peltier effect. The Peltier effect is due to the thermodynamically reversible conversion of energy as a current flows across the junction, in contrast to the irreversible dissipation of energy associated with I^2R losses. The Peltier heat is the quantity of heat in addition to the quantity I^2R that must be removed from the junction to maintain the junction at a constant temperature. This amount of energy is proportional to the current flowing through the junction; the proportionality constant is the Peltier coefficient π_{AB}, and the heat transfer required to maintain a constant temperature is

$$Q_\pi = \pi_{AB}\, I \qquad\qquad (7.12)$$

due to the Peltier effect alone. This behavior was discovered by Jean Charles Athanase Peltier (1785–1845) during experiments with Seebeck's thermocouple. He observed that passing a current through a thermocouple circuit having two junctions, as in Figure 7.11, raised the temperature at one junction, while lowering the temperature at the other junction. This effect forms the basis of a device known as a Peltier refrigerator, which provides cooling without moving parts.

In addition to the Seebeck effect and the Peltier effect, there is a third phenomenon that occurs in thermoelectric circuits. Consider the conductor shown in Figure 7.13, which is subject to a longitudinal temperature gradient, and also subject to a potential difference, such that there is a flow of current and heat in the conductor. Again, to maintain a constant temperature in the conductor it is found that a quantity of energy different than the Joule heat, I^2R, must be removed from the conductor. First noted by William Thomson (1824–1907, Lord Kelvin from 1892) in 1851, this energy is expressed in terms of the Thomson coefficient, σ as

$$Q_\sigma = \sigma I(T_1 - T_2) \qquad\qquad (7.13)$$

For a thermocouple circuit, all three of these effects may be present and may contribute to the overall emf of the circuit.

FIGURE 7.13   Thomson effect due to simultaneous flows of current and heat.

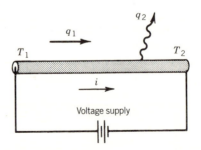

q_1 Energy flow due to temperature gradient
q_2 Heat transfer to maintain constant temperature

Fundamental Thermocouple Laws

The use of thermocouple circuits to measure temperature is based upon observed behaviors of carefully controlled thermocouple materials and circuits. The following laws provide the basis necessary for temperature measurement with thermocouples:

1. Law of Homogeneous Materials. **A thermoelectric current cannot be sustained in a circuit of a single homogeneous material by the application of heat alone, regardless of how it might vary in cross section.**

Simply stated, this law requires that at least two materials be used to construct a thermocouple circuit for the purpose of measuring temperature. It is interesting to note that a current may occur in a inhomogeneous wire that is nonuniformly heated; however, this is neither useful nor desirable in a thermocouple.

2. Law of Intermediate Materials. **The algebraic sum of the thermoelectric forces in a circuit composed of any number of dissimilar materials is zero if all of the circuit is at a uniform temperature.**

This law allows a material other than the thermocouple materials to be inserted into a thermocouple circuit without changing the output emf of the circuit. As an example, consider the thermocouple circuit shown in Figure 7.14 where the junctions of the measuring device are made of copper and material B is an alloy (not pure copper). The electrical connection between the measuring device and the thermocouple circuit forms yet another thermocouple junction. The law of intermediate metals, in this case, provides that the measured emf will be unchanged from the open-circuit emf which corresponds to the temperature difference between T_1 and T_2, if $T_3 = T_4$. Another practical consequence of this law is that copper extension wires may be used to transmit thermocouple emfs to a measuring device. This is very beneficial since most materials used in thermocouple wire can be significantly more expensive than copper.

3. Law of Successive or Intermediate Temperatures. **If two dissimilar homogeneous materials produce thermal emf$_1$ when the junctions are at T_1 and T_2 and produce thermal emf$_2$ when the junctions are at T_2 and T_3, the emf generated when the junctions are at T_1 and T_3 will be emf$_1$ + emf$_2$.**

The law of intermediate temperatures is of great importance in practical temperature measurements. This law allows a thermocouple calibrated for one reference temperature to be used at another reference temperature.

FIGURE 7.14   Typical thermocouple measuring circuit.

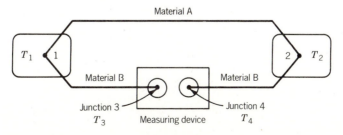

Basic Temperature Measurement with Thermocouples

The basic thermocouple circuit shown in Figure 7.14 can be used to measure the difference between the two temperatures T_1 and T_2. For practical temperature measurements, one of these junctions becomes a reference junction, and is maintained at some known, constant reference temperature. The other junction then becomes the measuring junction, and the emf existing in the circuit for any temperature T_1 provides a direct indication of the temperature of the measuring junction.

Figure 7.15 shows a basic thermocouple measuring system, using a Chromel–constantan thermocouple, an ice bath to create a reference temperature, copper

FIGURE 7.15   Thermocouple temperature measurement circuits.

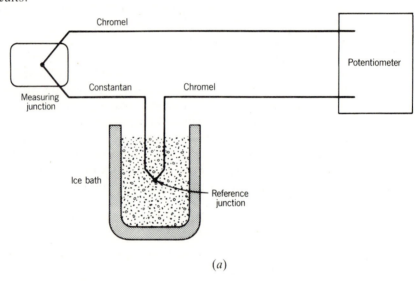

(a)

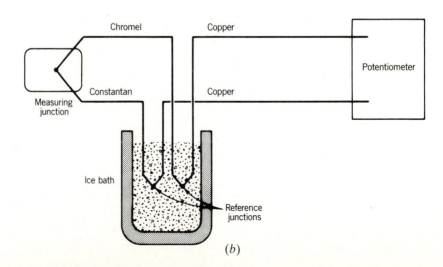

(b)

extension wires, and a potentiometer to measure the output voltage of the circuit. The law of intermediate materials ensures that neither the potentiometer nor the extension wires will change the emf of the circuit, as long as the connecting junctions are at the same temperature. All that is required to be able to measure temperature with this circuit is to know the relationship between the output emf and the temperature of the measuring junction, for the particular reference temperature. One method of determining this relationship is to calibrate the thermocouple. However, we shall see that for reasonable levels of uncertainty for temperature measurement, standard materials and procedures allow thermocouples to be accurate temperature measuring devices, without the necessity of calibration.

The provisions for a reference junction should provide a temperature that is accurately known, stable, and reproducible. A very common reference junction temperature is provided by the ice point, 0 °C, because of the ease with which it can be obtained. The creation of a reference junction temperature of 0 °C is accomplished in either of two basic ways. Prior to the development of an electronic means of creating a reference point in the electric circuit, an ice bath served to provide the reference junction temperature. An ice bath is typically made by filling a vacuum flask, or Dewar, with finely crushed ice, and adding just enough water to create a transparent slush. Surprising results are often obtained when the temperature of a mixture of ice and water is measured to verify the ice point is achieved. A few ice cubes floating in water does not create a 0 °C environment! Ice baths can be constructed to provide a reference junction temperature with an uncertainty to within ±0.01 °C.

Electronic reference junctions provide a convenient means of the measurement of temperature without the necessity to construct an ice-point reference. Numerous manufacturers produce commercial temperature measuring devices with built-in reference junction compensation. The electronics generally rely on a thermistor to determine the local environment temperature, as shown in Figure 7.16. Uncertainties for the reference junction temperature in this case are on the order of ±0.1 °C, with ±0.5 °C as typical.

Thermocouple Standards

The National Institute of Standards and Technology (NIST) provides specifications for the materials and construction of standard thermocouple circuits for temperature measurement. Many material combinations exist for thermocouples; these material combinations are identified by the thermocouple type and denoted by a letter. Table 7.4 shows the letter designations and the polarity

FIGURE 7.16   Basic thermistor circuit for thermocouple reference junction compensation.

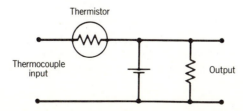

TABLE 7.4   Thermocouple Designations

| Type | Material Combination | | Applications |
	Positive	Negative	
E	Chromel(+)	Constantan(−)	Highest sensitivity (<1000 °C)
J	Iron(+)	Constantan(−)	Nonoxidizing environment (<760 °C)
K	Chromel(+)	Alumel(−)	High temperature (<1372 °C)
S	Platinum/ 10% rhodium	Platinum(−)	Long-term stability high temperature (<1768 °C)
T	Copper(+)	Constantan(−)	Reducing or vacuum environments (<400 °C)

of common thermocouples, along with some basic application information for each type. The choice of a type of thermocouple depends on the temperature range to be measured, the particular application, and the desired uncertainty level.

To determine the emf output of a particular material combination, a thermocouple is formed from a candidate material and a standard platinum alloy to form a thermocouple circuit having a 0 °C reference temperature. Figure 7.17 shows the output of various materials in combination with platinum-67. The notation indicates the thermocouple type. The law of intermediate temperatures then allows the emf of any two materials whose emf relative to platinum is known to be determined. Figure 7.18 shows a plot of the emf as a function of temperature for some common thermocouple material combinations. The slope of the curves in this figure corresponds to the static sensitivity of the thermocouple measuring circuit. Choices of a thermocouple type for a particular application should consider the nature of the measuring environment, the temperature range of interest, and the required uncertainty level. Significant cost differences exist for more exotic thermocouple materials.

Table 7.5 provides the standard composition of thermocouple materials, along with standard limits of error for the various material combinations. These limits specify the expected maximum errors resulting from the thermocouple materials. The NIST uses high-purity materials to establish the standard value of voltage output for a thermocouple composed of two specific materials. This results in standard tables or equations used to determine a measured temperature from a value of emf. An example of such a table is provided in Table 7.6 for an iron/constantan thermocouple, usually referred to as a J-type thermocouple. Because of the widespread need to measure temperature, an industry has grown up to supply high grade thermocouple wire. Manufacturers can also provide thermocouples having special tolerance limits relative to the NIST standard voltages ranging from ±1.0 °C to perhaps ±0.1 °C. Thermocouples constructed of standard thermocouple wire do not require calibration to provide measurement of temperature with accuracies within the tolerance limits of the thermocouple wire given in Table 7.5.

FIGURE 7.17   Thermal emf of thermocouple materials relative to platinum-67. *Note*: JP indicates the positive leg of a J thermocouple, or iron. [From R. P. Benedict, *Fundamentals of Temperature, Pressure and Flow Measurements*, 3d ed., copyright © 1984 by John Wiley & Sons, New York. Reprinted by permission.]

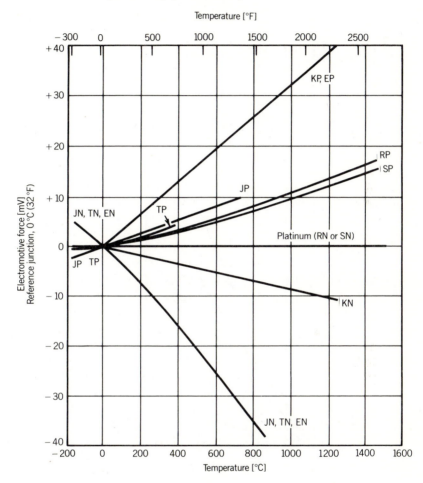

Thermocouple Voltage Measurement

The Seebeck voltage for a thermocouple circuit is measured with no current flow in the circuit. From our discussion of the Thomson and Peltier effects, it is clear that the emf will be slightly changed from the open-circuit value when there is a current flow in the thermocouple circuit. As such, the best method for the measurement of thermocouple voltages is a device that minimizes current flow, such as a potentiometer. A potentiometer has for many years been the laboratory standard for voltage measurement in thermocouple circuits. A potentiometer, as described in Chapter 6, achieves nearly zero loading error by inducing no current to flow at a balanced condition. However, in situations

FIGURE 7.18   Thermocouple voltage output as a
function of temperature for some common thermocouple
materials. Reference junction is at 0 °C. [From R. P.
Benedict, *Fundamentals of Temperature, Pressure and
Flow Measurements*, 3d ed., copyright © 1984 by John
Wiley & Sons, New York. Reprinted by permission.]

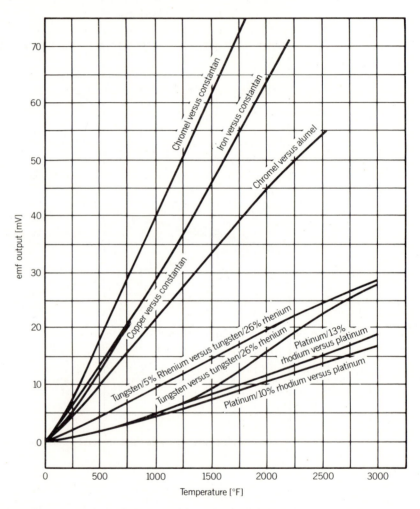

requiring the measurement of a time-varying temperature, a typical potentiometer is not practical. For such applications, voltage measuring devices that have
a very high input impedance can be used with minimal loading error. These
devices can also be used in either static or dynamic measuring situations where
the loading error created by the measurement device is acceptable for the particular application. For such needs, high-impedance voltmeters have been incorporated into commercially available temperature indicators, temperature
controllers, and digital data acquisition systems.

TABLE 7.5 Standard Thermocouple Compositions[a]

Type	Wire Positive	Wire Negative	Standard Limits of Error[b]
S	Platinum	Platinum/ 10% rhodium	+1.5 °C or 0.25%
R	Platinum	Platinum/ 13% rhodium	±1.5 °C
B	Platinum/ 30% rhodium	Platinum/ 6% rhodium	±0.5%
T	Copper	Constantan	±1.0 °C or 0.75%
J	Iron	Constantan	±2.2 °C or 0.75%
K	Chromel	Alumel	±2.2 °C or 0.75%
E	Chromel	Constantan	±1.7 °C or 0.5%

Alloy Designations

Constantan: 55% copper with 45% nickel
Chromel: 90% nickel with 10% chromium
Alumel: 94% nickel with 3% manganese, 2% aluminum, and 1% silicon

[a]From Temperature Measurements ANSI PTC 193-1974.
[b]Use greater value; these limits of error do not include installation errors.

EXAMPLE 7.6

The thermocouple circuit shown in Figure 7.19 is used to measure the temperature T_1. The thermocouple junction labeled 2 is at a temperature of 0 °C, maintained by an ice-point bath. The voltage output is measured using a potentiometer, and found to be 9.667 mV. What is T_1?

KNOWN

A thermocouple circuit having one junction at 0 °C and a second junction at an unknown temperature. The circuit produces an emf of 9.667 mV.

FIGURE 7.19 Thermocouple circuit for Example 7.6.

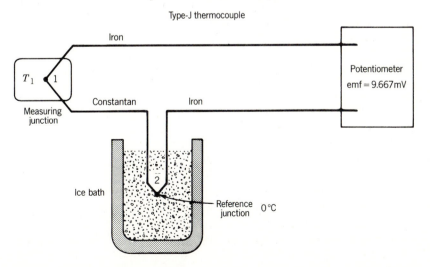

TABLE 7.6   Thermocouple Reference Table for Type J Thermocouple[a]

Temperatures in Degrees C									Reference Junction at 0 C
Deg C	0	−1	−2	−3	−4	−5	−6	−7	−8	−9
-210	-8.096									
-200	-7.890	-7.912	-7.934	-7.955	-7.976	-7.996	-8.017	-8.037	-8.057	-8.076
-190	-7.659	-7.683	-7.707	-7.731	-7.755	-7.778	-7.801	-7.824	-7.846	-7.868
-180	-7.402	-7.429	-7.455	-7.482	-7.508	-7.533	-7.559	-7.584	-7.609	-7.634
-170	-7.122	-7.151	-7.180	-7.209	-7.237	-7.265	-7.293	-7.321	-7.348	-7.375
-160	-6.821	-6.852	-6.883	-6.914	-6.944	-6.974	-7.004	-7.034	-7.064	-7.093
-150	-6.499	-6.532	-6.565	-6.598	-6.630	-6.663	-6.695	-6.727	-6.758	-6.790
-140	-6.159	-6.194	-6.228	-6.263	-6.297	-6.331	-6.365	-6.399	-6.433	-6.466
-130	-5.801	-5.837	-5.874	-5.910	-5.946	-5.982	-6.018	-6.053	-6.089	-6.124
-120	-5.426	-5.464	-5.502	-5.540	-5.578	-5.615	-5.653	-5.690	-5.727	-5.764
-110	-5.036	-5.076	-5.115	-5.155	-5.194	-5.233	-5.272	-5.311	-5.349	-5.388
-100	-4.632	-4.673	-4.714	-4.755	-4.795	-4.836	-4.876	-4.916	-4.956	-4.996
-90	-4.215	-4.257	-4.299	-4.341	-4.383	-4.425	-4.467	-4.508	-4.550	-4.591
-80	-3.785	-3.829	-3.872	-3.915	-3.958	-4.001	-4.044	-4.087	-4.130	-4.172
-70	-3.344	-3.389	-3.433	-3.478	-3.522	-3.566	-3.610	-3.654	-3.698	-3.742
-60	-2.892	-2.938	-2.984	-3.029	-3.074	-3.120	-3.165	-3.210	-3.255	-3.299
-50	-2.431	-2.478	-2.524	-2.570	-2.617	-2.663	-2.709	-2.755	-2.801	-2.847
-40	-1.960	-2.008	-2.055	-2.102	-2.150	-2.197	-2.244	-2.291	-2.338	-2.384
-30	-1.481	-1.530	-1.578	-1.626	-1.674	-1.722	-1.770	-1.818	-1.865	-1.913
-20	-0.995	-1.044	-1.093	-1.141	-1.190	-1.239	-1.288	-1.336	-1.385	-1.433
-10	-0.501	-0.550	-0.600	-0.650	-0.699	-0.748	-0.798	-0.847	-0.896	-0.945
0	0	-0.050	-0.101	-0.151	-0.201	-0.251	-0.301	-0.351	-0.401	-0.451

Deg C	0	1	2	3	4	5	6	7	8	9
0	0	0.050	0.101	0.151	0.202	0.253	0.303	0.354	0.405	0.456
10	0.507	0.558	0.609	0.660	0.711	0.762	0.813	0.865	0.916	0.967
20	1.019	1.070	1.122	1.174	1.225	1.277	1.329	1.381	1.432	1.484
30	1.536	1.588	1.640	1.693	1.745	1.797	1.849	1.901	1.954	2.006
40	2.058	2.111	2.163	2.216	2.268	2.321	2.374	2.426	2.479	2.532
50	2.585	2.638	2.691	2.743	2.796	2.849	2.902	2.956	3.009	3.062
60	3.115	3.168	3.221	3.275	3.328	3.381	3.435	3.488	3.542	3.595
70	3.649	3.702	3.756	3.809	3.863	3.917	3.971	4.024	4.078	4.132
80	4.186	4.239	4.293	4.347	4.401	4.455	4.509	4.563	4.617	4.671
90	4.725	4.780	4.834	4.888	4.942	4.996	5.050	5.105	5.159	5.213
100	5.268	5.322	5.376	5.431	5.485	5.540	5.594	5.649	5.703	5.758
110	5.812	5.867	5.921	5.976	6.031	6.085	6.140	6.195	6.249	6.304
120	6.359	6.414	6.468	6.523	6.578	6.633	6.688	6.742	6.797	6.852
130	6.907	6.962	7.017	7.072	7.127	7.182	7.237	7.292	7.347	7.402
140	7.457	7.512	7.567	7.622	7.677	7.732	7.787	7.843	7.898	7.953
150	8.008	8.063	8.118	8.174	8.229	8.284	8.339	8.394	8.450	8.505
160	8.560	8.616	8.671	8.726	8.781	8.837	8.892	8.947	9.003	9.058
170	9.113	9.169	9.224	9.279	9.335	9.390	9.446	9.501	9.556	9.612
180	9.667	9.723	9.778	9.834	9.889	9.944	10.000	10.055	10.111	10.166
190	10.222	10.277	10.333	10.388	10.444	10.499	10.555	10.610	10.666	10.721
200	10.777	10.832	10.888	10.943	10.999	11.054	11.110	11.165	11.221	11.276
210	11.332	11.387	11.443	11.498	11.554	11.609	11.665	11.720	11.776	11.831
220	11.887	11.943	11.998	12.054	12.109	12.165	12.220	12.276	12.331	12.387
230	12.442	12.498	12.553	12.609	12.664	12.720	12.776	12.831	12.887	12.942
240	12.998	13.053	13.109	13.164	13.220	13.275	13.331	13.386	13.442	13.497
250	13.553	13.608	13.664	13.719	13.775	13.830	13.886	13.941	13.997	14.052
260	14.108	14.163	14.219	14.274	14.330	14.385	14.441	14.496	14.552	14.607
270	14.663	14.718	14.774	14.829	14.885	14.940	14.995	15.051	15.106	15.162
280	15.217	15.273	15.328	15.383	15.439	15.494	15.550	15.605	15.661	15.716
290	15.771	15.827	15.882	15.938	15.993	16.048	16.104	16.159	16.214	16.270
300	16.325	16.380	16.436	16.491	16.547	16.602	16.657	16.713	16.768	16.823
310	16.879	16.934	16.989	17.044	17.100	17.155	17.210	17.266	17.321	17.376
320	17.432	17.487	17.542	17.597	17.653	17.708	17.763	17.818	17.874	17.929
330	17.984	18.039	18.095	18.150	18.205	18.260	18.316	18.371	18.426	18.481
340	18.537	18.592	18.647	18.702	18.757	18.813	18.868	18.923	18.978	19.033
350	19.089	19.144	19.199	19.254	19.309	19.364	19.420	19.475	19.530	19.585
360	19.640	19.695	19.751	19.806	19.861	19.916	19.971	20.026	20.081	20.137
370	20.192	20.247	20.302	20.357	20.412	20.467	20.523	20.578	20.633	20.688
380	20.743	20.798	20.853	20.909	20.964	21.019	21.074	21.129	21.184	21.239

TABLE 7.6   (*Continued*)

Temperatures in Degrees C									Reference Junction at 0 C
Deg C	0	1	2	3	4	5	6	7	8	9
390	21.295	21.350	21.405	21.460	21.515	21.570	21.625	21.680	21.736	21.791
400	21.846	21.901	21.956	22.011	22.066	22.122	22.177	22.232	22.287	22.342
410	22.397	22.453	22.508	22.563	22.618	22.673	22.728	22.784	22.839	22.894
420	22.949	23.004	23.060	23.115	23.170	23.225	23.280	23.336	23.391	23.446
430	23.501	23.556	23.612	23.667	23.722	23.777	23.833	23.888	23.943	23.999
440	24.054	24.109	24.164	24.220	24.275	24.330	24.386	24.441	24.496	24.552
450	24.607	24.662	24.718	24.773	24.829	24.884	24.939	24.995	25.050	25.106
460	25.161	25.217	25.272	25.327	25.383	25.438	25.494	25.549	25.605	25.661
470	25.716	25.772	25.827	25.883	25.938	25.994	26.050	26.105	26.161	26.216
480	26.272	26.328	26.383	26.439	26.495	26.551	26.606	26.662	26.718	26.774
490	26.829	26.885	26.941	26.997	27.053	27.109	27.165	27.220	27.276	27.332
500	27.388	27.444	27.500	27.556	27.612	27.668	27.724	27.780	27.836	27.893
510	27.949	28.005	28.061	28.117	28.173	28.230	28.286	28.342	28.398	28.455
520	28.511	28.567	28.624	28.680	28.736	28.793	28.849	28.906	28.962	29.019
530	29.075	29.132	29.188	29.245	29.301	29.358	29.415	29.471	29.528	29.585
540	29.642	29.698	29.755	29.812	29.869	29.926	29.983	30.039	30.096	30.153
550	30.210	30.267	30.324	30.381	30.439	30.496	30.553	30.610	30.667	30.724
560	30.782	30.839	30.896	30.954	31.011	31.068	31.126	31.183	31.241	31.298
570	31.356	31.413	31.471	31.528	31.586	31.644	31.702	31.759	31.817	31.875
580	31.933	31.991	32.048	32.106	32.164	32.222	32.280	32.338	32.396	32.455
590	32.513	32.571	32.629	32.687	32.746	32.804	32.862	32.921	32.979	33.038
600	33.096	33.155	33.213	33.272	33.330	33.389	33.448	33.506	33.565	33.624
610	33.683	33.742	33.800	33.859	33.918	33.977	34.036	34.095	34.155	34.214
620	34.273	34.332	34.391	34.451	34.510	34.569	34.629	34.688	34.748	34.807
630	34.867	34.926	34.986	35.046	35.105	35.165	35.225	35.285	35.344	35.404
640	35.464	35.524	35.584	35.644	35.704	35.764	35.825	35.885	35.945	36.005
650	36.066	36.126	36.186	36.247	36.307	36.368	36.428	36.489	36.549	36.610
660	36.671	36.732	36.792	36.853	36.914	36.975	37.036	37.097	37.158	37.219
670	37.280	37.341	37.402	37.463	37.525	37.586	37.647	37.709	37.770	37.831
680	37.893	37.954	38.016	38.078	38.139	38.201	38.262	38.324	38.386	38.448
690	38.510	38.572	38.633	38.695	38.757	38.819	38.882	38.944	39.006	39.068
700	39.130	39.192	39.255	39.317	39.379	39.442	39.504	39.567	39.629	39.692
710	39.754	39.817	39.880	39.942	40.005	40.068	40.131	40.193	40.256	40.319
720	40.382	40.445	40.508	40.571	40.634	40.697	40.760	40.823	40.886	40.950
730	41.013	41.076	41.139	41.203	41.266	41.329	41.393	41.456	41.520	41.583
740	41.647	41.710	41.774	41.837	41.901	41.965	42.028	42.092	42.156	42.219
750	42.283	42.347	42.411	42.475	42.538	42.602	42.666	42.730	42.794	42.858
760	42.922	42.986	43.050	43.114	43.178	43.242	43.306	43.370	43.434	43.498

[a]From NBS Monograph 125, Thermocouple Reference Tables Based on the IPTS-68, issued March 1974.

FIND

The temperature T_1.

ASSUMPTIONS

Thermocouple follows NIST standard.

SOLUTION

Standard thermocouple tables such as Table 7.6 are referenced to 0 °C. The temperature of the reference junction for this case is 0 °C. Therefore, the temperature corresponding to an output voltage may simply be determined from Table 7.6, in this case as 180 °C.

COMMENT

Because of the law of intermediate metals, the junctions formed at the potentiometer do not affect the voltage measured for the thermocouple circuit, and the voltage output reflects accurately the temperature difference between junctions 1 and 2.

EXAMPLE 7.7

Suppose the thermocouple circuit in the previous example (Example 7.6) now has junction 2 maintained at a temperature of 30 °C, and produces an output voltage of 8.131 mV. What temperature is sensed by the measuring junction?

KNOWN

T_2 is 30 °C, and the output emf is 8.131 mV.

ASSUMPTIONS

Thermocouple follows NIST standard.

FIND

The temperature of the measuring junction.

SOLUTION

By the law of intermediate temperatures the output emf for a thermocouple circuit having two junctions, one at 0 °C and the other at T_1, would be the sum of the emfs for a thermocouple circuit between 0 and 30 °C and between 30 °C and T_1. Thus,

$$\text{emf}_{0\text{-}30} + \text{emf}_{30\text{-}T_1} = \text{emf}_{0\text{-}T_1}$$

This relationship allows the voltage reading from the nonstandard reference temperature to be converted to a 0 °C reference temperature by adding $\text{emf}_{0\text{-}30} = 1.536$ to the existing reading. This results in an equivalent output voltage, referenced to 0 °C as

$$1.536 + 8.131 = 9.667 \text{ mV}$$

Clearly, this thermocouple is sensing the same temperature as in the previous example, 180 °C. This value is determined from Table 7.6.

COMMENT

Note that the effect of raising the reference junction temperature is to lower the output voltage of the thermocouple circuit. Negative values of voltage, as compared with the polarity listed in Table 7.4, indicate that the measured temperature is less than the reference junction temperature.

EXAMPLE 7.8

A J-type thermocouple measures a temperature of 212 °F, and is referenced to 32 °F. The thermocouple is AWG 30 (30-gauge or 0.010-in. wire diameter), and is arranged in a circuit as shown in Figure 7.15a. The length of the thermocouple wire is 10 ft, in order to run from the measurement point to the ice bath and to a potentiometer. The resolution of the potentiometer is 0.005 mV. If the thermocouple wire has a resistance per unit length, as specified by the manufacturer, of 5.6 Ω/ft, estimate the residual current in the thermocouple circuit at balanced conditions.

KNOWN

A potentiometer having a resolution of 0.005 mV is used to measure the emf of a J-type thermocouple which is 10 ft long.

FIND

The residual current in the thermocouple circuit.

SOLUTION

The total resistance of the thermocouple circuit is 56 Ω, for 10 ft of thermocouple wire. The residual current is then found from Ohm's law as

$$I = \frac{E}{R} = \frac{0.005 \text{ mV}}{56 \ \Omega} = 8.9 \times 10^{-8} \text{ A}$$

COMMENT

The loading error due to this current flow is about 0.005 mV/54.3 mV/°C $\approx$ 0.1 °C.

EXAMPLE 7.9

Suppose a high-impedance voltmeter is used in place of the potentiometer in Example 7.8. Determine the minimum input impedance required for the voltmeter that will limit the loading error to the same level as the potentiometer.

KNOWN

Loading error should be less than 8.9 $\times$ 10^{-8} A.

FIND

Input impedance for a voltmeter that would produce the same current flow or loading error.

SOLUTION

At 212 °F a J-type thermocouple referenced to 32 °F will have a Seebeck voltage of $E_s = 5.268$ mV. At this temperature, the required voltmeter impedance to limit the current flow to 8.9×10^{-8} A is found from Ohm's law:

$$\frac{E_s}{I} = 5.268 \times 10^{-3} \text{ V}/8.9 \times 10^{-8} \text{ A}$$

$$= 59.2 \text{ k}\Omega$$

COMMENT

This input impedance is not unusually high for microvoltmeters, and indicates that such a voltmeter would be a reasonable choice for the measurement of the thermocouple emf in this situation. As always, the allowable loading error should be determined based on the required uncertainty in the measured temperature.

Multiple-Junction Thermocouple Circuits

A thermocouple circuit composed of two junctions of dissimilar metals produces an open-circuit emf that is related to the temperature difference between the two junctions. More than two junctions can be employed in a thermocouple circuit, and thermocouple circuits can be devised to measure temperature differences or average temperature, or to amplify the output voltage of a thermocouple circuit.

Thermopiles

Thermopile is a term used to describe a multiple-junction thermocouple circuit that is designed to amplify the output of the circuit. Since thermocouple voltage outputs are typically in the millivolt range, increasing the voltage output may be a key element in reducing the uncertainty in the temperature measurement, or may be necessary to allow transmission of the thermocouple signal to the recording device. Figure 7.20 shows a thermopile for providing an amplified output signal; in this case the output voltage would be N times the single thermocouple output, where N is the number of junctions in the circuit. The average output voltage corresponds to the average temperature level sensed by the N junctions. This thermopile arrangement can be used to measure a spatially averaged temperature, or to measure a single value of temperature. The measurement of a single value of temperature entails considerations of the physical size of a thermopile, as compared to a single thermocouple. In transient measurements, a thermopile may have a more limited frequency range than a single thermocouple, due to its increased thermal capacitance. Thermopiles are particularly useful for reducing the uncertainty in measuring small temperature differences between the measuring and reference junctions. The principle has also been used to generate small amounts of power in spacecraft.

Figure 7.21 shows a series arrangement of thermocouple junctions designed to measure the average temperature difference between junctions. This thermocouple circuit could be used in an environment where a uniform temperature was desired. In that case, a voltage output would indicate that a temperature

FIGURE 7.20   Thermopile arrangement. [From R. P.
Benedict, *Fundamentals of Temperature, Pressure and
Flow Measurements*, 3d ed., copyright © 1984 by John
Wiley & Sons, New York. Reprinted by permission.]

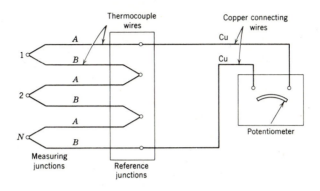

Note: In a thermopile the measuring junctions are usually located at the same physical
location, to measure one temperature.

difference existed between two of the thermocouple junctions. It should be
noted, however, that a zero voltage output could also occur if the temperature
differences that occurred in the circuit summed to a zero emf. Alternatively,
junctions 1, 2, . . ., *N* could be located at one physical location, while junctions
1', 2', . . ., *N'* could be located at another physical location. Applications for
such a circuit might be the measurement of heat flux through a solid.

FIGURE 7.21   Thermocouples arranged to sense
temperature differences. [From R. P. Benedict,
*Fundamentals of Temperature, Pressure and Flow
Measurements*, 3d ed., copyright © 1984 by John Wiley &
Sons, New York. Reprinted by permission.]

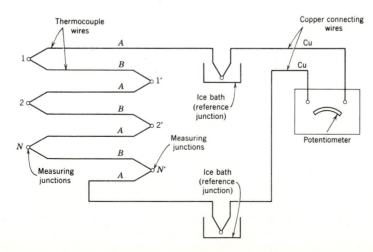

FIGURE 7.22   Parallel arrangement of thermocouples for
sensing the average temperature of the measuring
junctions. [From R. P. Benedict, *Fundamentals of
Temperature, Pressure and Flow Measurements*, 3d ed.,
copyright © 1984 by John Wiley & Sons, New York.
Reprinted by permission.]

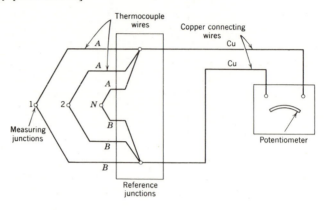

Thermocouples in Parallel

When a spatially averaged temperature is desired, multiple thermocouple
junctions can be arranged as shown in Figure 7.22. In such an arrangement of
N junctions, a mean emf is produced, given by

$$\overline{\text{emf}} = \frac{1}{N} \sum_{i=1}^{N} (\text{emf})_i \tag{7.14}$$

The mean emf is indicative of a mean temperature,

$$\overline{T} = \frac{1}{N} \sum_{i=1}^{N} T_i \tag{7.15}$$

7.6 RADIATIVE
TEMPERATURE MEASUREMENTS

The measurement of temperature through the detection of thermal radiation
presents some unique advantages. The sensor for thermal radiation need not be
in contact with the surface to be measured, making this method attractive for
a wide variety of applications. The basic operation of a radiation thermometer
is predicated upon some knowledge of the radiation characteristics of the surface
whose temperature is being measured, relative to the calibration of the thermometer. The spectral characteristics of radiative measurements of temperature
is beyond the scope of the present discussion; an excellent source for further
information is found in [8].

Radiation Fundamentals

Radiation refers to the emission of electromagnetic waves from the surface of an object. This radiation has characteristics of both waves and particles, which leads to a description of the radiation as being composed of photons. The photons generally travel in straight lines from points of emission to another surface, where they are absorbed, reflected or transmitted. This radiation exists over a large range of wavelengths that includes x rays, ultraviolet radiation, visible light, and thermal radiation, as shown in Figure 7.23. The thermal radiation emitted from an object is related to its temperature, and has wavelengths ranging from approximately 10^{-7} to 10^{-3} m. It is necessary to understand two key aspects of radiative heat transfer in relation to temperature measurements. First, the radiation emitted by an object is proportional to the fourth power of its temperature. In the ideal case, this may be expressed as

$$E_b = \sigma T^4 \qquad (7.16)$$

where E_b is the flux of energy radiating from an ideal surface, or the blackbody emissive power. The emissive power of a body is the energy emitted per unit area and per unit time. The term blackbody implies a surface that absorbs all incident radiation, and as a result emits radiation in an "ideal" manner.

The emissive power is a direct measure of the total radiation emitted by an object. However, energy is emitted by an ideal radiator over a range of wavelengths, and at any given temperature the distribution of the energy emitted as a function of wavelength is unique. Max Planck (1858–1947) developed the basis for the theory of quantum mechanics in 1900 as a result of examining the wavelength distribution of radiation. He proposed the following equation to describe

FIGURE 7.23 The electromagnetic spectrum. [From F. P. Incropera and D. P. DeWitt, *Fundamentals of Heat and Mass Transfer*, 2d ed., copyright © 1985 by John Wiley & Sons, New York. Reprinted by permission.]

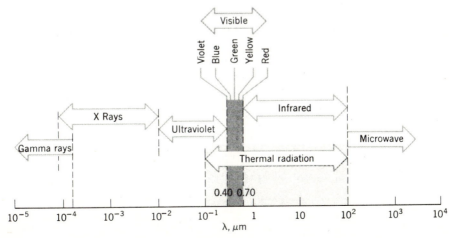

the wavelength distribution of thermal radiation for an ideal or blackbody radiator:

$$E_{b\lambda} = \frac{2\pi h_p c^2}{\lambda^5 \left[\exp(h_p c/k_B \lambda T) - 1\right]}$$ (7.17)

where

$E_{b\lambda}$ = total emissive power at the wavelength
λ = wavelength
c = speed of light
h_p = Planck's constant = 6.6256×10^{-34} J-s
k_B = Boltzmann's constant = 1.3805×10^{-23} J/K

Figure 7.24 is a plot of this wavelength distribution for various temperatures. For the purposes of radiative temperature measurements, it is crucial to note that the maximum energy emission shifts to shorter wavelengths at higher tem-

FIGURE 7.24   Planck distribution of blackbody emissive power as a function of wavelength. [From *Fundamentals of Heat and Mass Transfer*, 2d ed., copyright © 1985 by John Wiley & Sons, New York. Reprinted by permission.]

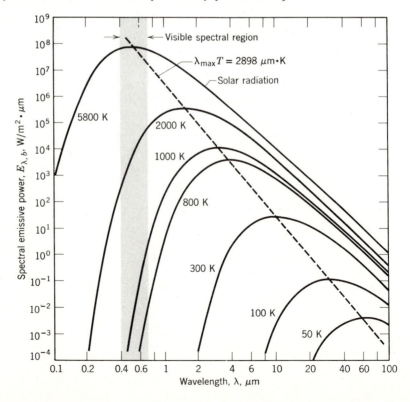

peratures. Our experiences confirm this behavior through observation of color changes as a surface is heated.

Consider an electrical heating element, as can be found in an electric oven. With no electric current flow through the element, it appears almost black, its room temperature color. With a current flow, the element temperature rises and it appears to change color to a dull red, and perhaps to a reddish orange. If its temperature continued to increase, eventually it would appear white. This change in color signifies a shift in the maximum intensity of the emitted radiation to shorter wavelengths, out of the infrared and into the visible. There is also an increase in total emitted energy. The Planck distribution provides a basis for the measurement of temperature through color comparison.

Radiation Detectors

Radiative energy flux can be detected in a sensor by two basic techniques. The detector is subject to radiant energy from the source whose temperature is to be measured. The first technique involves a thermal detector in which absorbed radiative energy elevates the detector temperature, as shown in Figure 7.25. These thermal detectors are certainly the oldest sensors for radiation, and the first such detector can probably be credited to Sir William Herschel, who verified the presence of infrared radiation using a thermometer and a prism. The equilibrium temperature of the detector is a direct measure of the amount of radiation absorbed. The resulting rise in temperature must then be measured. Thermopile detectors provide a thermoelectric power resulting from a change in temperature. A thermistor can also be used as the detector, and results in a change in resistance with temperature.

A second basic type of detector relies on the interaction of a photon with an electron, resulting in an electric current. In a photomultiplier tube, the emitted electrons are accelerated and used to create an amplified current, which is measured. Photovoltaic cells may be employed as radiation detectors. The photovoltaic effect results from the generation of a potential across a p-n junction in a semiconductor when it is subject to a flux of photons. Electron–hole pairs are formed if the incident photon has an energy level of sufficient magnitude. This process will result in the direct conversion of radiation into electrical energy, and results in high sensitivity and a fast response time when used as a detector.

FIGURE 7.25   Schematic diagram of a basic radiometer: 1, lens; 2, focusing mirror; 3, detector (thermopile or thermistor).

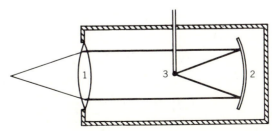

In general, the photon detectors tend to be spectrally selective, so that the relative sensitivity of the detector tends to change with the wavelength of the measured radiation.

Many considerations enter into the choice of a detector for radiative measurements. If time response is important, photon detectors are significantly faster than thermopile or thermistor detectors, and therefore have a much wider frequency response. Photodetectors saturate, while thermopile sensors may slowly change their characteristics over time. Some instruments have variations of sensitivity with the incident angle of the incoming radiation; this factor may be important for solar insolation measurements. Other considerations include wavelength sensitivity, cost, and allowable operating temperatures.

Radiative Temperature Measurements

Commercially applicable radiation thermometers vary widely in their complexity and the accuracy of the resultant measurements. We will consider only the basic techniques that allow measurement of temperature. Perhaps the simplest form, a radiometer measures a source temperature by measuring the voltage output from a thermopile detector. A schematic of such a device is shown in Figure 7.25. The increase in temperature of the thermopile is a direct indication of the temperature of the radiation source. One application of this principle is in the measurement of total solar radiation incident upon a surface. Figure 7.26 shows a schematic of a pyranometer, used to measure global solar irradiance. It would have a hemispherical field of view, and measures both the direct or beam radiation, and diffuse radiation. The diffuse and beam components of radiation can be separated by shading the pyranometer from the direct solar radiation, thereby measuring the diffuse component.

Optical pyrometry identifies the temperature of a surface by its color, or more precisely the color of the radiation it emits. A schematic of an optical pyrometer is shown in Figure 7.27. A standard lamp is calibrated so that the current flow through its filament is controlled and calibrated in terms of the filament temperature. Comparison is made optically between the color of this filament and the surface of the object whose temperature is being measured. The comparator can be the human eye. Uncertainties in the measurement may be reduced by appropriately filtering the incoming light. Corrections must be applied for surface

FIGURE 7.26   Pyranometer construction.

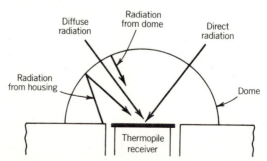

FIGURE 7.27   Schematic diagram of a disappearing
filament optical pyrometer.

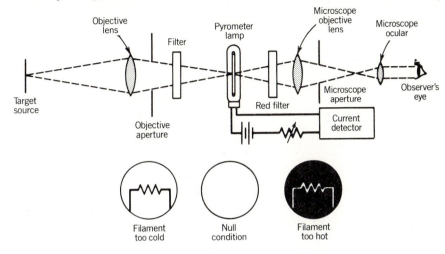

Appearance of lamp filament in eyepiece of optical pyrometer.

emissivity associated with the measured radiation; uncertainties vary with the
skill of the user, and generally are on the order of 5 °C. Replacing the human
eye with a different detector extends the range of useful temperature measurement, and improves the precision.

The major advantage of an optical pyrometer lies in its ability to measure
high temperatures remotely. For example, it could be used to measure the
temperature of a furnace without having any sensor in the furnace itself. For
many applications this provides a safe and economical means of measuring high
temperatures.

Optical Fiber Thermometers

The optical fiber thermometer is based on the creation of an ideal radiator
that is optically coupled to a fiber optic transmission system [9,10], as shown in
Figure 7.28. The temperature sensor in this system is a thin, single-crystal aluminum oxide (sapphire) fiber; a metallic coating on the tip of the fiber forms a

FIGURE 7.28   Optical fiber thermometer: 1, blackbody
cavity (iridium film); 2, sapphire fiber (single crystal); 3,
protective coating (Al_2O_3).

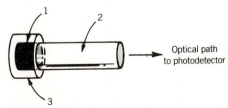

blackbody radiating cavity, which radiates directly along the sapphire crystal fiber. The single-crystal sapphire fiber is necessary because of the high-temperature operation of the thermometer. The operating range of this thermometer is 300 to 1900 °C. Signal transmission is accomplished using standard, low-temperature fiber optics. A specific wavelength band of the transmitted radiation is detected and measured, and these raw data are reduced to yield the temperature of the blackbody sensor.

The absence of electrical signals associated with the sensor signal provides excellent immunity from electromagnetic and radio-frequency interference. The measurement system has superior frequency response and sensitivity. The system has been employed for measurement in combustion applications. Temperature resolution of 0.0001 °C or better is possible.

7.7   PHYSICAL ERRORS IN TEMPERATURE MEASUREMENT

In general, errors in temperature measurement derive from two fundamental sources. The first source of errors derives from uncertain information about the temperature of the sensor itself, such as results from random interpolation errors, calibration bias errors, or other effects. Instrument and procedure uncertainty in the sensor temperature can be reduced by improved calibration or by changes in the measuring and recording instruments to reduce error components. However, a second source of errors in temperature measurement could occur even if the temperature of the probe itself could be measured exactly. This would occur if the probe was not properly sensing the temperature that it was intended to measure.

In this section we discuss the errors in temperature measurements that affect the sensor temperature such that it will not accurately represent the intended measured temperature. While the sensor temperature itself may be very accurately measured, heat transfer between the sensor and its total environment affects the equilibrium temperature of any sensor. These errors are called *insertion errors*.

A list of typical errors associated with the use of temperature sensors is provided in Table 7.7. Precision errors in temperature measurements are a result of limits of precision of measuring and recording equipment, and time variations in the measuring environment. Thermocouples have some characteristics that can lead to bias and precision errors, such as the effect of extension wires and electrical connection plugs. Another major source of error for thermocouples involves the accuracy of the reference junction.

Suppose it is desired to measure the outdoor temperature. This measurement could employ a large dial thermometer which might be placed on a football field or a tennis court in the direct sunlight, and assumed to represent "the temperature," perhaps as high as 125 °F. But what temperature is being indicated by this thermometer? Certainly the thermometer is not measuring the air temperature, nor is it measuring the temperature of the field or the court. The thermometer is subject to the very error sources that we wish to describe and analyze. **The thermometer**, very simply, **indicates its own temperature**! The temperature

TABLE 7.7   Measuring Errors Associated with
Temperature Sensors

Precision Errors

1. Imprecision of readings
2. Time and spatial variations

Bias Errors

1. Insertion errors, heating or cooling of junctions

    a. Conduction errors
    b. Radiation errors
    c. Recovery errors

2. Effects of plugs and extension wires

    a. Non-isothermal connections
    b. Loading errors

3. Ignorance of materials or material changes during measurements

    a. Aging following calibration
    b. Annealing effects
    c. Cold work hardening

4. Ground loops
5. Magnetic field effects
6. Galvanic error
7. Reference junction inaccuracies

of the thermometer is the thermodynamic equilibrium temperature that results from the radiant energy from the sun, convective exchange with the air, and conduction heat transfer with the surface on which it is resting. Considering the fact that these thermometers typically have a glass cover, which acts as a solar collector, it is very likely that the thermometer temperature is significantly higher than the air temperature.

The physical mechanisms that may cause a temperature probe to indicate a temperature different than that intended include conduction, radiation, and recovery errors. In any real measurement system, their effects could be coupled, and therefore should not be considered independently. However, for simplicity, each of these error sources will be considered separately. Our purpose is to provide only approximate analyses of the errors, and not to provide predictive techniques for correcting measured temperatures. These analyses should be used for selection and design of temperature measuring systems. The goal of the measurement engineer should be to minimize these errors, as far as is possible, through the careful installation and design of temperature probes.

EXAMPLE 7.10

An effective method of evaluating data acquisition and reduction errors associated with the use of multiple temperature sensors within a test rig is to provide a known temperature point at which the sensor outputs can be compared.

Suppose the outputs from M similar thermocouple sensors (e.g., all T-type) are to be measured and stored on an M-channel data acquisition system. Each sensor is referenced to the same reference junction temperature (e.g., ice point) and operated in the normal manner. The sensors are exposed to a known and uniform temperature. N (say 30) readings for each of the M thermocouples are recorded. What information can be obtained from the data?

KNOWN

$\quad\quad M\ (j = 1, 2, \ldots, M)$ thermocouples

$\quad\quad N\ (i = 1, 2, \ldots, N)$ readings measured for each thermocouple

SOLUTION

The mean value for all readings of the ith thermocouple is given as

$$\overline{T}_j = \frac{1}{N} \sum_{i=1}^{N} T_{ij}$$

The pooled mean for all the thermocouples is given as

$$\langle \overline{T} \rangle = \frac{1}{M} \sum_{j=1}^{M} \overline{T}_j$$

The difference between the pooled mean temperature and the known temperature would provide an estimate of the bias limit that can be expected from any channel during data acquisition. On the other hand, the differences between each $\overline{T}_j$ and $\langle \overline{T} \rangle$ must reflect the precision among the M channels. The precision index for the data acquisition and reduction instrumentation system is then

$$\langle S_{\mathrm{T}} \rangle = \sqrt{\frac{\displaystyle\sum_{j=1}^{M} \sum_{i=1}^{N} (T_{ij} - \overline{T}_j)^2}{M(N-1)}}$$

with degrees of freedom, $\nu = M(N-1)$

COMMENT

Elemental errors accounted for in these estimates include:

- Reference junction precision errors
- Precision errors in the known temperature
- Data acquisition system precision errors
- Extension cable and connecting plug bias errors
- Thermocouple emf–T correlation bias errors

These estimates would not include instrument calibration errors or probe insertion errors.

FIGURE 7.29   Temperature probe inserted into a measuring environment.

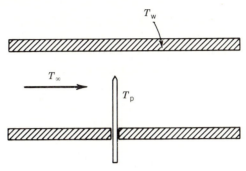

Conduction Errors

Errors that result from conduction heat transfer between the measuring environment and the ambient are often called *immersion errors*. Consider the temperature probe shown in Figure 7.29. In many circumstances, a temperature probe extends from the measuring environment through a wall into the ambient environment, where indicating or recording systems are located. The probe and the electrical leads form a path for the conduction of energy from the measuring environment to the ambient. The fundamental nature of the error created by conduction in measured temperatures can be illustrated by the model of a temperature probe shown in Figure 7.30, where it is assumed that the measured temperature is higher than the ambient temperature. The essential physics of immersion errors associated with conduction can be discerned by modeling the temperature probe as a fin. If we consider a differential element of the fin, as shown in Figure 7.30b, at steady state, there is energy conducted along the fin, and transferred by convection from the surface. The surface area for convection

FIGURE 7.30   Model of a temperature probe as a one-dimensional fin.

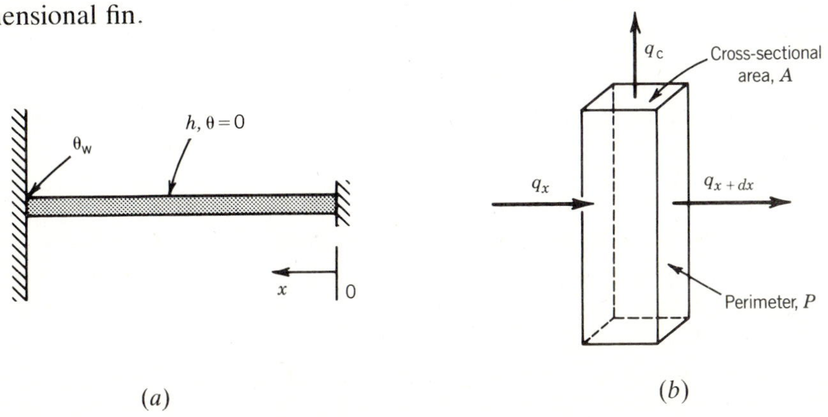

(a) (b)

is $P\ dx$, where P is the perimeter or circumference. Applying the first law of thermodynamics to this differential element yields

$$q_{x+dx} - q_x = hP\ dx\ [T(x) - T_\infty] \tag{7.18}$$

where h is the convection coefficient. If q is expanded in a Taylor series about the point x, and the substitutions

$$\theta = T - T_\infty \qquad q = -kA\frac{dT}{dx} \qquad m = \sqrt{\frac{hP}{kA}} \tag{7.19}$$

are made, then the governing differential equation becomes

$$\frac{d^2\theta}{dx^2} - m^2\theta = 0 \tag{7.20}$$

Here k is the effective thermal conductivity of the temperature probe. The solution to this differential equation for the boundary conditions that the wall has a temperature T_w, or a normalized value $\theta_w = T_w - T_\infty$, and the end of the fin is small in surface area, is

$$\frac{\theta(x)}{\theta_w} = \frac{\cosh mx}{\cosh mL} \tag{7.21}$$

The point $x = 0$ is the location where the temperature is being assumed to be measured, and therefore the solution is evaluated at $x = 0$ as

$$\frac{\theta(0)}{\theta_w} = \frac{T(0) - T_\infty}{T_w - T_\infty} = \frac{1}{\cosh mL} \tag{7.22}$$

From this analysis, the error due to conduction, e_c can be estimated. An ideal sensor would indicate the fluid temperature, T_∞; therefore, if the sensor temperature is $T_p = T(0)$, then the conduction error is

$$e_c = T_p - T_\infty = \frac{T_w - T_\infty}{\cosh mL} \tag{7.23}$$

The purpose of this analysis is to gain some physical understanding of ways to minimize conduction errors (not to correct inaccurate measurements). The behavior of this solution is such that the ideal temperature probe would have $T_p = T_\infty$, or $\theta(0) = 0$, implying that $e_c = 0$. Equation 7.22 shows that a value of $\theta(0)$ different from zero results from a nonzero value of θ_w, and a finite value of cosh (mL). The difference between the fluid temperature being measured and the wall temperature should be as small as possible; clearly, this implies that the wall should be insulated to minimize this temperature difference, and the resulting conduction error.

The term $\cosh(mL)$ should be as large as possible. The behavior of the cosh function is shown in Figure 7.31. Since the hyperbolic cosine monotonically

FIGURE 7.31   Behavior of the hyperbolic cosine.

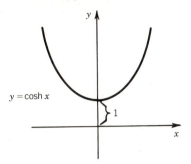

increases for increasing values of the argument, the goal of a probe design should be to maximize the value of the product mL, or $(hP/kA)^{1/2}L$. Thus, two important conclusions are that the probe should be as small in diameter as possible, and should be inserted as far as possible into the measuring environment, away from the bounding surface, producing a large value of L. A small diameter increases the ratio of the perimeter, P, to the cross-sectional area, A. For a circular cross section, this ratio is $4/D$, where D is the diameter. A good rule of thumb based on (7.23) is to have an $L/D > 50$ for negligible conduction error. In general, the thermal conductivity of a temperature probe and the convection coefficient are not design parameters.

Although this analysis clearly indicates the fundamental aspects of conduction errors in temperature measurements, it does not provide the capability to correct measured temperatures for conduction errors. Conduction errors should be minimized through appropriate design and installation of temperature probes. Usually, the physical situation is sufficiently complex to preclude accurate mathematical description of the measurement errors. Additional information on modeling conduction errors may be found in [11].

Radiation Errors

Consider a temperature probe used to measure a gas temperature. In the presence of significant radiation heat transfer, the equilibrium temperature of a temperature probe may be different than the fluid temperature being measured. Because radiation heat transfer is proportional to the fourth power of temperature, the importance of radiation effects increase as the absolute temperature of the measuring environment increases. The error due to radiation can be estimated by considering steady-state thermodynamic equilibrium conditions for a temperature sensor. Consider the case where energy is transferred to/from a sensor by convection from the environment and from/to the sensor by radiation to a body at a different temperature, such as a pipe or furnace wall. For this analysis conduction will be neglected. A first law analysis of a system containing the probe, at steady-state conditions, yields

$$q_c + q_r = 0$$

where

$$q_c = \text{convective heat transfer}$$
$$q_r = \text{radiative heat transfer}$$

The heat-transfer components can then be expressed in terms of the appropriate fundamental relations as

$$q_c = hA_s(T_\infty - T)$$
$$q_r = \sigma\epsilon(T_w^4 - T^4)$$

Assuming that the surroundings may be treated as a blackbody, the first law for a system consisting of the temperature probe is

$$hA_s(T_\infty - T_p) = FA_s\epsilon\sigma(T_p^4 - T_s^4) \tag{7.24}$$

A temperature probe is generally small compared to its surroundings, which justifies the assumption that the surroundings may be treated as black. The radiation error, e_r, is estimated by

$$e_r = (T_p - T_\infty) = \frac{F\epsilon\sigma}{h}(T_s^4 - T_p^4) \tag{7.25}$$

where

$$\sigma = \text{the Stefan–Boltzmann constant } (\sigma = 5.669 \times 10^{-8} \text{ W/m}^2 \cdot \text{K}^4)$$
$$\epsilon = \text{emissivity of the sensor}$$
$$F = \text{radiation view factor}$$
$$T_p = \text{probe temperature}$$
$$T_s = \text{temperature of the surroundings}$$

Again, if the sensor is small compared to the scale of the surroundings, the view factor from the sensor to the surroundings may be taken as 1.

EXAMPLE 7.11

A typical situation where radiation would be important occurs in measuring the temperature of a furnace. Figure 7.32 shows a small temperature probe for which conduction errors are negligible, which is placed in a high-temperature enclosure, where the fluid temperature is T_∞ and the walls of the enclosure are at T_w. Convection and radiation are assumed to be the only contributing heat-transfer modes at steady state. Develop an expression for the equilibrium temperature of the probe. Also determine the equilibrium temperature of the probe and the radiation error in the case where $T_\infty = 800$ °C, $T_w = 500$ °C, and the emissivity of the probe is 0.8. The convective heat transfer coefficient is 100 W/m²-°C.

KNOWN

Temperatures $T_\infty = 800$ °C and $T_w = 500$ °C, with $h = 100$ W/m²-°C. The emissivity of the probe is 0.8.

FIGURE 7.32   Analysis of a temperature probe in a radiative and convective environment.

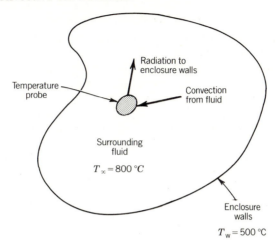

FIND

An expression for the equilibrium temperature of the probe, and the resulting probe temperature for the stated conditions.

ASSUMPTIONS

The surroundings may be treated as black, and conduction heat transfer is neglected.

SOLUTION

The probe is modeled as a small spherical body within the enclosed furnace; the radiation view factor from the probe to the furnace is 1.0. At steady state, an equilibrium temperature may be found from an energy balance. The first law for a system consisting of the temperature probe, from equation 7.24, is

$$h(T_\infty - T_p) + \sigma\epsilon(T_w^4 - T_p^4) = 0 \qquad (7.26)$$

The probe is attempting to measure T_∞. However, equation 7.26 may be solved by trial and error to yield an equilibrium temperature for the probe of 642.5 °C. The radiation error in this case is -157.5 °C. This result indicates that radiative heat transfer can create significant measurement errors at elevated temperatures.

COMMENT

Equation 7.25 does not allow direct solution for the radiation error, since it contains the probe temperature. A trial and error approach, as followed here, provides the simplest solution.

Radiation shielding is a key concept in controlling radiative heat transfer; shielding for radiation is analogous to insulation to reduce conduction heat transfer. A *radiation shield* is an opaque surface interposed between a temperature sensor and its radiative surroundings so as to reduce electromagnetic wave interchange. In principle, the shield attains an equilibrium temperature closer to the fluid temperature than the surroundings. Because the probe can no longer "see" the surroundings, with the radiation shield in place, the probe temperature is closer to the fluid temperature. Additional information on radiation error in temperature measurements may be found in [11]. The following example serves to demonstrate radiation errors and the effect of shielding.

EXAMPLE 7.12

Consider again the oven in example 7.11 where the oven is maintained at a temperature of 800 °C. Because of energy losses the walls of the oven are cooler, having a temperature of 500 °C. For the present case, consider the temperature probe as a small spherical object located in the oven, having no thermal conduction path to the ambient. (All energy exchange is through convection and radiation.) Under these conditions, the probe temperature is 642.5 °C, as found in Example 7.11.

Suppose a radiation shield is placed between the temperature probe and the walls of the furnace, which blocks the path for radiative energy transfer. Examine the effect of adding a radiation shield on the probe temperature.

KNOWN

A radiation shield is added to a temperature probe in an environment with $T_\infty = 800$ °C and $T_w = 500$ °C.

FIND

The radiation error in the presence of the shield.

ASSUMPTIONS

The radiation shield completely surrounds the probe, and the surroundings may be treated as a blackbody.

SOLUTION

The shield equilibrium temperature is higher than the wall temperature by virtue of convection with the fluid. As a result, the probe "sees" a higher temperature surface, and the probe temperature is closer to the fluid temperature, resulting in less measurement error.

For a single radiation shield placed so that it completely surrounds the probe, which is small compared to the size of the enclosure and has an emissivity of 1, the equilibrium temperature of the shield can be determined from equation 7.24. The temperature of the shield is found by trial and error solution to be 628 °C. Because the sensor now "sees" the shield, rather than the wall, the temperature measured by the probe will be 697 °C, which is also determined by trial and error from equation 7.24.

One shield with an emissivity of 1 provides for an improvement over the case of no shields, but a better choice of the surface characteristics of the shield material can result in much better performance. If the shield has an emissivity of 0.1, the shield temperature rises to 756 °C, and the probe temperature to 771 °C.

COMMENT

Although shielding provides improved temperature measurements by reducing radiative heat transfer, the primary area for improvement in this temperature measurement could be the elevation of the wall temperature through insulation.

This discussion of radiation shielding serves to demonstrate the usefulness of shielding as a means of improving temperature measurements in radiative environments. As with conduction errors, the development should be used to guide the design and installation of temperature sensors, not to correct measured temperatures. Further information on radiation errors may be found in [4].

Recovery Errors in Temperature Measurement

The kinetic energy of a gas moving at high velocity can be converted to sensible energy by reversibly and adiabatically bringing the flow to rest at a point. The temperature resulting from this process is called the *stagnation* or *total temperature*, T_t. On the other hand, the *static temperature* of the gas, T_∞, is the temperature that would be measured by an instrument moving at the local stream velocity. From a molecular point of view, the static temperature measures the magnitude of the random kinetic energy of the molecules that comprise the gas, while the stagnation temperature includes both the directed and random components of kinetic energy. Generally, the engineer would be content with knowledge of either temperature, but in high-speed gas flows neither temperature is indicated by the sensor.

For negligible changes in potential energy, and in the absence of heat transfer or work, the energy equation for a flow may be written

$$h_1 + \frac{U^2}{2g_c} = h_2 \tag{7.27}$$

where state 2 refers to the stagnation condition, and state 1 to a condition where the gas is flowing with the velocity U. Assuming ideal gas behavior, the enthalpy difference $h_2 - h_1$ may be expressed as $c_p(T_2 - T_1)$, or in terms of static and stagnation temperatures

$$\frac{U^2}{2g_c} = c_p(T_t - T_\infty) \tag{7.28}$$

The term $U^2/2g_c c_p$ is called the *dynamic temperature*.

What implication does this have for the measurement of temperature in a flowing gas stream? The physical nature of gases at normal pressures and temperatures is such that the velocity of the gas on a solid surface is zero, because of the effects of viscosity. Thus, when a temperature probe is placed in a moving fluid, the fluid is brought to rest on the surface of the probe. However, this process may not be thermodynamically reversible, and the gas in question may not behave as an ideal gas. Deceleration of the flow by the probe converts some portion of the directed kinetic energy of the flow to thermal energy, and elevates the temperature of the probe above the static temperature of the gas. The fraction of the kinetic energy recovered as thermal energy is called the *recovery factor*, r, defined as

$$r \equiv \frac{T_p - T_\infty}{\dfrac{U^2}{2g_c c_p}} \tag{7.29}$$

where T_p represents the equilibrium temperature of the stationary (with respect to the flow) real temperature probe. In general, r may be a function of the velocity of the flow, or more precisely, the Mach number and Reynolds number of the flow, and the shape and orientation of the temperature probe. For thermocouple junctions of round wire, Moffat [12] reports values for r of

$$0.68 \pm 0.07 \ (95\%) \qquad \text{for wires normal to the flow}$$
$$0.86 \pm 0.09 \ (95\%) \qquad \text{for wires parallel to the flow}$$

These recovery factor values tend to be constant at velocities for which temperature errors are significant, usually flows where the Mach number is greater than 0.1. For thermocouples having a welded junction, a spherical weld bead significantly larger than the wire diameter tends to a value of the recovery factor of 0.75, for the wires parallel or normal to the flow. The relationships between temperature and velocity for temperature probes with known recovery factors are

$$T_p = T_\infty + \frac{rU^2}{2g_c c_p} \tag{7.30}$$

or in terms of the recovery error, e_U,

$$e_U = T_\infty - T_p = -\frac{rU^2}{2g_c c_p} \tag{7.31}$$

The probe temperature is related to the stagnation temperature by

$$T_p = T_t - \frac{(1 - r)U^2}{2g_c c_p} \tag{7.32}$$

Fundamentally, in liquids the stagnation and static temperatures are essentially equal [4], and the recovery error may generally be taken as zero for liquid flows. In any case, high-velocity flows are rarely encountered in liquids.

EXAMPLE 7.13

A temperature probe having a recovery factor of 0.86 is to be used to measure a flow of air at velocities up to the sonic velocity, at a pressure of 1 atm and a static temperature of 30 °C. Calculate the value of the recovery error in the temperature measurement as the velocity of the air flow increases, from 0 to the speed of sound, using equation 7.31.

KNOWN

$r = 0.86$ $p_\infty = 1$ atm abs $= 101$ kPa abs
$M \leq 1$ $T_\infty = 30$ °C $= 303$ K

FIND

The recovery error as a function of air velocity.

ASSUMPTIONS

Air behaves as an ideal gas.

SOLUTION

Assuming that air behaves as an ideal gas, the speed of sound is expressed as

$$c = \sqrt{kRTg_c} \qquad (7.33)$$

For air at 101 kPa and 303 K, with $R = 0.287$ kJ/kg-K, the speed of sound is approximately 349 m/s.

The *Mach number* is defined as the ratio of the flow velocity to the speed of sound. Figure 7.33 shows the error in temperature measurement as a function of Mach number for this temperature probe.

FIGURE 7.33   Behavior of recovery error as a function of Mach number.

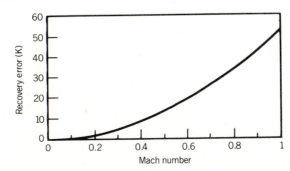

COMMENT

Typically, the static and total temperatures of the flowing fluid stream are to be determined from the measured probe temperature. In this case a second independent measurement of the velocity is necessary.

7.8   SUMMARY

Temperature is a fundamentally important quantity in science and engineering, both in concept and practice. As such, temperature is one of the most widely measured engineering variables, providing the basis for a variety of control and safety systems. This chapter provides the basis for the selection and installation of temperature sensors.

Temperature is defined for practical purposes through the establishment of a temperature scale, such as the Kelvin scale, that encompasses fixed reference points and interpolation standards. The International Temperature Scale 1990 is the accepted standard for establishing a universal means of temperature measurement.

The two most common methods of temperature measurement employ thermocouples and resistance temperature detectors. Standards for the construction and use of these temperature measuring devices have been established and provide the basis for selection and installation of commercially available sensors and measuring systems.

Installation effects on the accuracy of temperature measurements are a direct result of the influence of radiation, conduction, and convection heat transfer on the equilibrium temperature of a temperature sensor. The installation of a temperature probe into a measuring environment can be accomplished in such a way as to minimize the uncertainty in the resulting temperature measurement.

REFERENCES

1. Patterson, E. C., Eponyms: Why Celsius? *American Scientist* 77(4): 413, 1989.
2. Committee Report, The International Temperature Scale of 1990, *Metrologia* 27(3), 1990. (The text of the superceded International Practical Temperature Scale of 1968 appears as an appendix in the National Bureau of Standards monograph 124. An amended version was adopted in 1975, and the English text published: *Metrologia* 12: 7–17, 1976.)
3. Quinn, T. J., News from BIPM, *Metrologia* 26: 69–74, 1989.
4. Benedict, R. P., *Fundamentals of Temperature, Pressure and Flow Measurements,* 3d ed., Wiley, New York, 1984.
5. McGee, T. D., *Principles and Methods of Temperature Measurement,* Wiley-Interscience, New York, 1988.
6. Diehl, W., Thin-film PRTD, *Measurements and Control,* 155–159, Dec. 1982.
7. Thermistor Definitions and Test Methods, Electronic Industries Standard RS-275-A (ANSI Standard C83.68-1972), June 1971.

8. Dewitt, D. P., and G. D. Nutter, *Theory and Practice of Radiation Thermometry*, Wiley-Interscience, New York, 1988.

9. Dils, R. R., High-temperature optical fiber thermometer, *Journal of Applied Physics*, 54(3): 1198–1201, 1983.

10. Optical fiber thermometer, *Measurements and Control*, April 1987.

11. Sparrow, E. M., Error estimates in temperature measurement, in E. R. G. Eckert and R. J. Goldstein (Eds.), *Measurements in Heat Transfer*, 2d ed., Hemisphere, Washington, DC, 1976.

12. Moffat, R. J., Gas temperature measurements, in *Temperature—Its Measurement and Control in Science and Industry*, Vol. 3, Part 2, Reinhold, New York, 1962.

NOMENCLATURE

c speed of light in a vacuum $[l-t]$

c_p specific heat $[l^2 - t^{-2}/°]$

d thickness of bimetallic strip $[l]$

e_c conduction temperature error $[°]$

e_U recovery temperature error $[°]$

e_r radiation temperature error $[°]$

emf   electromotive force

h convective heat transfer coefficient $[m - t^{-3}/°]$

k thermal conductivity $[m - l - t^{-3}/°]$

k ratio of specific heats (equation for the speed of sound)

l length $[l]$

m $\sqrt{hp/kA}$ (equation 7.15) $[l^{-2}]$

q heat flux $[m - t^3]$

r recovery factor

r_c radius of curvature $[l]$

u uncertainty

u_d design-stage uncertainty

A_c cross-sectional area $[l^2]$

A_s surface area $[l^2]$

B bias limit

C_α coefficient of thermal expansion $[l - l^{-1}/°]$

D diameter $[l]$

E_b blackbody emissive power $[m - t^{-3}]$

$E_{b\lambda}$ spectral emissive power $[m - t^{-3}]$

E_i input voltage $[V]$

E_1 voltage drop across R_1 $[V]$

F radiation view factor

I current $[A]$

L length of temperature probe $[l]$

P precision index

Q heat transfer $[m - l^2 - t^{-3}]$

R resistance $[\Omega]$

R gas constant $[l^2 - t^{-2}/°]$

R_0 reference resistance $[\Omega]$

R_T thermistor resistance $[\Omega]$

$S_{\bar{x}}$ standard deviation of the means for the variable x

T temperature $[°]$

T_0 reference temperature $[°]$

T_p probe temperature $[°]$

T_w wall or boundary temperature $[°]$

T_∞ fluid temperature $[°]$

U fluid velocity $[l - t^{-1}]$

α temperature coefficient of resistivity $[\Omega/°]$

α_{AB} Seebeck coefficient

β material constant for thermistor resistance $[°]$

β constant in polynomial expansion

π_{AB} Peltier coefficient $[m - l^2 - t^{-3} - A^{-1}]$

σ Thomson coefficient $[m - l^2 - t^{-3} - A^{-1}]$

σ Stefan–Boltzmann constant for radiation $[m - t^{-3}/(°)^4]$

θ nondimensional temperature
θ_x sensitivity index for variable x
       (uncertainty analysis)
ρ_c resistivity [Ω-l]

δ thermistor dissipation
       constant [$m - l^2 - t^{-3}/°$]
ϵ emissivity

PROBLEMS

7.1 Define and discuss the significance of the following terms, as they apply to temperature and temperature measurements:
   a.   temperature scale
   b.   temperature standards
   c.   fixed points
   d.   interpolation

7.2 Fixed temperature points in the International Temperature Scale are phase equilibrium states for a variety of pure substances. Discuss the conditions necessary within an experimental apparatus to accurately reproduce these fixed temperature points. How would elevation, weather, and material purity affect the uncertainty in these fixed points?

7.3 Calculate the resistance of a platinum wire that is 2 m in length and has a diameter of 0.1 cm. The resistivity of platinum at 25 °C is 9.83×10^{-6} Ω-cm. What implications does this result have for the construction of a resistance thermometer using platinum?

7.4 An RTD forms one arm of a Wheatstone bridge, as shown in Figure 7.34. The RTD is used to measure a constant temperature, with the bridge operated in a balanced mode. The RTD has a resistance of 25 Ω at a temperature of 0 °C, and a thermal coefficient of resistance, $\alpha = 0.003925$ °C^{-1}. The value of the variable resistance R_1 must be set to 41.485 Ω to balance the bridge circuit, with the RTD in thermal equilibrium with the measuring environment.
   a.   Determine the temperature of the RTD.
   b.   Compare this circuit to the equal-arm bridge in Example 7.2. Which circuit provides the greater static sensitivity?

FIGURE 7.34   Wheatstone bridge circuit for Problem 7.4

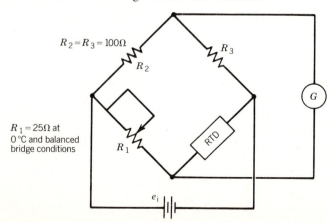

7.5 A thermistor is placed in a 100 °C environment, and its resistance measured as 20 000 Ω. The material constant, β, for this thermistor is 3650 °C. If the thermistor is then used to measure a particular temperature, and its resistance is measured as 500 Ω, determine the thermistor temperature.

7.6 Estimate the required level of uncertainty in the measurement of resistance for a platinum RTD if the RTD is to serve as a local standard for the calibration of a temperature measurement system for an uncertainty of ±0.005 °C. Assume $R(0 °C) = 100$ Ω.

7.7 Define and discuss the following terms related to thermocouple circuits:
a. thermocouple junction
b. thermocouple laws
c. reference junction
d. Peltier effect
e. Seebeck coefficient

7.8 The output emf from the thermocouple circuit shown in Figure 7.35 is 10.721 mV. What is the measuring junction temperature?

FIGURE 7.35   Thermocouple circuit for Problem 7.8: 1, measuring junction; 2, reference junction.

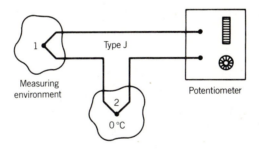

7.9 a. The thermocouple shown in Figure 7.36a yields an output voltage of 7.947 mV. What is the temperature of the measuring junction?

FIGURE 7.36   Schematic diagram for Problem 7.9.

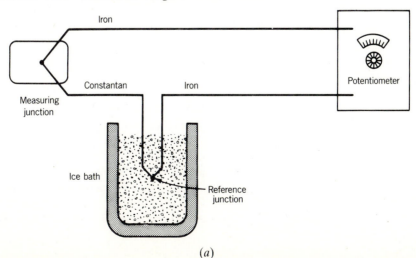

(a)

b. The ice bath that maintains the reference junction temperature melts, allowing the reference junction to reach a temperature of 25 °C. If the measuring junction of part (a) remains at the same temperature, what voltage would be measured by the potentiometer?

c. Copper extension leads are installed as shown in Figure 7.36b. For an output voltage of 7.947 mV, what is the temperature of the measuring junction?

FIGURE 7.36 Schematic diagram for Problem 7.9.

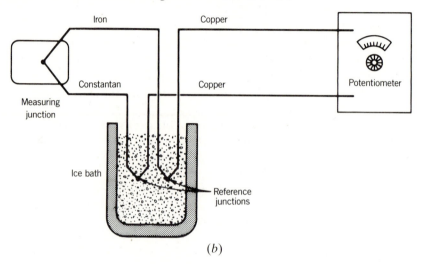

(b)

7.10 A J-type thermocouple referenced to 70 °F has a measured output emf of 2.878 mV. What is the temperature of the measuring junction?

7.11 A J-type thermocouple referenced to 0 °C indicates 4.115 mV. What is the temperature of the measuring junction?

7.12 A temperature measurement requires an uncertainty of ±2 °C at a temperature of 200 °C. A standard T-type thermocouple is to be used with a readout device that provides electronic ice-point reference junction compensation, and has a stated accuracy of ±0.5 °C with 0.1 °C resolution. Determine if the uncertainty constraint is met at the design stage.

7.13 A temperature difference of 3.0 °C is measured using a thermopile having three pairs of measuring junctions, arranged as shown in Figure 7.21.

a. Determine the output of the thermopile for J-type thermocouple wire, if all pairs of junctions sense the 3.0 °C temperature difference. The average temperature of the junctions is 80 °C.

b. If the thermopile is constructed of wire that has a maximum emf variation from the NIST standard values of ±0.8%, and the voltage measuring capabilities in the system are such that the uncertainty is ±0.0005 V, perform an uncertainty analysis to estimate the uncertainty in the measured temperature difference at the design stage.

7.14 Complete the following table for a J-type thermocouple:

Temperature [°C]		emf
Measured	Reference	[mV]
100	0	
	0	−0.5
100	50	
	50	2.5

7.15 A thermopile is constructed as shown in Figure 7.20, to measure a single temperature. For a four-junction thermopile, referenced to 0 °C, what would be the emf produced at a temperature of 125 °C? If a voltage measuring device was available that had a total uncertainty of ±0.0001 V, how many junctions would be required in the thermopile to reduce the uncertainty in the measured temperature to 0.1 °C?

7.16 You are employed as a heating, ventilating, and air conditioning engineer. Your task is to decide where in a residence to place a thermostat, and how it is to be mounted on the wall. A thermostat contains a bimetallic temperature measuring device that serves as the sensor for the control logic of the heating and air conditioning system for the house. Consider the heating season. When the temperature of the sensor falls 1 °C below the set-point temperature of the thermostat, the furnace is activated; when the temperature rises 1 °C above the set point, the furnace is turned off. Discuss where the thermostat should be placed in the house, what factors could cause the temperature of the sensor in the thermostat to be different from the air temperature, and possible causes of discomfort for the occupants of the house. How does the thermal capacitance of the temperature sensor affect the operation of the thermostat? Why are thermostats typically set 5 °C higher in the air conditioning season?

7.17 A J-type thermocouple for use at temperatures between 0 and 100 °C was calibrated at the steam point in a device called a hypsometer. A hypsometer creates a constant temperature environment at the saturation temperature of water, at the local barometric pressure. The steam-point temperature is strongly affected by barometric pressure variations. Atmospheric pressure on the day of this calibration was 30.1 in Hg. The steam-point temperature as a function of barometric pressure may be expressed as

$$T_{st} = 212 + 50.422 \left(\frac{p}{p_0} - 1 \right) - 20.95 \left(\frac{p}{p_0} - 1 \right)^2 \quad [°F]$$

where $p_0 = 29.921$ in Hg. At the steam point, the emf produced by the thermocouple, referenced to 0 °C, is measured as 5.310 mV. Construct a calibration curve for this thermocouple by plotting the difference between the thermocouple reference table value and the measured value ($emf_{ref} - emf_{meas}$) versus temperature. What is this difference at 0 °C? Suggest a means for measuring temperatures between 0 and 100 °C using this calibration, and estimate the contribution to the total uncertainty.

7.18 A J-type thermocouple is calibrated against an RTD standard within ±0.01 °C between 0 and 200 °C. The emf is measured with a potentiometer having 0.001-mV resolution and less than 0.015-mV bias. The reference junction temperature is provided by an ice bath. The calibration procedure yields the following results:

T_{RTD} [°C]	0.00	20.50	40.00	60.43	80.25	100.65
emf [mV]	0.010	1.038	2.096	3.207	4.231	5.336

    a. Determine a polynomial to describe the relation between the temperature and thermocouple emf.
    b. Estimate the uncertainty in temperature using this thermocouple and potentiometer.
    c. Suppose the thermocouple is connected to a digital temperature indicator having a resolution of 0.1 °C and better than 0.3 °C accuracy. Estimate the uncertainty in indicated temperature.

7.19 A beaded thermocouple is placed in a duct in a moving gas stream having a velocity of 200 ft/s. The thermocouple indicates a temperature of 1400 °R.
    a. Determine the true static temperature of the fluid, based on correcting the reading for velocity errors. Take the specific heat of the fluid to be 0.6 Btu/lb$_m$°R, and the recovery factor to be 0.22.
    b. Estimate the error in the thermocouple reading due to radiation if the walls of the duct are at 1200 °R. The view factor from the probe to the duct walls is 1, the convective heat-transfer coefficient, h, is 30 Btu/h-ft²-°R, and the emissivity of the temperature probe is 1.

7.20 It is desired to measure the static temperature of the air outside of an aircraft flying at 20 000 ft with a speed of 300 miles per hour, or 438.3 ft/s. A temperature probe is used that has a recovery factor, r, of 0.75. If the static temperature of the air is 413 °R, and the specific heat is 0.24 Btu/lb$_m$°R, what is the temperature indicated by the probe? Local atmospheric pressure at 20 000 ft. is approximately 970 lb/ft², which results in an air density of 0.0442 lb$_m$/ft³. Discuss additional factors that might affect the accuracy of the static temperature reading.

7.21 Consider the typical construction of a sheathed thermocouple, as shown in Figure 7.37. Analysis of this geometry to determine conduction errors in temperature measurement is difficult. Suggest a method for placing a

FIGURE 7.37   Typical construction of a sheathed thermocouple.

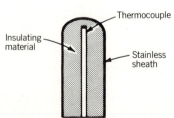

realistic upper limit on the conduction error for such a probe, for a specified immersion depth into a convective environment.

7.22 An iron–constantan thermocouple is placed in a moving air stream in
a duct, as shown in Figure 7.38. The thermocouple reference junction
is maintained at 100 °C. The emf output from the thermocouple is
14.143 mV.
 a. Determine the thermocouple junction temperature.
 b. By considering recovery and radiation errors, estimate the total error
     in the indicated temperature. Discuss whether this estimate of the
     measurement error is conservative, and why or why not. The heat-
     transfer coefficient may be taken as 70 Btu/h-ft²-°F.

Air Properties	*Thermocouple Properties*
$c_p = 0.24 \text{ Btu/lb}_m°F$	$r = 0.7$
$u = 200 \text{ ft/s}$	$\epsilon = 0.25$

FIGURE 7.38   Schematic diagram for Problem 7.22.

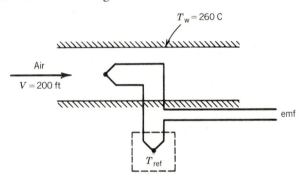

7.23 In Example 7.5, an uncertainty value for R_T was determined at 125 °C as
$B_{R_T} = \pm247 \ \Omega$. Show that this value is correct by performing an uncertainty analysis on R_T. In addition, determine the value of B_{R_T} at the
temperatures 150 and 100 °C. What error is introduced into the uncertainty analysis for β by using the value of B_{R_T} at 125 °C?

7.24 The thermocouple circuit shown in Figure 7.39 measures the temperature
T_1. The potentiometer limits of error are given as:

Limits of error: ±0.05% of reading + 15 μV at 25 °C

Resolution: 5 μV

Give a best estimate for the temperature T_1, if the output emf is 9 mV.

FIGURE 7.39   Thermocouple circuit for Problem 7.24.

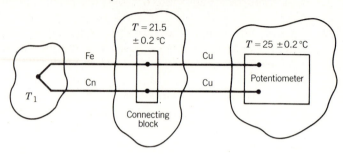

7.25   A concentration of salt of 600 ppm in tap water will cause a 0.1 °F change in the freezing point of water. For an ice bath prepared using tap water and ice cubes from tap water at a local laboratory, and having 1500 ppm of salts, determine the error in the ice-point reference. Upon repeated measurements, is this error manifested as a bias or precision error? Explain.

7.26   A platinum RTD ($\alpha = 0.00392$ °C^{-1}) is to be calibrated in a fixed-point environment. The probe itself is used in a balanced mode with a Wheatstone bridge, as shown in Figure 7.40. The bridge resistances are known to an uncertainty of ± 0.001 Ω (95%). At 0 °C the bridge balances when $R_c = 100.000$ Ω. At 100 °C the bridge balances when $R_c = 139.200$ Ω.

FIGURE 7.40   Bridge circuit for Problem 7.26.

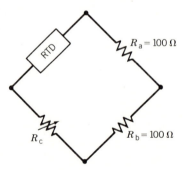

a.   Find the RTD resistance corresponding to 0 and 100 °C, and the uncertainty in each value.

b.   Calculate the uncertainty in determining a temperature using this RTD–bridge system for a measured temperature that results in $R_c = 300$ Ω. Assume $u_\alpha = \pm 1 \times 10^{-5}$ (95%).

CHAPTER 8
PRESSURE AND VELOCITY MEASUREMENTS

8.1   INTRODUCTION

In this chapter, methods to quantify and to measure the pressure within a stationary or moving fluid are presented. Instruments and procedures for establishing known values of pressure for calibration purposes, as well as various types of transducers for pressure measurement are discussed. Three well-established methods for the measurement of the local velocity within a moving fluid are also discussed. Practical considerations, including common error sources, for pressure and velocity measurements are presented.

8.2   PRESSURE CONCEPTS

Pressure represents a contact force per unit area. It acts inward and normal to the surface of any physical boundary which a fluid contacts. A basic understanding of the origin of a pressure involves the consideration of the forces acting between the fluid molecules and the solid boundaries containing the fluid. For example, consider the measurement of pressure at the wall of a vessel containing a perfect gas. As a molecule with some amount of kinetic energy collides with the solid boundary, it will rebound off in a different direction. From Newton's second law, we know that the change in linear momentum of the molecule produces an equal but opposite force on the boundary. It is the net effect of these collisions that yields the pressure sensed at the boundary surface. Factors that affect the magnitude or frequency of the collisions, such as fluid temperature and fluid density, will affect the pressure. In fact, this reasoning is the basis of the kinetic theory from which the ideal gas equation of state may be derived.

A pressure scale can be related to molecular activity, as well, since a lack of any molecular activity must form the limit of absolute zero pressure. A pure vacuum, which contains no molecules, would form the primary standard for absolute zero pressure. As shown in Figure 8.1, the *absolute pressure scale* is quantified relative to this absolute zero pressure. The pressure under standard atmospheric conditions is defined [1] as 1.0132×10^5 N/m² absolute, or more

FIGURE 8.1   Relative pressure scales.

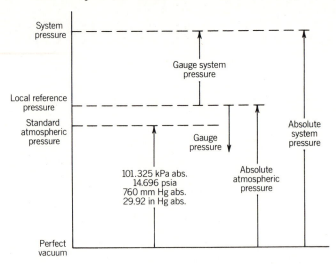

commonly, in pascals as 101 320 Pa absolute ($1 \text{ N/m}^2 = 1$ Pa). This is equivalent to

1 atm absolute

14.696 lb/in.^{2} absolute (psia)

Also indicated in Figure 8.1 is a gauge pressure scale. The *gauge pressure scale* is measured relative to some absolute reference pressure which is defined in a manner convenient to the measurement. The relation between an absolute pressure, p_{abs}, and its corresponding gauge pressure, p_g, is given by

$$p_g = p_{abs} - p_0 \qquad (8.1)$$

where p_0 is a reference pressure. A commonly used reference pressure is the local absolute atmospheric pressure existing at the time of the measurement. For example, if, as in Figure 8.1, we choose the reference pressure as 1 atm absolute, then the pressure at one atmosphere can be stated as being either 1 atm absolute or 0 atm. As a convention, an absolute pressure will always be noted as such by using the term absolute or the abbreviations "abs" or "a" following its units. Absolute pressure will be a positive number. Gauge pressure can be positive or negative depending on the value of measured pressure relative to the reference pressure. A *differential pressure*, such as $p_1 - p_2$, is a relative measure and cannot be written as an absolute pressure.

Pressure can also be described in terms of the pressure exerted on a surface submerged in a column of fluid at a depth, h, as depicted in Figure 8.2. From hydrostatics, the pressure at any depth within a fluid of specific weight γ can be written as

$$p_{abs}(h) = p_0(h_0) + \gamma h \qquad (8.2)$$

FIGURE 8.2   Hydrostatic head and pressure.

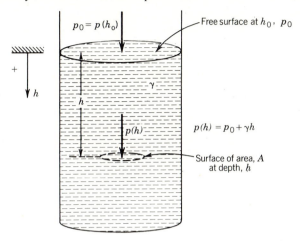

In (8.2), p_0 is determined at and h measured from an arbitrary datum line, h_0. The fluid specific weight is given by $\gamma = \rho g / g_c$. When (8.2) is rearranged, the equivalent head of fluid of depth, h, becomes

$$h = \frac{p_{abs} - p_0}{\gamma} \tag{8.3}$$

The equivalent pressure head at one standard atmosphere is defined to be

$$760 \text{ mm Hg abs} = 760 \text{ torr abs} = 1 \text{ atm abs}$$
$$= 10\ 350.8 \text{ mm } H_2O \text{ abs}$$
$$= 29.92 \text{ in. Hg abs}$$
$$= 407.513 \text{ in. } H_2O \text{ abs}$$

The standard is based on mercury with a density of 0.0135951 kg/cm^3 at 0 °C and water at 0.000998207 kg/cm^3 at 20 °C [1].

EXAMPLE 8.1

   Determine the absolute and gauge pressures and the equivalent pressure head at a depth of 10 m below the free surface of a pool of water at 20 °C.

KNOWN

   $h = 10$ m, where $h = 0$ is the free surface
   $T = 21$ °C

$\rho_{H_2O} = 998.207$ kg/m^3

Specific gravity of mercury, $S_{Hg} = 13.57$

ASSUMPTIONS

Water density constant

$p(h = 0) = 1.0132 \times 10^5$ N/m^2 abs

FIND

p_{abs}, p_g, and h

SOLUTION

The absolute pressure can be determined directly from equation 8.2. Using the pressure at the free surface as the reference pressure and the datum line for h_0, the absolute pressure must be

$$p_{abs} = 1.0132 \times 10^5 \text{ N/m}^2 + \frac{(997.4 \text{ kg/m}^3)(9.8 \text{ m/s}^2)(10 \text{ m})}{1 \text{ kg} - \text{m/N} - \text{s}^2}$$

$$= 1.9906 \times 10^5 \text{ N/m}^2 \text{ abs}$$

$$= 199.06 \text{ kPa abs} = 1.96 \text{ atm abs} = 28.88 \text{ lb/in.}^2 \text{ abs}$$

The gauge pressure is found from equation 8.1 to be

$$p_g = p_{abs} - p_0 = \gamma h$$

$$= 9.7745 \times 10^5 \text{ N/m}^2$$

$$= 97.7 \text{ kPa} = 0.96 \text{ atm} = 14.18 \text{ lb/in.}^2$$

For example, then, the pressure could be stated as

$$p = 1.96 \text{ atm abs} = 0.96 \text{ atm}$$

The pressure in terms of equivalent head is found from equation 8.3:

$$h_{abs} = \frac{1.9906 \times 10^5}{(997.4 \text{ kg/m}^3)(9.8 \text{ m/s}^2)}$$

$$= 20.36 \text{ m H}_2\text{O abs} = 1.50 \text{ m Hg abs}$$

or in terms of gauge pressure relative to 760 mm Hg abs:

$$h = \frac{(1.9906 \times 10^5) - (1.0132 \times 10^5) \text{ N/m}^2}{(998.207 \text{ kg/m})(9.8 \text{ m/s}^2)(1 \text{ N} - \text{s}^2/\text{kg} - \text{m})}$$

$$= 10 \text{ m H}_2\text{O} = 0.73 \text{ m Hg}$$

8.3 PRESSURE REFERENCE INSTRUMENTS

The units of pressure are defined by the standards of the fundamental dimensions of mass, length, and time. However, a primary standard for pressure per se does not exist. Pressure transducers are generally calibrated by comparison against certain reference instruments from which pressure can be determined through methods intrinsic to the reference instrument operating principle. In this section, several basic reference instruments are discussed that are used to establish pressure for the purposes of calibration by comparison, as well as for general measurement.

McLeod Gauge

The McLeod gauge [2] is a pressure-measuring instrument and laboratory reference standard used to establish gas pressures in the subatmospheric range of 1 mm Hg abs down to 0.1 μm Hg abs. One variation of this instrument is sketched in Figure 8.3a in which the gauge is connected directly to the low-pressure source. The glass tubing is arranged so that a sample of the gas at the low pressure to be determined can be trapped by inverting the gauge from the sensing position in Figure 8.3a to that of the measuring position of Figure 8.3b. In this way, the trapped gas within the capillary is isothermally compressed by a rising column of mercury. Boyles law is then used to relate the two pressures on either side of the mercury to the distance of travel of the mercury within the

---

FIGURE 8.3 McLeod gauge.

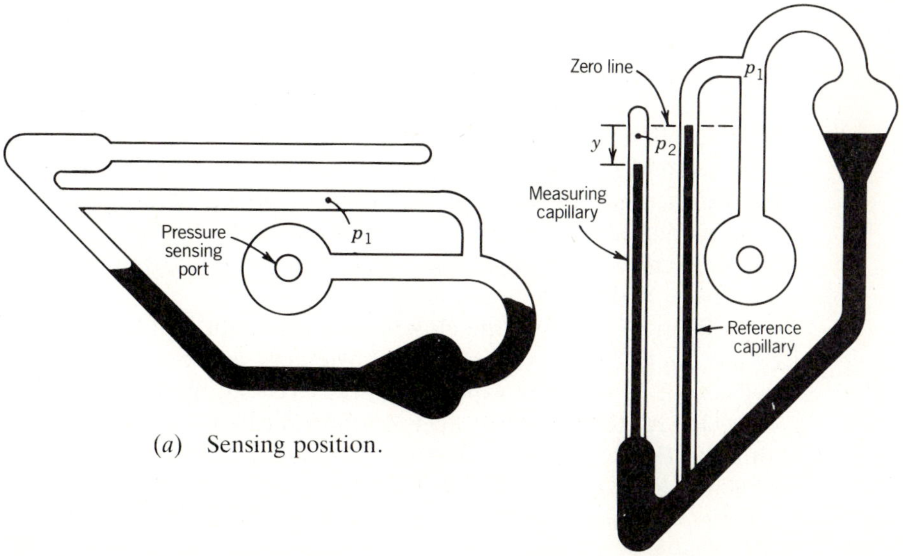

(a) Sensing position.

(b) Indicating position.

capillary. Mercury is the preferred working fluid because of its high density and very low vapor pressure.

At the equilibrium and measuring position in Figure 8.3b, the capillary pressure, p_2, is related to the unknown gas pressure to be determined, p_1, by

$$p_2 = p_1 \frac{V_1}{V_2}$$

where V_1 is the gas volume of the gauge in Figure 8.3a (a constant for a gauge at any pressure) and V_2 is the capillary volume in Figure 8.3b. But $V_2 = Ay$, where A is the known cross-sectional area of the capillary and y is the vertical length of the capillary occupied by the gas (which equals the height of the reference column of mercury). Letting γ be the specific weight of the mercury, the difference in pressures is related by

$$p_2 - p_1 = \gamma y$$

such that the unknown gas pressure is just a function of y:

$$p_1 = \frac{Ay^2}{V_1 - Ay} \tag{8.4}$$

In practice, a commercial McLeod gauge will have the capillary etched and calibrated to indicate pressure, p_1, or equivalent head, p_1/γ, directly.

The McLeod gauge indication generally does not require correction. The reference stem is provided to offset capillary forces acting in the measuring capillary. Instrument bias error in indicated pressure will be on the order of 0.5% (95%) at 1 μm Hg abs and increasing to 3% (95%) at 0.1 μm Hg abs.

Barometer

A barometer consists of an initially evacuated tube that is closed on one end. The open end is inverted and immersed within a liquid-filled reservoir as shown in the illustration of the Fortin barometer in Figure 8.4. The reservoir is open to atmospheric pressure which forces the liquid to rise up the tube. From equations 8.2 and 8.3, the resulting height of the liquid column above the reservoir free surface is a measure of the absolute atmospheric pressure in equivalent head (equation 8.3). Evangelista Torricelli (1608–1647), a colleague of Galileo, can be credited with developing and interpreting the working principles of the barometer in 1644.

As Figure 8.4 shows, the closed end of the tube will be at the vapor pressure of the liquid at room temperature. Mercury is the most common liquid used because of its very low vapor pressure. Even so, the barometer will need to be corrected for temperature effects on the indicated pressure, for temperature and altitude effects on the weight of mercury, and for deviations from standard gravity (9.80665 m/s² or 32.17405 ft/s²). Correction curves are usually provided by instrument manufacturers.

Barometers are used as local standards for the measurement of atmospheric pressure. Under standard conditions for pressure temperature and gravity, the

FIGURE 8.4 Fortin barometer.

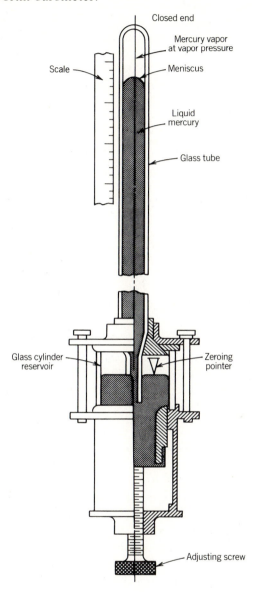

mercury will rise 760 mm (29.92 in.) above the reservoir surface. The U.S. National Weather Service always reports a barometric pressure that has been corrected to sea level elevation.

Manometers

A manometer is an instrument that utilizes the hydrostatic equation relating pressure to the hydrostatic equivalent head of fluid. Several variations in manometer design are available but all are basically pressure differential instruments used to measure differential pressures ranging on an order of magnitude of equivalent head of between 0.01 mm of manometer fluid to several meters.

The U-tube manometer in Figure 8.5 consists of a transparent tube filled with an indicating liquid of specific weight, γ_m. This forms two free surfaces of the manometer liquid. The difference in pressures p_1 and p_2 applied across the two free surfaces brings about a deflection, H, in the level of the manometer liquid. For a fluid of specific weight γ the hydrostatic equation can be applied to the manometer of Figure 8.5 as

$$p_1 = p_2 + \gamma x + \gamma_m H - \gamma(H + x)$$

which yields the relation between the manometer deflection and applied differential pressure,

$$p_1 - p_2 = (\gamma_m - \gamma)H \qquad (8.5)$$

For an input of a differential pressure with an output of a manometer deflection, equation 8.5 indicates that the static sensitivity of the U-tube manometer is given by

$$K = \frac{1}{\gamma_m - \gamma}$$

For the maximum manometer sensitivity then, liquids giving as small a value of $(\gamma_m - \gamma)$ as possible should be used as manometeric fluids. From a practical standpoint, however, the choice of fluid for the manometer must typically have a greater specific weight than and must not be soluble with the working fluid.

FIGURE 8.5   U-tube manometer.

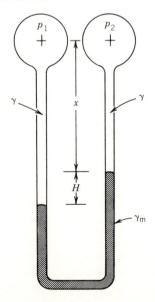

FIGURE 8.6 Micromanometer.

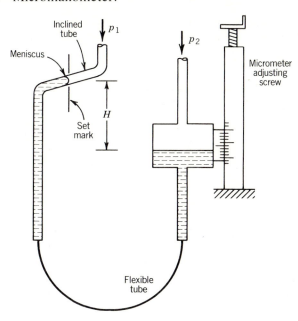

The manometer fluid should be selected to provide a deflection that is measurable yet not so great that it becomes awkward to observe.

A variation in the U-tube manometer is the micromanometer shown in Figure 8.6. These special purpose instruments are used to measure very small differential pressures, down to 0.005 mm H_2O (0.0002 in. H_2O). In the micromanometer, the manometer reservoir is moved up or down until the level of the manometer fluid within the reservoir is at the same level as a set mark within a magnifying sight glass. At that point the manometer meniscus will be at the set mark, and this serves as a reference position. Changes in pressure bring about fluid displacement so that the reservoir must be moved up or down to bring the meniscus back to the set mark. The amount of this repositioning is equal to the change in equivalent pressure head. The position of the reservoir is controlled by a micrometer or other calibrated displacement measuring device so that relative changes in pressure can be measured with a high resolution.

The inclined tube manometer is also used to measure small changes in pressure. It is essentially a U-tube manometer with one tube leg inclined at an angle, typically from 10 to 30° relative to the horizontal. As indicated in Figure 8.7, a change in pressure equivalent to a deflection of height H in a U-tube manometer would bring about a change in position of the meniscus in the inclined leg of $H/\sin \theta$. This provides increased sensitivity over the conventional U-tube by a factor of $1/\sin \theta$.

A number of elemental errors affect the instrument uncertainty of all types of manometers. These include scale and alignment errors, zero error, temperature error, gravity error, and capillary and meniscus errors. The specific weight of the manometer fluid will vary with temperature but can be corrected. For

FIGURE 8.7   Inclined tube manometer.

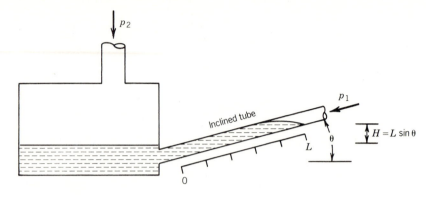

example, the common manometer fluid of mercury has a temperature dependence approximated by

$$\gamma_{Hg} = \frac{848.707}{1 + 0.000101(T - 32)} \quad [\text{lb/ft}^3]$$

A gravity correction for elevation, z, and latitude, ϕ, can be applied to correct for gravity error effects,

$$e_1 = -(2.637 \times 10^{-3} \cos 2\phi + 9.6 \times 10^{-8}z + 5 \times 10^{-5}) \quad (8.6a)$$

$$= -(2.637 \times 10^{-3} \cos 2\phi + 2.9 \times 10^{-8}z + 5 \times 10^{-5}) \quad (8.6b)$$

where ϕ is in degrees and z is in feet for (8.6a) and meters in (8.6b). Surface tension effects between the manometer and measured fluids give rise to capillary forces on the manometer column and lead to the development of a meniscus. Although the actual effect varies with purity of the manometer liquid, these effects can be minimized by using manometer tube bores of greater than about 6 mm (0.25 in.). In general, the instrument uncertainty in measuring pressure can be as low as 0.02 to 0.2% of the reading.

EXAMPLE 8.2

An inclined manometer with indicating leg at 30° is to be used at 70°F to measure a gas pressure of nominal magnitude of 100 N/m² relative to ambient. "Unity" oil ($S = 1$) is to be used. The specific weight of the oil is judged to be known at 9770 ± 0.5% N/m² (95%) at 70 °F, the angle of inclination can be set to within 1° using a bubble level, and the manometer resolution is 1 mm with a manometer zero error equal to its interpolation error. Estimate the uncertainty in indicated differential pressure at the design stage. $\gamma_{gas} = 11.5 \pm 0.5\%$ N/m³ (95%).

KNOWN

$p \approx 100 \text{ N/m}^2$

Manometer

Resolution: 1 mm

Zero error: 0.5 mm

$\theta = 30 \pm 1° \text{ (95% assumed)}$

$\gamma_m = 9770 \pm 0.5\% \text{ N/m}^2 \text{ (95%)}$

ASSUMPTIONS

Temperature and capillary effects in manometer and gravity error in the specific weights of the fluids are negligible.

FIND

u_d

SOLUTION

The relation between pressure and manometer deflection is given by equation 8.5 with $H = L \sin \theta$:

$$p_1 - p_2 = L(\gamma_m - \gamma) \sin \theta$$

where p_2 is ambient pressure. Hence, the indicated gauge pressure can be written as

$$p = L(\gamma_m - \gamma) \sin \theta$$

For a nominal pressure of 100 N/m², the nominal manometer rise L would be

$$L = \frac{p}{(\gamma_m - \gamma) \sin \theta} \approx 21 \text{ mm}$$

For the design stage analysis, $p = f(\gamma, \gamma_m, L, \theta)$, so that the uncertainty in pressure, p, is estimated by

$$(u_d)_p = \pm \sqrt{\left[\frac{\partial p}{\partial \gamma_m} (u_d)_{\gamma_m}\right]^2 + \left[\frac{\partial p}{\partial L} (u_d)_L\right]^2 + \left[\frac{\partial p}{\partial \theta} (u_d)_\theta\right]^2}$$

where the uncertainty in γ is assumed to be negligible ($\gamma_m \gg \gamma$). At assumed 95% confidence levels, the manometer specific weight uncertainty and angle uncertainty are set by judgment at

$$(u_d)_{\gamma_m} = (9770 \text{ N/m}^2)(0.005) \approx 49 \text{ N/m}^2$$

$$(u_d)_\theta = 0.0175 \text{ rad}$$

The uncertainty in estimating the pressure from the indicated deflection is due both to the manometer resolution and zero point bias:

$$(u_d)_L = \sqrt{(u_0)^2 + (u_c)^2}$$

$$= \sqrt{(0.5 \text{ mm})^2 + (0.5 \text{ mm})^2} = 0.7 \text{ mm}$$

This gives a design-stage uncertainty of,

$$(u_d)_p = \pm \sqrt{(0.26)^2 + (3.42)^2 + (3.10)^2}$$

$$= \pm 4.6 \text{ N/m}^2 \quad (95\%)$$

COMMENT

At a 30° inclination and for this pressure, the uncertainty in pressure is affected almost equally by the instrument inclination and deflection uncertainties. As the inclination approaches a vertical orientation, that is, the U-tube manometer, inclination uncertainty becomes less important. However, for a U-tube manometer, the deflection is reduced to less than 11 mm, a 50% reduction in manometer sensitivity, with an associated design-stage uncertainty of 6.8 N/m^2.

Deadweight Testers

The deadweight tester is used as a laboratory standard for the calibration of pressure-measuring devices over the pressure range from 69 to 7×10^7 N/m^2 (0.01 to 10 000 psi). This device determines pressure directly through its fundamental definition of a force per unit area. A deadweight tester, such as that shown in Figure 8.8, consists of an internal chamber filled with a liquid, and a close-fitting piston and cylinder. A pressure is produced by the compression of the liquid, usually oil. This pressure acts on the end of the carefully machined piston. A static equilibrium will exist when the external pressure exerted by the piston on the fluid balances with the chamber pressure. This external piston

FIGURE 8.8   Deadweight tester.

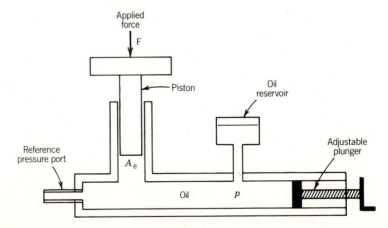

pressure is created by a downward force acting over the equivalent area, A_e, of the piston. The weight of the piston plus the additional weight of calibrated masses are used to provide this external force, F. At static equilibrium the piston will float and the chamber pressure can be deduced as

$$p = \frac{F}{A_e} + \sum \text{errors} \tag{8.7}$$

A pressure-measuring device, such as a pressure transducer, can be connected to the tester's reference port and calibrated by comparison to the chamber pressure.

The instrument uncertainty in the chamber pressure using a deadweight tester can be as low as 0.05 to 0.01% of the reading. A number of elemental errors contribute, including air buoyancy effects, variations in local gravity, uncertainty in the known mass of the piston and added masses, shear effects, thermal expansion of the piston area, and elastic deformation of the piston.

An indicated pressure, p_i can be corrected for gravity effects, e_1, and air buoyancy effects, e_2, by

$$p = p_i(1 + e_1 + e_2) \tag{8.8}$$

The error introduced into the indicated tester pressure can be corrected for gravity effects by (8.6) and for the air buoyancy effects by

$$e_2 = -\frac{\gamma_{air}}{\gamma_{masses}} \tag{8.9}$$

The tester fluid lubricates the piston and the piston will be partially supported by the shear forces of the oil between the cylinder and the piston. Error is introduced which will vary inversely with the tester fluid viscosity [3]. Hence, high-viscosity oils are preferred. In a typical tester, this error is less than 0.01% of the reading. At high pressures, elastic deformation of the piston will affect the actual piston area. For the above two reasons, the effective area is usually based on the average of the piston and cylinder diameters another source of uncertainty in the measurement.

EXAMPLE 8.3

A deadweight tester indicates 100.00 lb/in.^{2}, or 100.00 psi, at 70°F in Clemson, SC ($\phi = 34°$, $z = 841$ ft). Manufacturer specifications for the effective piston area were stated at 72 °F so that thermal expansion effects remain negligible. Assume $\gamma_{air} = 0.076$ lb/ft^3 and $\gamma_{mass} = 496$ lb/ft^3. Estimate the corrected pressure.

KNOWN

$p_i = 100.00$ psi

$z = 841$ ft

$\phi = 34°$

ASSUMPTIONS

Bias corrections for altitude and latitude apply.

FIND

p

SOLUTION

From equation 8.9, the indicated pressure can be corrected for buoyancy effects as

$$e_2 = -\frac{\gamma_{air}}{\gamma_{masses}} = -0.000154$$

and for gravity effects by (8.6a)

$$e_1 = [2.637 \times 10^{-3} \cos(2 \times 34) + 9.6 \times 10^{-8} \times 841 + 5 \times 10^{-5}]$$

$$= -0.001119$$

From (8.8), the corrected pressure becomes

$$p = 100 \times (1 - 0.000154 - 0.001119) \text{ lb/in.}^2 = 99.87 \text{ lb/in.}^2$$

COMMENT

This amounts to the correction of an indicated signal for known bias. The result, p, is not necessarily the true value for pressure since other precision and bias errors may be present.

8.4   PRESSURE TRANSDUCERS

A pressure transducer converts a measured pressure into a mechanical or electrical signal. The transducer is actually a hybrid sensor–transducer. The primary sensor is usually an elastic element that deforms or deflects under pressure. Several common elastic elements used, shown in Figure 8.9, include the Bourdon tube, bellows, capsule, and diaphragm. A secondary transducer element converts the elastic element deflection into a readily measurable signal such as an electrical voltage or mechanical rotation of a pointer. There are myriads of methods available to perform this secondary function. But electrical transducers will require additional external signal conditioning equipment and external power supplies to drive their electrical output signals.

Pressure transducers are subject to some or all of the following elemental errors: resolution, zero shift error, linearity error, sensitivity error, hysteresis, and drift due to environmental temperature changes. Electrical transducers are also subject to loading error between the transducer output and its indicating device (Chapter 6). This error increases transducer nonlinearity over its oper-

FIGURE 8.9   Elastic elements used as pressure sensors.

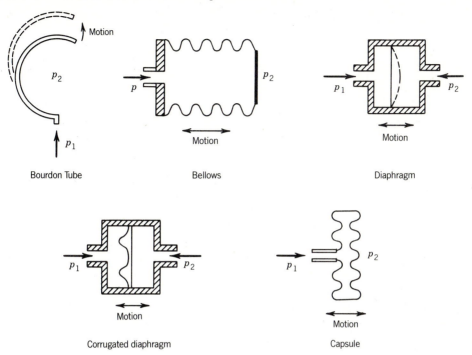

ating range. A voltage follower can be inserted at the output of the transducer to isolate transducer load.

Bourdon Tube

The Bourdon tube is a curved metal tube having an elliptical cross section that mechanically deforms under pressure. It is used as the primary sensor in a large class of pressure gauges. In practice, one tube end is held fixed and the input pressure applied internally. A pressure difference between the outside of the tube and the inside of the tube will bring about tube deformation and a deflection of the tube free end. This action of the tube under pressure can be likened to the action of a deflated balloon that is inflated slightly. The magnitude of the deflection is proportional to the magnitude of the pressure difference. Several variations exist, such as the C shape (Figure 8.9), the spiral, and the twisted tube. The exterior of the tube is usually open to atmosphere (hence, the origin of the term "gauge" pressure referring to pressure referenced to atmospheric pressure), but in some variations the tube may be placed within a sealed housing and the tube exterior exposed to some other reference pressure.

The Bourdon tube mechanical dial gauge is perhaps the most commonly used pressure transducer. A typical design is shown in Figure 8.10, in which the secondary element is a mechanical linkage that converts the tube displacement into a rotation of a pointer. The instrument has a range in which pressure will be linearly related to pointer rotation range of the instrument and this is usually

FIGURE 8.10   Bourdon tube gauge.

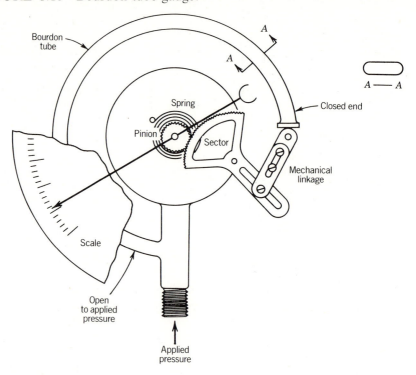

the range specified by its manufacturer. Designs exist that can be used for low or high pressures. Various gauges exist that are suitable within a pressure range of 0.1 to 100 000 psi. Some can resolve pressure to within 0.02 psi. The best Bourdon tube gauges have instrument uncertainties as low as 0.1% of the full-scale deflection of the gauge, but values of 0.5 to 2% are more common.

Bellows and Capsule

A bellows sensing element is a thin-walled, flexible metal tube formed into deep convolutions and sealed at one end (Figure 8.9). One end is held fixed and pressure is applied internally. A difference between the internal and external pressures will cause the bellows to change length. The bellows is housed within a chamber that can be sealed and evacuated for absolute measurements, vented through a reference pressure port for differential measurements, or opened to atmosphere for gauge pressure measurements. A capsule sensing element is a thin-walled, flexible metal tube similar to the bellows but tending to have a wider diameter and shorter length (Figure 8.9).

A mechanical linkage is used to convert the translational displacement of the bellows or capsule sensors into a measurable form. A common secondary trans-ducer is the sliding arm potentiometer (voltage-divider, Chapter 6) found in the potentiometric pressure transducer shown in Figure 8.11. Another type uses a linear variable displacement transducer (Chapter 10) to measure bellows or

FIGURE 8.11   Potentiometer transducer

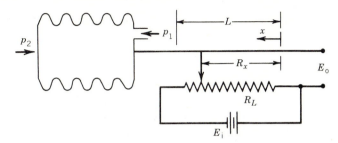

capsule displacement. The LVDT design has a high sensitivity and is commonly found in pressure transducers rated for low pressures and small pressure ranges, such as zero to several inches of water absolute, gauge or differential.

EXAMPLE 8.4

An engineer finds a desirable potentiometric bellows gauge with a stated range of 0 to 100 psi in a catalog. It uses a sliding contact potentiometer having a terminal resistance of 50 Ω–10 kΩ $\pm$ 10% over full scale. The following manufacturer information is available:

Calibration linearity:	<0.5% full scale
Repeatability:	$\pm$0.1% full scale
Excitation voltage:	5–10 V dc

A voltmeter having an input impedance of 100 kΩ is available to measure output. For an applied transducer excitation voltage of 5 V, estimate the expected instrument uncertainty in measuring a 50-psi pressure.

KNOWN

$E_i = 5$ V

$R_L = 50 \ \Omega$–10 kΩ $\pm$ 10% (95% assumed)

ASSUMPTIONS

Manufacturer specifications are reasonably accurate.

FIND

u_c

SOLUTION

The electrical arrangement of a potentiometric gauge is a voltage divider circuit (Chapter 6). The circuit is illustrated in Figure 8.12. For an infinite measuring

FIGURE 8.12   Potentiometer transducer circuit for
Example 8.4.

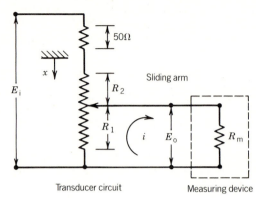

Transducer circuit   Measuring device

instrument input impedance and assuming that the terminal resistance is at its
minimum at $p = 0$ psi and at its maximum at $p = 100$ psi, then

$$E_o(p) = \frac{R_L(p)E_i}{10 \text{ k}\Omega}$$

so that $E_o(0) = 0.0250$ V, $E_o(50) = 2.4875$ V, and $E_o(100) = 5$ V. The output
is nearly linear, so the static sensitivity will be approximately

$$K = \frac{E_o(100) - E_o(0)}{100 - 0 \text{ psi}} = 0.04975 \text{ V/psi}$$

The transducer instrument uncertainty will be affected by the elemental errors
due to linearity, e_1, repeatability, e_2, and loading, e_3. For a full-scale voltage of
5 V, the specifications indicate

$$e_1 = 5 \text{ V} \times 0.005 = 0.0250 \text{ V} \qquad e_2 = 5 \text{ V} \times 0.001 = 0.0050 \text{ V}$$

Loading error will depend on the ratio of terminal resistance to measuring
instrument input impedance, R_m. Using equation 6.37 with Figure 6.19, at 50
psi, $R_1 = 4975$ Ω, $R_2 = 5025$ Ω, and $R_m = 100\,000$ Ω. So with interstage loading,
the output voltage becomes

$$E_o(50) = 2.4268 \text{ V}$$

The error due to loading is $e_3 = 2.4875 - 2.4268 = 0.0607$ V.
    At 50 psi, the instrument uncertainty, u_c, is estimated to be

$$u_c = \left[\sum_{i=1}^{3} e_i^2 \right]^{1/2}$$

$$= \sqrt{0.0250^2 + 0.0050^2 + 0.0607^2} = 0.0652 \text{ V}$$

For a static sensitivity of 0.04975 V/psi, this equates to

$$u_c = 1.311 \text{ psi}$$

Diaphragms

An effective primary element is a diaphragm, which is a thin elastic circular plate supported about its circumference. The action of a diaphragm within a pressure transducer is similar to the action of a trampoline and a pressure differential across the surface of the diaphragm acts to deform it. The magnitude of the deformation is proportional to the pressure difference. Both membrane and corrugated designs are used. Membranes are made of metal or nonmetallic material, such as plastic or neoprene. The material chosen depends on the pressure range anticipated and the fluid in contact with it. Corrugated diaphragms contain a number of corrugations that serve to increase diaphragm stiffness and to increase the diaphragm effective surface area.

Pressure transducers that use a diaphragm sensor are well suited for either static or dynamic pressure measurements. They have good linearity and resolution over their useful range. An advantage of the diaphragm sensor is that the very low mass and relative stiffness of the thin diaphragm gives the sensor a very high natural frequency with a small damping ratio. Hence, these transducers can have a very wide frequency response and very short 90% rise and settling times. The natural frequency of a circular diaphragm can be estimated by [4]

$$\omega_n = 64.15 \sqrt{\frac{E_m t^2 g_c}{12(1 - v_p^2) \, r^4 \rho}} \tag{8.10}$$

where E_m is the bulk modulus [psi or N/m^2], t the thickness [in. or m], r the radius [in. or m], ρ the material density [lb/in.^{3} or kg/m^3], and v_p Poisson's ratio for the diaphragm material with $g_c = 386 \text{ lb}_m\text{-in./lb-s}^2 = 1 \text{ kg-m/N-s}^2$. The maximum elastic deflection of a uniformly loaded, circular diaphragm supported about its circumference occurs at its center and can be estimated by

$$y_{max} = \frac{3(p_1 - p_2)(1 - v_p^2)r^4}{16 E_m t^3} \tag{8.11}$$

provided that the deflection does not exceed one-third the diaphragm thickness. Diaphragms should be selected so as to not exceed this maximum deflection over the anticipated operating range.

Various secondary elements are available to translate this displacement of the diaphragm into a measurable signal. Several methods are discussed below.

Strain Gauge Elements

The most common method for converting diaphragm displacement into a measurable signal is to sense the strain induced on the diaphragm surface as it is displaced. Strain gauges, devices whose measurable resistance is proportional to their sensed strain (Chapter 11), can be bonded directly onto the diaphragm or onto a deforming element (such as a thin beam) attached to the diaphragm so as to deform with the diaphragm and sense strain. Strain gauge resistance is reasonably linear over a wide range of strain and can be directly related to the diaphragm sensed pressure [5]. A diaphragm transducer using strain gauge detection is depicted in Figure 8.13.

The use of semiconductor technology in pressure transducer construction has led to the development of a variety of very fast, very small, highly sensitive strain gauge diaphragm transducers. Silicone piezoresistive strain gauges can be diffused into a single crystal of silicone wafer which forms the diaphragm. Semiconductor strain gauges have a sensitivity that is 50 times greater than conventional metallic strain gauges. And because the piezoresistive gauges are integral to the diaphragm, they are relatively immune to the thermoelastic strains prevalent in conventional metallic strain gauge–diaphragm constructions. Furthermore, a silicone diaphragm will not creep with age (as will a metallic gauge), thus minimizing calibration drift over time. However, gauge failure is catastrophic and silicone is not well suited to wet environments.

Capacitance Elements

When one or more fixed metal plates are placed directly above or below a metallic diaphragm, a capacitor is created that forms an effective secondary element. Such a transducer using this method is depicted in Figure 8.14. It is known as a capacitance transducer. The capacitance, C, developed between two parallel plates separated by a distance, t, is determined by

$$C = \frac{c\epsilon A}{t} \tag{8.12}$$

FIGURE 8.13   Diaphragm pressure transducer.

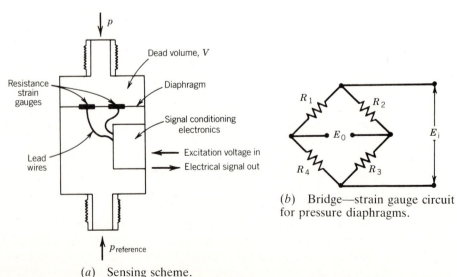

(b)   Bridge—strain gauge circuit for pressure diaphragms.

(a)   Sensing scheme.

FIGURE 8.14 Capacitance pressure transducer.

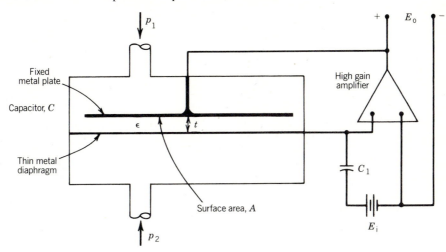

where ϵ is the dielectric constant of the material between the plates (for air, $\epsilon = 1$), A is the overlapping area of the two plates, and c is the proportionality constant given by 0.225 (when A is measured in [in.^{2}] and t in [in.]) or 0.0885 (when A is measured in [cm^2] and t in [cm]). Displacement of the diaphragm changes the average gap separation. In the circuit shown, the measured voltage will be essentially linear with developed capacitance

$$E_o = \frac{C_1}{C} E_i \qquad (8.13)$$

and pressure can be inferred. The capacitance pressure transducer possesses the attractive features of other diaphragm transducers, including small size and a very wide operating range. However, it is sensitive to temperature changes and has a relatively high impedance output.

Piezoelectric Crystal Elements

Piezoelectric crystals form effective secondary elements for dynamic (transient) pressure measurements. Under the action of compression, tension, or shear, a piezoelectric crystal will deform and develop a surface charge, q, which is proportional to the force acting to bring about the deformation. In a piezoelectric pressure transducer, a preloaded crystal is mounted to the diaphragm sensor as indicated in Figure 8.15. Pressure acts normal to the crystal axis and changes the crystal thickness, t. This sets up a charge

$$q = K_q p A$$

FIGURE 8.15　Piezoelectric pressure transducer.

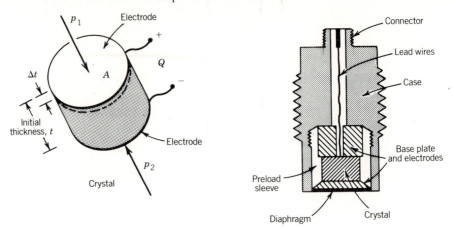

where p is the pressure acting over the electrode area A and K_q is the crystal charge sensitivity, a material property. The voltage developed across the electrodes is given by

$$E_o = \frac{q}{C}$$

where C is the capacitance of the crystal–electrode combination. The capacitance can be obtained from equation 8.12 to yield

$$E_o = \frac{K_q pt}{c\epsilon} = K_E pt \qquad (8.14)$$

where K_E is the voltage sensitivity of the transducer. The crystal sensitivities for quartz, the most common material used, are $K_q = 2.2 \times 10^{-9}$ coulombs/N and $K_E = 0.055$ V-m/N.

EXAMPLE 8.5

A common method to estimate the data acquisition and reduction errors present in the pressure-measuring instrumentation of a test rig is to apply a series of replication tests on the M calibrated pressure transducers used on that rig (when the number of transducers installed is large, a few transducers can be selected at random and tested). In such a test each transducer is interconnected through a common manifold to the output of a single deadweight tester (or suitable standard) so that each transducer is exposed to exactly the same static pressure. Each transducer is operated at its normal excitation voltage. The test proceeds as follows: record applied test pressure (from the deadweight tester), record each transducer output N (say, $N \geq 10$) times, vent manifold to local atmosphere, close manifold to reapply test pressure, repeat procedure at least K (say, $K \geq 5$) times. What information is available in such test data?

KNOWN

M $(j = 1, 2, \ldots, M)$ transducers
$N(i = 1, 2, \ldots, N)$ repetitions at a test pressure
$K(k = 1, 2, \ldots, K)$ replications of test pressure

SOLUTION

The mean value for each transducer for any replication is given by the mean of the repetitions for that transducer,

$$\bar{p}_{ij} = \frac{1}{N} \sum_{i=1}^{N} p_{ijk}$$

The mean value for the replications of the jth transducer is given by

$$<\bar{p}_j> = \frac{1}{K} \sum_{k=1}^{K} \bar{p}_{jk}$$

The difference between the pooled mean pressure and the known applied pressure would provide an estimate of the bias limit to be expected from any transducer during data acquisition. On the other hand, differences between $\bar{p}_{jk}$ and $<\bar{p}_j>$ must be due to the precision error in the data acquisition and reduction procedure. This precision error is estimated by considering the variation of the test pressure mean, $\bar{p}_{jk}$, about the pooled mean, $<\bar{p}_j>$, for each transducer:

$$<S_j> = \sqrt{\frac{\sum_{k=1}^{K} (\bar{p}_{jk} - <\bar{p}_j>)^2}{K - 1}}$$

The pooled standard deviation of the mean for M transducers provides the estimate of the precision index of the data acquisition and reduction procedure:

$$<S_{\bar{p}}> = \sqrt{\frac{\sum_{j=1}^{M} \sum_{k=1}^{K} (\bar{p}_{jk} - <p_j>)^2}{M(K - 1)}}$$

with degrees of freedom, $\nu = M(K - 1)$.

COMMENT

The statistical estimate contains the effects on precision due to:

- Pressure standard (applied pressure repeatability)
- Pressure transducer repeatability (repetition and replication)
- Excitation voltages
- Recording system

It does not contain the effects of instrument calibration errors, large deviations in environmental conditions, pressure tap design errors, or dynamic pressure effects. Bias limits for these must be set by the engineer based on other information.

Note that if the above procedure were to be repeated over a range of different applied known pressures, then this would be a calibration. This would allow instrument calibration errors and data reduction curve fit errors over the range to be entered into the analysis.

8.5 PRESSURE MEASUREMENTS IN MOVING FLUIDS

Pressure measurements in moving fluids deserve special consideration. Consider the flow over the bluff body shown in Figure 8.16. Assume that the upstream flow is uniform and steady. Points along the two streamlines labeled as A and B are to be studied. Along streamline A, the flow moves with a velocity, U_1, such as at point 1 upstream of the body. As the flow approaches point 2 it must slow down and finally stop at the front end of the body. Point 2 is known as the stagnation point and streamline A the stagnation streamline for this flow. Along streamline B, the velocity at point 3 will be U_3 and because the upstream flow is considered to be uniform it follows that $U_1 = U_3$. As the flow along B approaches the body, it is deflected around the body. From conservation of mass principles, $U_4 > U_3$. Application of conservation of energy between points 1 and 2 and between 3 and 4 yields

$$p_1 + \frac{\rho U_1^2}{2g_c} = p_2 + \frac{\rho U_2^2}{2g_c}$$

$$p_3 + \frac{\rho U_3^2}{2g_c} = p_4 + \frac{\rho U_4^2}{2g_c}$$

(8.15)

FIGURE 8.16   Streamline flow over a bluff body.

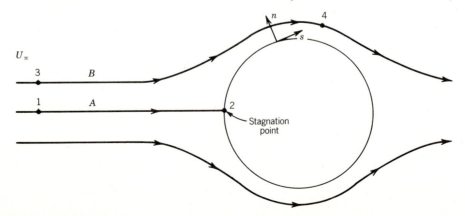

FIGURE 8.17   Total pressure measurement devices.

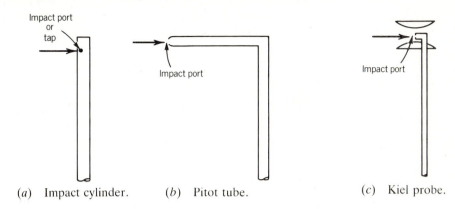

(a)   Impact cylinder.   (b)   Pitot tube.   (c)   Kiel probe.

However, because point 2 is the stagnation point, $U_2 = 0$, and

$$p_2 = p_t = p_1 + \frac{\rho U_1^2}{2g_c} \tag{8.16}$$

Hence, it follows that $p_2 > p_1$ by an amount equal to $U_1^2/2g_c$, an amount equivalent to the kinetic energy of the flow as it moves along the streamline. If the flow is brought to rest in an isentropic manner (i.e., no energy is lost through irreversible processes such as through a transfer of heat[1]), this translational kinetic energy will be transferred completely into p_2. The value of p_2 is known as the stagnation or the total pressure and will be noted as p_t. The *total pressure* can be determined by bringing the flow to rest at a point in an isentropic manner.

The pressures at 1, 3, and 4 are known as the stream or static pressures[2] of the flow. Because the flow is uniform, $U_1 = U_3$, so that $p_3 = p_1$. The static pressure and velocity at points 1 and 3 are known as the *freestream pressure* and *freestream velocity*. However, as the flow accelerates around the body its velocity increases such that, from (8.15), $p_4 \neq p_3$. The pressure at point 4 is called a *local static pressure*. The *static pressure* is that pressure sensed by a fluid particle as it moves with the same velocity as the local flow.

In practice, the total pressure is measured using an impact probe, such as those depicted in Figure 8.17. A small hole in the impact probe is aligned with the flow so as to cause the flow to come to rest at the hole. The sensed pressure is transferred through the impact probe to a pressure transducer or other pressure sensing device such as a manometer. Alignment with the flow is somewhat critical, although the probes in Figure 8.17a and b are relatively insensitive (within about 1% error in indicated reading) to misalignment within a $\pm 7°$ angle.

---

[1]This is a realistic assumption for subsonic flows. In supersonic flows, the assumption will not be valid across a shock wave.

[2]The term static pressure is somewhat of a misnomer in moving fluids but its use here conforms to common expression.

A special type of impact probe shown in Figure 8.17c, known as a Kiel probe, uses a shroud around the impact port. The effect of the shroud is to force the local flow to align itself with the shroud axis so as to impact directly onto the impact port. This effectively decreases total pressure misalignment sensitivity to negligible levels up to about ±40°.

As it is defined the static pressure is more awkward to measure than the total pressure since it requires the measurement probe to physically move through the fluid at the flow velocity. However, consider the coordinate system, s-n, fixed to the body as shown in Figure 8.16. As the flow passes over any real object, a boundary layer will be formed. Within a boundary layer, the pressure gradient in the direction normal to the streamwise direction s will be [6] $dp/dn \approx 0$. This suggests that the local value of static pressure can be sensed in a direction that is oriented normal to the flow streamline. Within ducted flows, static pressure is sensed by wall taps, small holes drilled into the duct wall perpendicular to the flow direction at the measurement point. The tap hole diameter is typically between 1 and 10% of the pipe diameter, with the smaller size preferred [5].

From equation 8.15 and Figure 8.16, we see that curvature of the flow streamlines must affect the local static pressure, since $p_1 = p_3$ but $p_3 \neq p_4$. Curvature of the flow streamlines introduces acceleration forces that change the local velocity. The freestream static pressure given by p_1 and p_3 can be measured only where there is no curvature of the streamlines. This brings about several important considerations. If pressure at the wall is to be measured, the wall taps should not disturb the flow in any way, for a disturbance would cause streamline curvature. The tap should be square with the wall with no drilling burrs [8]. If a pressure probe is inserted into the flow for the purpose of measuring static pressure then it should be a streamlined design to minimize the disturbance of the flow. It should be physically small enough so as not to cause more than a negligible increase in velocity in the vicinity of measurement. The static pressure sensing port should be located well downstream of the leading edge of the probe so as to allow the streamlines to realign themselves parallel with the probe.

Such a probe design, the improved Prandtl tube, is shown in Figure 8.18a. The probe consists of 8 holes arranged about the probe circumference and positioned 8 to 16 probe diameters downstream of the probe leading edge and 16 probe diameters upstream of its support stem. These positions are chosen to minimize static pressure error in a measurement. This is illustrated in Figure 8.18b where the relative static error, $p_e/p_v = [p_i - p]/\frac{1}{2}\rho U^2$, as a function of tap location along the probe body is plotted with p_i the indicated pressure. A pressure transducer or manometer is connectd to the probe stem to measure the sensed pressure. Real viscous effects around the static probe cause a slight discrepancy between the actual static pressure and the indicated static pressure. To account for this a correction factor, C_0, is used with $p = C_0 p_i$ where $0.99 < C_0 < 0.995$.

8.6 DESIGN AND INSTALLATION: TRANSMISSION EFFECTS

The size of pressure tap diameter and the length of tubing between a pressure tap and the connecting transducer form a pressure-measuring system that can

FIGURE 8.18   Improved Prandtl tube.

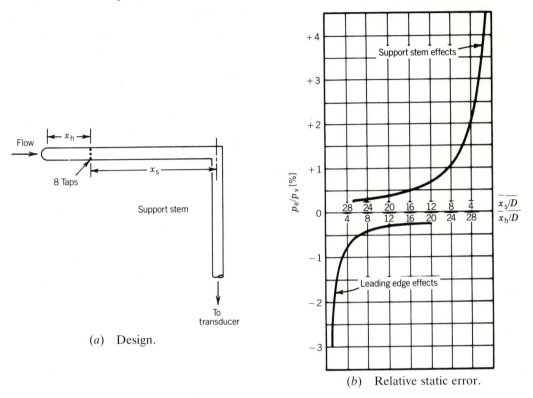

(a)   Design.

(b)   Relative static error.

have dynamic response characteristics very different from the pressure transducer itself. In the discussions to follow it is assumed that the transducer has frequency response and rise time characteristics that exceed those of the entire pressure measuring system (tubing plus transducer). Consider the configuration depicted in Figure 8.19 in which a rigid tube of length L and diameter d is used to connect a pressure tap to a pressure transducer of internal dead volume V (e.g., Figure 8.13). Initially, we can assume that under static conditions the input pressure at the tap will be indicated by the pressure transducer. But if the pressure tap is exposed to a time-dependent pressure, $p_a(t)$, the response behavior of the tubing

FIGURE 8.19   Wall tap–pressure transducer connection.

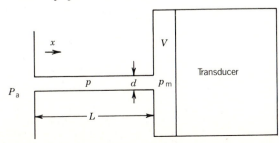

will cause the system output from the transducer, $p(t)$, to lag the input. By considering the one-dimensional pressure forces acting on a lumped mass of fluid within the connecting tube, a model for the pressure system can be developed. Be aware that this model is based on several very simplifying assumptions and should be used only as a guide in the design of a time-dependent pressure measurement system, not as an accurate indicator of exact system performance or as a correction method.

Gases

Compressibility of a gas can be accounted for through the fluid bulk modulus of elasticity, E_m. Pressure changes will act on the fluid slug in an effort to move it back and forth by a distance x within the tube. At any instant, we can expect the fluid to be acted upon by the driving pressure force, $p_a \pi d^2/4$, a damping force due to fluid shear forces, $8\pi\mu L\dot{x}$, and a compression-restoring force, $\pi^2 E_m d^2 x/16V$. Summing the forces in Newton's second law yields the response equation,

$$\frac{4L\rho V}{\pi E_m d^2} \ddot{p}_m + \frac{128\mu LV}{\pi E_m d^4} \dot{p}_m + p_m = p_a(t) \tag{8.17}$$

in which p_m is the measured pressure and p_a the applied pressure. Using equation 3.12, this gives

$$\omega_n = \frac{d\sqrt{\pi E_m/LV\rho}}{2}$$

$$\zeta = \frac{32\mu\sqrt{VL/\pi E_m \rho}}{d^3}$$

If we define $a = \sqrt{kRTg_c}$ as the acoustic wave velocity of a perfect gas, where T is the absolute temperature of the gas, the natural frequency and damping ratio of the system become

$$\omega_n = \frac{d\sqrt{\pi a^2/LV}}{2} \tag{8.18}$$

$$\zeta = \frac{32\mu\sqrt{VL/\pi}}{a\rho d^3} \tag{8.19}$$

When the tube volume, $V_t = \pi d^2 L/4 \gg V$, then the entire pressure-measuring system will behave in a manner similar to an operating organ pipe. A series of standing pressure waves occur that alter the model to the extent that a better predictor is given by [9]

$$\omega_n = \frac{a}{L[(0.5 + 4V)/V_t]^{1/2}} \tag{8.20}$$

$$\zeta = \frac{16\mu L\sqrt{(0.5 + 4V)/V_t}}{d^2\rho a} \tag{8.21}$$

One should note that the natural frequency of these systems is proportional to the tube diameter to length ratio and that damping is inversely related to this ratio. Larger diameters and shorter tubes improve pressure system response.

EXAMPLE 8.6

A pressure transducer with a natural frequency of 100 kHz is connected to a 0.10-in. static wall pressure tap using a 0.10-in i.d. rigid tube that is 5 in. long. The transducer has a dead volume of 1 in.^{3}. Determine the pressure transmission system magnitude ratio response to fluctuating pressures of air at 72 °F if the fluctuations are about 1 atm abs mean pressure. $R_{air} = 53.3$ ft-lb/lb$_m$-^{0}R, and $\mu = 4 \times 10^{-7}$ lb-s/ft^2.

KNOWN

$L = 5$ in. $k = 1.4$
$d = 0.1$ in. $T = 72 °F = 532 °R$
$V = 1$ in.^{3} $\rho = p/RT = 0.075$ lb$_m$/ft^3

ASSUMPTIONS

Air behaves as a perfect gas.

FIND

Find $M(\omega)$

SOLUTION

The magnitude ratio is given by equation 3.18:

$$M(\omega) = \frac{1}{\{[1 - (\omega/\omega_n)^2]^2 + [2\zeta(\omega/\omega_n)]^2\}^{1/2}}$$

We need to find ω_n and ζ and solve $M(\omega)$ for various frequencies.
For this geometry,

$$V_t = \frac{\pi d^2 L}{4} = 0.04 \text{ in.}^3$$

Therefore, $V_t \ll V$. Using $a = \sqrt{kRTg_c} = 1130$ ft/s, then from equation 8.18

$$\omega_n = \frac{d\sqrt{\pi a^2/LV}}{2}$$

$$= 537.4 \text{ rad/s}$$

and from (8.19)

$$\zeta = \frac{32\mu\sqrt{VL/\pi}}{a\rho d^3}$$

$$= 0.08$$

It becomes clear that the system is lightly damped. Computation of $M(\omega)$ yields the following representative values:

ω [rad/s]	$M(\omega)$
63	1.01
315	1.50
535	6.27
3150	0.03

Liquids

In liquid flows, sudden pressure changes will be transported through the pressure system more readily. Still the analysis leading to equation 8.17 will hold. However, a momentum correction factor or equivalent mass is introduced to account for the inertial effects not included in a one-dimensional analysis [e.g., 10]. This in effects increases the inertial force by 1.33 to yield

$$\omega_n = \frac{d\sqrt{3\pi E_m/L\rho V}}{4} \tag{8.22}$$

$$\zeta = \frac{16\mu\sqrt{3VL/\pi E_m\rho}}{d^3} \tag{8.23}$$

Heavily Damped Systems

In systems in which equation 8.19, 8.21, or 8.23 indicates a damping ratio greater than 1.5, the system can be considered as heavily damped. The behavior of the pressure-measuring system will closely follow that of a first-order system. A typical pressure transducer will have a rated compliance, C_{vp}, which is a measure of the transducer volume change relative to an applied pressure change. The response of the first-order system is indicated through its time constant which can be approximated by [11]

$$\tau = \frac{128\mu L C_{vp}}{\pi d^4} \tag{8.24}$$

An important aspect of (8.24) is that the time constant is proportional to L/d^4. Long and small diameter connecting tubes will result in very sluggish measurement system response to changes in pressure.

8.7   FLUID VELOCITY MEASURING SYSTEMS

Velocity measuring systems are used when information about the velocity of a moving fluid within a localized portion of the flow is needed. Desirable information can consist of the mean velocity, as well as any of the dynamic components of the velocity. Dynamic components are found in pulsating or oscillating flows, or in turbulent flows. For most general engineering applications, information about the mean flow velocity is usually sufficient. The dynamic velocity information is often sought during applied and basic fluid mechanics research and development, such as in attempting to study airplane wing response to air turbulence, a complex periodic waveform as the wing sees it.

In general, the instantaneous velocity can be written as

$$U(t) = \overline{U} + u \tag{8.25}$$

where $\overline{U}$ is the mean velocity and u is the time-dependent dynamic component of the velocity. The instantaneous velocity can also be expressed in terms of a Fourier series

$$U(t) = \overline{U} + \sum C_i \sin(\omega_i t + \phi_i') \tag{8.26}$$

so that the mean velocity and the amplitude and frequency information concerning the dynamic velocity component can be found through a Fourier analysis of the time-dependent velocity signal.

Pitot–Static Pressure Probe

For a steady, incompressible, isentropic flow, equation 8.15 can be written at any arbitrary point x in the flow field as

$$p_t = p_x + \frac{1}{2g_c} \rho U_x^2 \tag{8.27}$$

or as

$$p_v = p_t - p_x = \frac{1}{2g_c} \rho U_x^2 \tag{8.28}$$

where p_v, the difference between the total static pressures at any point in the flow, is called the *dynamic pressure*. The determination of the dynamic pressure of a moving fluid at point x would provide a method for the estimation of the local velocity existing at point x. From (8.28),

$$U_x = \sqrt{\frac{2g_c p_v}{\rho}} = \sqrt{\frac{2g_c(p_t - p_x)}{\rho}} \tag{8.29}$$

In practice, equation 8.29 is utilized through a device known as a pitot–static pressure probe. Such an instrument has an outward appearance similar to that of an improved Prandtl static pressure probe except that the pitot–static probe contains an interior pressure tube attached to an impact port at the leading edge of the probe, as shown in Figure 8.20. This creates two coaxial internal cavities within the probe, one exposed to the total pressure and the second exposed to

FIGURE 8.20   Pitot–static pressure probe.

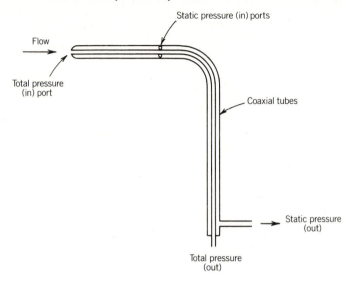

the static pressure. The two pressures are typically measured using a differential pressure transducer so as to indicate p_v directly.

The pitot–static pressure probe is relatively insensitive to misalignment over the yaw angle range of $\pm 15°$. When possible, the probe can be rotated until a maximum signal is measured, a condition that is indicative of alignment with the mean flow direction. However, the probes have a lower velocity limit of use which is brought about by strong viscous effects in the entry regions of the pressure ports. In general, viscous effects should not be a concern, provided that the Reynolds number based on the probe radius, $\mathrm{Re}_r = Ur/\nu > 500$ where ν is the kinematic viscosity of the fluid. For $10 < \mathrm{Re}_r < 500$, a correction to the dynamic pressure should be applied, $p_v = C_v p_i$, where

$$C_v = 1 + \frac{4}{\mathrm{Re}_r} \qquad (8.30)$$

However, even with this correction the uncertainty (bias limit) in measured dynamic pressure will be on the order of 40% at $\mathrm{Re}_r \approx 10$ but decreases to 1% at $\mathrm{Re}_r \approx 500$.

In high-speed gas flows, compressibility effects near the probe leading edge necessitate a closer inspection of the governing equation for a pitot-static pressure probe. Recalling equation 7.28, which states the energy balance for a perfect gas between the freestream and a stagnation point,

$$\frac{U^2}{2g_c} = c_p(T_t - T) \qquad (7.28)$$

For an isentropic process, the relationship between temperature and pressure can be stated as

$$\frac{T}{T_t} = \left(\frac{p}{p_t}\right)^{(k-1)/k}$$ (8.31)

where k is the ratio of specific heats for the gas, $k = c_p/c_v$. The Mach number of a moving fluid relates its local velocity to the local speed of sound,

$$M = \frac{U}{a}$$ (8.32)

where the acoustic wave velocity, or speed of sound, is defined for a perfect gas as

$$a = \sqrt{kRTg_c}$$ (8.33)

where T is the absolute temperature of the gas. Combining equations 7.28 with 8.31 through 8.33 and using a binomial expansion yields the relationship between total pressure and static pressure in a moving compressible flow,

$$p_v = p_t - p = \frac{1}{2g_c}\rho U^2[1 + M^2/4 + (2 - k)M^4/24) + \cdots]$$ (8.34)

It is apparent that equation 8.34 reduces to equation 8.28 when $M \ll 1$. The error in the estimate of p_v based on use of equation 8.28 relative to the true dynamic pressure becomes significant for $M > 0.2$ as shown in Figure 8.21.

FIGURE 8.21   Relative error between equations 8.28 and 8.34.

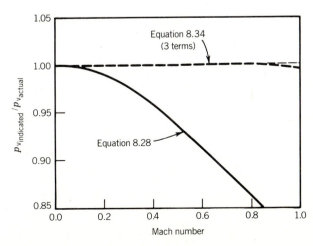

For $M > 1$, the local velocity can be estimated by the Rayleigh relation

$$U = \sqrt{2g_c[k/(k - 1)](p/\rho)[(p_t/p)^{(k-1)/k} - 1]}$$ (8.35)

where p and p_t must be measured by independent means.

Thermal Anemometry

The rate at which energy, $\dot{Q}$, is transferred between a warm body at T_s and a cooler moving fluid at T_f is proportional both to the temperature difference between them and to the thermal conductance of the heat transfer path, hA. This thermal conductance increases with fluid velocity, thereby increasing the rate of heat transfer at any given temperature difference. Hence, a relationship between the rate of heat transfer and velocity will exist and this forms the working basis of a *thermal anemometer*.

A thermal anemometer utilizes a sensor, a metallic resistance temperature device (RTD) element, which makes up one active leg of a Wheatstone bridge circuit, as indicated in Figure 8.22. The resistance–temperature relation for such a sensor was shown in Chapter 7 to be well represented by

$$R_s = R_0[1 + \alpha(T_s - T_0)]$$ (8.36)

so that sensor temperature can be inferred through a resistance measurement. A current is passed through the sensor to heat it to some desired temperature above that of the host fluid. The relationship between the rate of heat transfer from the sensor and the cooling fluid velocity is given by King's law [12]:

$$\dot{Q} = I^2 R_s = A + BU^n$$ (8.37)

where A and B are constants that depend on the fluid and sensor physical properties and operating temperatures, and n is a constant that depends on sensor dimensions [13]. Typically, $0.45 \leq n \leq 0.52$ [14]. A, B, and n are found through calibration.

Two types of sensors are common: the hot-wire and the hot-film. As shown in Figure 8.23, the hot-wire sensor consists of a tungsten or platinum wire ranging from 1 to 4 mm in length and from 5 to 15 µm in diameter. The wire is supported

FIGURE 8.22   Thermal anemometer circuit, constant resistance mode.

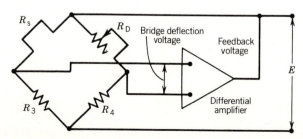

FIGURE 8.23   Schematic of hot-wire sensor.

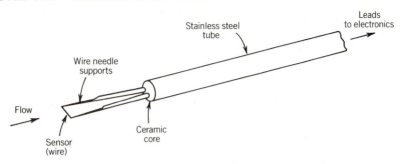

between two rigid needles that protrude from a ceramic tube that houses the lead wires. A hot-film sensor usually consists of a thin (2-μm) platinum or gold film deposited onto a glass substrate and covered with a high thermal conductivity coating. The coating acts to electrically insulate the film and offers some mechanical protection. Hot wires are generally used in electrically nonconducting fluids, while hot films are used both in nonconducting and conducting fluids and where a more rugged sensor is needed.

Two anemometer bridge operating modes are possible: (1) constant current and (2) constant resistance. In constant current operation, a fixed current is passed through the sensor to heat it. The sensor resistance, and therefore its temperature, are permitted to vary with the rate of heat transfer between the sensor and its environment. Bridge deflection voltage provides a measure of the cooling velocity. The more common mode of operation for velocity measurements is constant resistance. In constant resistance operation, the sensor resistance, and therefore its temperature, is originally set by adjustment of the bridge balance. The sensor resistance is then maintained constant by using a differential feedback amplifier to sense small changes in bridge balance, which would be equivalent to sensing changes in the sensor set point resistance. The feedback amplifier rapidly readjusts the bridge applied voltage, thereby adjusting the sensor current to bring the sensor back to its set point resistance and corresponding temperature. Since the current through the sensor will vary with changes in the velocity, the instantaneous power ($I^2 R_s$) required to maintain this constant temperature is equivalent to the instantaneous rate of heat transfer from the sensor ($\dot{Q}$). In terms of the instantaneous applied bridge voltage, E, required to maintain a constant sensor resistance, the velocity is found by the correlation

$$E^2 = C + DU^n \qquad (8.38)$$

where constants C, D and n are found by calibration under a fixed sensor and fluid temperature condition. An electronic or digital linearizing scheme is usually employed to condition the signal by performing the transformation,

$$E_1 = K \left(\frac{E^2 - C}{D} \right)^{1/n}$$

such that the measured output from the linearizer, E_1, is

$$E_1 = KU \qquad (8.39)$$

where K is found through a static calibration.

For simple mean velocity measurements the thermal anemometer is a simple device to use. It has a better usable sensitivity than the pitot–static tube at lower velocities. Multiple velocity components can be measured by using multiple sensors, each sensor aligned differently to the mean flow direction and operated by independent anemometer circuits [13,15]. In highly turbulent flows with fluctuations of $\sqrt{\overline{u^2}} \geq 0.1\overline{U}$, signal interpretation can become complicated but has been well investigated [e.g., 15]. Low-frequency fluid temperature fluctuations can be compensated for by placing resistor R_3 directly adjacent to the sensor and exposed to the flow. An extensive bibliography of thermal anemometry theory and signal interpretation exists [16].

In constant temperature mode using a fast responding differential feedback amplifier, a hot-wire system can attain a frequency response that is flat up to 100 000 Hz, which makes it particularly useful in fluid mechanics turbulence research. However, less expensive and more rugged systems are commonly used for industrial flow monitoring where a fast dynamic response is desirable. An upper frequency limit on a cylindrical sensor of diameter d is brought about by the natural oscillation in the flow immediately downstream of a body which vibrates the sensor. The frequency of this oscillation, known as the Strouhal frequency, occurs at approximately

$$f \approx 0.22 \frac{\overline{U}}{(2\pi d)} \qquad (10^2 < \mathrm{Re}_d < 10^7) \qquad (8.40)$$

The heated sensor will warm the fluid within its proximity. Under flowing conditions this will not cause any measurable problems so long as the condition,

$$\mathrm{Re}_d \geq Gr^{1/3} \qquad (8.41)$$

is met where $\mathrm{Re}_d = Ud/\nu$, $Gr = d^3 g\beta(T_s - T_t)/\nu$, and β is the coefficient of thermal expansion of the fluid. Equation 8.41 ensures that the inertial forces of the moving fluid dominate over the buoyant forces brought on by the heated sensor. For air, this forms a lower velocity limit on the order of 2 ft/s (0.6 m/s).

Doppler Anemometry

The Doppler effect describes the phenomenon experienced by an observer whereby the frequency of light or sound waves emitted from a source that is traveling away from or toward the observer will be shifted from its original value. Most readers are familiar with the change in pitch of a train heard by an observer as the train changes from approaching to receding. Any radiant energy wave, such as a sound or light wave, will experience a Doppler effect. The effect was recognized and modeled by Johann Doppler (1803–1853). The observed shift in frequency, called the Doppler shift, is directly related to the speed of the emitter relative to the observer. To an independent observer, the frequency of emission is perceived to be higher than actual if the emitter is moving toward the observer and lower if moving away, since the arrival of the emission at the observer location will be affected by the relative velocity of the emission source. The Doppler effect is used in astrophysics to measure the velocity of distant objects by monitoring the frequency of light emitted from a particular gas, usually hydrogen. Since in the visible light spectrum frequency is related to color, the common terms of red shift or blue shift refer to frequency shifts toward the red side of the spectrum or toward the blue side.

Doppler anemometry refers to a class of techniques that utilize the Doppler effect to measure local velocity in a moving fluid. In these techniques, the emission source and the observer remain stationary. However, small scattering particles suspended in the moving fluid can be used to generate the Doppler effect. The emission source is a coherent narrow incident wave. Either acoustic waves or light waves are used.

When a laser beam is used as the incident wave source, the velocity measuring device is called a *laser Doppler anemometer* (LDA). The first practical laser Doppler anemometer system was discussed by Yeh and Cummins in 1964 [17]. A laser beam provides a ready emission source that is monochromatic and remains coherent over long distances. As a moving particle suspended in the fluid passes through the laser beam it scatters light in all directions. An observer viewing this encounter between the particle and the beam will perceive the scattered light at a frequency, f_s:

$$f_s = f_i \pm f_D \tag{8.42}$$

where f_i is the frequency of the incident laser beam and f_D is the Doppler shift. Using visible light, an incident laser beam frequency will be on the order of 10^{14} Hz. For most engineering applications, the velocities are such that the Doppler shift frequency, f_D, will be on the order of 10^3 to 10^7 Hz. Such a small shift in the incident frequency can be difficult to detect in a practical instrument. An operating mode that overcomes this difficulty is the dual-beam mode shown in Figure 8.24. In this mode, a single laser beam is divided into two coherent beams of equal intensity using an optical beam splitter. These incident beams are passed through a focusing lens which focuses the beams to a point in the flow. The focal point forms the effective measuring volume (sensor) of the instrument. Particles suspended in and moving with the fluid will scatter light

FIGURE 8.24   Laser Doppler anemometer in dual-beam mode.

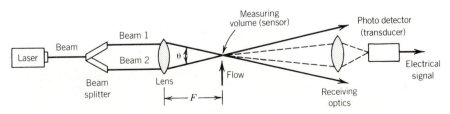

as they pass through the beams. The frequency of the scattered light will be that given by (8.42) everywhere but at the measuring volume. There, the two beams cross and the incident information from the two beams mixes, a process known as optical heterodyne. The outcome of this mixing is a cancellation of the incident frequency in the scattered light. A stationary observer, such as an optical photodiode, focused on the measuring volume will see two distinct frequencies, the Doppler shift frequency and the unshifted incident frequency instead of their sum. It is a simple matter to separate the much smaller Doppler frequency from the incident frequency by filtering.

For the setup shown in Figure 8.24, the velocity is related directly to the Doppler shift by

$$U = \frac{\lambda}{2 \sin \theta/2} f_D = d_f f_D \qquad (8.43)$$

where the component of the velocity measured is that which is in the plane of and bisector to the crossing beams. In theory, by using beams of different color or polarization, different velocity components can be measured simultaneously. However, the dependence of the lens focal length on color will cause a small displacement between the different measuring volumes formed by the different colors. For most applications this can be corrected. The LDA technique requires no direct calibration beyond determination of the parameters in d_f and the ability to measure f_D.

In dual-beam mode, the output from the photodiode transducer is a current of a magnitude proportional to the square of the amplitude of the scattered light seen and of a frequency equal to f_D, as shown in the oscilloscope trace of Figure 8.25. If the instantaneous velocity of a dynamic flow varies with time, the Doppler shift from successive scatterers will vary with time. This time-dependent fre-

FIGURE 8.25   Oscilloscope trace of photodiode output showing a Doppler signal.

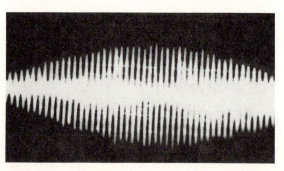

quency information can be extracted by any of a variety of processing equipment that can interpret the signal current. Several methods include:

- Frequency analyzers
- Frequency trackers
- Frequency counters

Frequency analyzers extract frequency information by performing a Fourier analysis (see Chapter 2) on the input signal. These are best suited for a statistical representation (mean velocity and standard deviation or rms velocity) of the velocity field. *Frequency trackers* act much like a radio tuner and contain a phase-locked loop that tunes to the input signal Doppler frequency and follows it in time. These require an almost continuous Doppler signal to maintain the phase loop lock and so are used in fluids containing a large number of particles. *Frequency counters* count the number of zero crossings in the periodic Doppler signal and are suited to flows containing relatively few particles. Both trackers and counters output a voltage that is proportional to the instantaneous velocity, which makes their signal easy to process. Many contain a digital output path for direct output to a digital computer for signal analysis. At very low light levels and very few scattering particles, the signal level to noise level can be very low. In such cases, photon correlation techniques are successful [18,19].

The LDA technique is particularly useful where probe blockage effects render other methods unsuitable, where fluid density and temperature fluctuations occur, or where environments hostile to physical sensors exist. An extended discussion of LDA techniques can be found in references [19,20].

EXAMPLE 8.7

A laser Doppler anemometer made up of a He–Ne laser ($\lambda = 632.8$ nm) is used to measure the velocity of water at a point in a flow. A 150-mm lens having a $\theta = 11°$ is used to operate the LDA in a dual-beam mode. If an average Doppler frequency of 1.41 MHz is measured, estimate the velocity of water.

KNOWN

$\bar{f}_D = 1.41$ MHz $F = 150$ mm

$\lambda = 632.8$ nm $\theta = 11°$

ASSUMPTIONS

Scattering particles follow the water exactly.

FIND

$\overline{U}$

SOLUTION

Using equation 8.43,

$$\overline{U} = \frac{\lambda}{2 \sin \theta/2} \overline{f}_D$$

$$= \left(\frac{632.8 \times 10^{-9} \text{ m}}{2 \sin 5.5°} \right) (1.41 \times 10^6 \text{ Hz})$$

$$= 4.655 \text{ m/s}$$

Selection of Velocity Measuring Methods

The selection of the best velocity measuring system for a particular application will depend on a number of factors which should be weighted accordingly. These factors include:

1. Required spatial resolution
2. Required velocity range
3. Sensitivity to velocity changes only
4. Required need to quantify dynamic velocity
5. Acceptable probe blockage of flow
6. Ability to be used in hostile environments
7. Calibration requirements
8. Low cost and ease of use.

When used under appropriate conditions, the uncertainty in velocity determined by any of the discussed methods can be as low as 1% of the measured velocity, although under special conditions LDA methods can have an uncertainty one order of magnitude lower [21].

Pitot–Static Pressure Methods

The pressure probe methods are best suited for the determination of mean velocity in fluids of constant density. Relative to other methods, they are the simplest and cheapest method available to measure velocity at a point. Probe blockage of the flow is not a problem in large ducts and away from walls. Fluid particulate will block the impact ports but aspirating models are available for such situations. They are subject to mean flow misalignment errors. They require no calibration and are well used in the field and laboratory alike.

Thermal Anemometers

Thermal anemometers are best suited for use in clean fluids of constant temperature and density. They offer a small spatial resolution and possess a high-frequency response, making them well suited for measuring dynamic velocities. However, signal interpretation in strongly dynamic flows can be complicated [15,22]. Hot-film sensors are less fragile and less susceptible to contamination than hot-wire sensors. Probe blockage is not significant in large ducts and away from walls. Thermal anemometers are 180° directionally insensitive, an important factor in flows that may contain flow reversal regions. An industrial grade system can be built rather inexpensively. The thermal anemometer is usually calibrated against either pressure probes or an LDA.

Laser Doppler Anemometers

The laser anemometer is a relatively expensive and technically complicated technique best suited for measurements in flows in which other methods are unsuited. The technique is particularly well suited to hostile, combusting, or highly dynamic flow environments. It can offer good frequency response, small spatial resolution, no probe blockage, and simple signal interpretation, but requires optical access and the presence of scattering particles. The method measures the velocity of particles suspended in the moving fluid, not the fluid velocity. Careful planning is required in particle selection to ensure that the particle velocities represent the fluid velocity exactly. The size and concentration of the particles will govern the system frequency response [23,24]. The technique requires no velocity calibration.

8.8 SUMMARY

Several reference pressure instruments have been presented which form the working standards for pressure transducer calibration. Pressure transducers convert sensed pressure into an output form that is readily quantifiable. These transducers come in many forms but tend to operate on either hydrostatic principles, expansion techniques, or force-displacement methods.

In moving fluids, special care must be taken in the measurement of the pressure to delineate between static and total pressure. Methods for the separate measurement of static and total pressure or for the measurement of the dynamic pressure are readily available and well documented.

The measurement of the local velocity within a moving fluid can be accomplished in a number of ways. Specifically, dynamic pressure, thermal anemometry, and Doppler anemometry methods have been presented. As discussed, each method offers advantages over the other and the best technique must be carefully weighed against the needs and constraints of a particular application.

REFERENCES

1. Brombacher, W. G., D. P. Johnson, and J. L. Cross, *NBS Monograph* 8, 1960.
2. McLeod, H., *Philos. Mag.* 48, 1874.
3. Sweeney, R. J., *Measurement Techniques in Mechanical Engineering*, Wiley, New York, 1953.
4. *Handbook of Experimental Stress Analysis*, M. Hetenyi, Ed., Wiley, New York, 1950.
5. Way, S., Bending of circular plates with large deflection, *Transactions of the ASME* 56, 1934.
6. Schlicting, H., *Boundary Layer Theory*, McGraw-Hill, New York, 1969.

7. Franklin, R. E., and J. M. Wallace, Absolute measurements of static-hole error using flush transducers, *Journal of Fluid Mechanics* 42, 1970.

8. Rayle, R. E., Influence of orifice geometry on static pressure measurements, ASME Paper No. 59-A-234, 1959.

9. Iberall, A. S., Attenuation of oscillatory pressures in instrument lines, *Transactions of the ASME* 72, 1950.

10. Streeter, V. L., and E. B. Wylie, *Fluid Mechanics*, 7th ed., McGraw-Hill, New York, 1979.

11. Doebelin, E. O., *Measurement Systems: Application and Design*, 2d ed., McGraw-Hill, New York, 1975.

12. King, L. V., On the convection from small cylinders in a stream of fluid: Determination of the convection constants of small platinum wires with application to hot-wire anemometry, *Proceedings of the Royal Society*, *London* 90, 1914.

13. Hinze, J. O., *Turbulence*, McGraw-Hill, New York, 1959.

14. Collis, D. C., and M. J. Williams, Two-dimensional convection from heated wires at low Reynolds numbers, *Journal of Fluid Mechanics* 6, 1959.

15. Rodi, W., A new method for analyzing hot-wire signals in a highly turbulent flow and its evaluation in a round jet, *DISA Information*, 17, Dantek Electronics, Denmark, 1975. See also, Bruun, H. H., Interpretation of X-wire signals, *DISA Information*, 18, Dantek Electronics, Denmark, 1975.

16. Freymuth, P., A bibliography of thermal anemometry, *TSI Quarterly* 4, 1978.

17. Yeh, Y., and H. Cummins, Localized fluid flow measurement with a He–Ne laser spectrometer, *Applied Physic Letters* 4, 1964.

18. *Photon Correlation and Light Beating Spectroscopy*, H. Z. Cummins and E. R. Pike, eds., Proc. NATO ASI, Plenum, New York, 1973.

19. Durst, F., A. Melling, and J. H. Whitelaw, *Principles and Practice of Laser Doppler Anemometry*, Academic, New York, 1976.

20. *Fluid Mechanics Measurements*, R. J. Goldstein, ed., Hemisphere, New York, 1983.

21. Goldstein, R. J., and D. K. Kried, Measurement of laminar flow development in a square duct using a laser doppler flowmeter, *Journal of Applied Mechanics* 34, 1967.

22. Yavuzkurt, S., A guide to uncertainty analysis of hot-wire data, *Transactions of the ASME*, *Journal of Fluids Engineering* 106, 1984.

23. Maxwell, B. R., and R. G. Seaholtz, Velocity lag of solid particles in oscillating gases and in gases passing through normal shock waves, NASA-TN-D-7490.

24. Dring, R. P., and Suo, M. Particle trajectories in swirling flows, *Transactions of the ASME*, *Journal of Fluids Engineering* 104, 1982.

NOMENCLATURE

d	diameter $[l]$	k	ratio of specific heats
e_i	elemental errors	p	pressure $[m\text{-}l^{-1}\text{-}t^{-2}]$
h	depth $[l]$	p_a	applied pressure $[m\text{-}l^{-1}\text{-}t^{-2}]$
h_0	reference depth ($h = 0$) $[l]$	p_{abs}	absolute pressure $[m\text{-}l^{-1}\text{-}t^{-2}]$

p_e	relative static pressure error $[m\text{-}l^{-1}\text{-}t^{-2}]$	K_c	charge sensitivity $[C\text{-}m^{-1}\text{-}l^{-1}\text{-}t^2]$
p_i	indicated pressure $[m\text{-}l^{-1}\text{-}t^{-2}]$	K_E	voltage sensitivity $[V\text{-}m^{-1}\text{-}t^2]$
p_m	measured pressure $[m\text{-}l^{-1}\text{-}t^{-2}]$	M	Mach number
p_o	reference pressure $[m\text{-}l^{-1}\text{-}t^{-2}]$	Re_d	Reynolds number, $Re_d = Ud/\nu$
p_t	total or stagnation pressure $[m\text{-}l^{-1}\text{-}t^{-2}]$	S	specific gravity
p_v	dynamic pressure $[m\text{-}l^{-1}\text{-}t^{-2}]$	U	velocity $[l\text{-}t^{-1}]$
q	charge	V	volume $[l^3]$
r	radius $[l]$	γ	specific weight $[m\text{-}l^{-2}\text{-}t^{-2}]$
t	thickness $[l]$	ϵ	dielectric constant
y	displacement $[l]$	λ	wavelength $[l]$
z	altitude $[l]$	ρ	density $[m\text{-}l^{-3}]$
C	capacitance $[C]$	τ	time constant $[t]$
E	voltage $[V]$	ϕ	latitude
E_m	bulk modulus of elasticity $[m\text{-}l^{-1}\text{-}t^{-2}]$	ω_n	natural frequency $[t^{-1}]$
Gr	Grashof number	μ	absolute viscosity $[m\text{-}t^{-1}\text{-}l^{-1}]$
H	manometer deflection $[l]$	ν	kinematic viscosity $[l^2/t]$
K	static sensitivity	ν_p	Poisson ratio

PROBLEMS

8.1 Convert the following absolute pressures to units of N/m^2, kPa, psi, mm Hg, and inches H_2O:
    a. 10.8 psia
    b. 1.75 atm abs
    c. 30.36 inches H_2O absolute
    d. 791 mm Hg abs

8.2 Convert the following gauge pressures into absolute pressure relative to one standard atmosphere:
    a. -0.55 psi
    b. 100 μm Hg
    c. 98.6 kPa
    d. 3 in. H_2O

8.3 A water-filled manometer is used to measure the pressure in an air-filled tank. One leg of the manometer is open to atmosphere. For a measured manometer deflection of 20 in. water, determine the tank static pressure. Barometric pressure is 14.59 psia.

8.4 A deadweight tester is used to provide a reference pressure during the calibration of a pressure transducer. A combination of 11.5 lb of 3-in.-diameter stainless steel disks is found to be necessary to balance the tester piston against its internal pressure. For an effective piston area of 0.785 in.^{2} and a piston mass of 2.43 lb, determine the reference pressure in psi, psia, and N/m^2. Barometric pressure is 30.15 in. Hg abs, elevation is 59 ft, and latitude is 42°.

8.5 An inclined tube manometer indicates a change in pressure of 2.2 in. H_2O

when switched from a null mode (both legs at atmospheric pressure) to deflection mode (one leg measuring, one leg at atmospheric pressure). For an inclination of 30° relative to horizontal, determine the pressure change indicated.

8.6   Determine the maximum deflection and the natural frequency of a 0.1-in. thick diaphragm made of steel (E_m = 30 Mpsi, ν_p = 0.32, ρ = 0.28 $lb_m/in.^3$), if the diaphragm must be 0.75 in. in diameter. Determine its differential pressure limit.

8.7   A strain gauge, diaphragm pressure transducer (accuracy: <0.1% reading) is subjected to a pressure differential of 1 psi. If the output is measured using a voltmeter having a resolution of 10 μV and accuracy of better than 0.1% of the reading, estimate the uncertainty in pressure at the design stage. How does this change at 10 and 100 psi?

8.8   Select a practical manometric fluid to measure pressures up to 10 psi of an inert gas (γ = 0.066 lb/ft³), if water (γ = 62.4 lb/ft³), oil (S = 0.82), and mercury (S = 13.57) are available.

8.9   An air pressure in the range of 200 to 400 N/m² is to be measured relative to atmosphere using a U-tube manometer with mercury (S = 13.57). Manometer resolution will be 1 mm with a zero error of 0.5 mm. Estimate the design-stage uncertainty in gauge pressure based on the manometer indication at 20 °C. Would an inclined manometer (θ = 30°) be a better choice if the inclination can be set to within 0.5°?

8.10   Plot the design-stage uncertainty in estimating a nominal pressure of 10 000 N/m² using an inclined manometer (resolution: 1 mm; zero error: 0.5 mm) with water at 20 °C for inclination angles of 10–90° (using 10° increments). The inclination angle can be set to within 1°.

8.11   A capacitance pressure transducer, such as shown in Figure 8.14, uses a C_1 of 0.01 ± 0.005 μf and an excitation voltage of 5 ± 1% V. The plates have an overlap area of 8 ± 0.01 mm² and are separated by an air gap of 1.5 ± 0.1 mm. If the plates move apart by 0.2 mm, estimate the change in capacitance and the output voltage.

8.12   The pressure fluctuations in a pipe filled with air at 20 °C is to be measured using a static wall tap, rigid connecting tubing, and a diaphragm pressure transducer. The transducer has a natural frequency of 100 000 Hz. For a tap and tubing diameter of 3.5 mm, a tube length of 0.25 m, and a transducer dead volume of 1600 mm³, estimate the resonance frequency of the system. What is the maximum frequency that this system can measure with no more than a 10% dynamic error? Plot the frequency response of the system.

8.13   A small tube is to connect a pressure tap to a transducer. The transducer has a dead volume of 0.3 cm³. Air at 1 atm and 20 °C is the transmitting fluid. Determine the length of tubing necessary to attenuate pressure fluctuations at 20 Hz by 50%.

8.14   A pitot–static pressure probe inserted within a large duct indicates a differential pressure of 8 in. H₂O. Determine the velocity measured.

8.15   Determine the resolution of a manometer required to measure the velocity of air from 10 to 100 ft/s using a pitot–static tube and a manometric fluid of mercury ($S = 13.57$) to a zero-order uncertainty of 5 and 1%.

8.16 A long cylinder is placed into a wind tunnel and aligned perpendicular to an oncoming freestream. Static wall pressure taps are located circumferentially about the centerline of the cylinder at 45° increments with 0° at the impact (stagnation) position. Each tap is connected to a separate manometer referenced to atmosphere. A pitot–static tube indicates an upstream dynamic pressure of 8 in. H_2O which is used to determine the freestream velocity. The following static pressures are measured:

Tap	p [in. H_2O]	Tap	p [in. H_2O]
0°	0.0	135°	9.1
45°	16.3	180°	9.4
90°	32.0		

Compute the local velocities around the cylinder if the total pressure in the flow remains constant. $p_{atm} = 14.7$ psia, $T_{atm} = 60$ °F

8.17 A pitot–static tube is used as a working standard to calibrate a hot-wire anemometer in 20 °C air. If dynamic pressure is measured using a water-filled micromanometer, determine the smallest manometer deflection for which the pitot–static tube can be considered as accurate without correction for viscous effects.

8.18 For the thermal anemometer in Figures 8.22 and 8.23, determine the decade resistance setting required to set a platinum sensor at 40 °C above ambient if the sensor ambient resistance is 110 Ω and $R_3 = 500$ Ω and $R_4 = 500$ Ω. $\alpha = 0.00395$ °C^{-1}

8.19 Determine the static sensitivity of the output from a constant resistance anemometer as a function of velocity. Is it more sensitive at high or at low velocities?

8.20 A laser Doppler anemometer setup in a dual-beam mode uses a 600-mm focal length lens ($\theta = 5.5°$) and an argon-ion laser ($\lambda = 514.4$ nm). Compute the Doppler shift frequency expected at 1, 10 and 100 m/s. Recompute for a 300-mm lens ($\theta = 7.3°$).

8.21 A set of 5000 measurements of velocity at a point in a flow using a dual-beam LDA give the following results:

$$\overline{U} = 21.37 \text{ m/s} \qquad S_U = 0.43 \text{ ft/s}$$

If the Doppler shift can be measured with an uncertainty of better than 0.9%, the optical angle can be measured to within 0.25°, and the laser can be tuned to $\lambda = 623.8 \pm 0.5\%$ nm, determine the best estimate of the velocity.

CHAPTER 9
FLOW MEASUREMENTS

9.1 INTRODUCTION

Flow rate can be stated in terms of a volume per unit time, known as the *volume flow rate*, or as a mass per unit time, known as the *mass flow rate*. Not only is this property useful in flow metering, but many engineering systems require flow measurement information for proper process control. For example, in heat exchange processes the rate at which energy can be removed from or added to a moving fluid is directly proportional to the mass flow rate of the fluid. Properly selected and designed flow measurement equipment can be a very high engineering priority because from the flow of water into our homes to the flow of petroleum at the oil well, flow measurements are vitally linked to the economy.

This chapter discusses some of the most common and accepted methods for flow quantification. Size, accuracy, cost, pressure drop, pressure losses, and compatibility with the fluid are important engineering design considerations for flow metering devices. All methods have both desirable and undesirable features that necessitate some compromise in the selection of the best method for the particular application and most of the more important of these considerations are included in this chapter. Inherent uncertainties in fluid properties, such as density, viscosity, or specific heat, can affect the accuracy of a flow measurement made using some metering methods. Novel techniques that preclude knowledge of fluid properties are being introduced for use in the more demanding of these applications. The chapter objective is to present both an overview of basic flow metering techniques for proper meter selection, as well as those design considerations important in the integration of a flow metering system with the process system it will meter.

9.2 HISTORICAL COMMENTS

Their importance in engineering systems gives flow measurement methods a rich history. The earliest available accounts of flow metering were recorded by Hero of Alexandria (ca. 150 B.C.) who proposed a scheme to regulate water flow using a siphon pipe attached to a constant head reservoir. And the early Romans developed elaborate water systems to supply public baths and private homes. In fact, Sextus Frontinius (A.D. 40–103), Commissioner of Water Works for Rome, prepared a treatise on Roman methods of water distribution. Evidence suggests that the Romans understood that a relation between volume flow rate and pipe flow area existed, although the role of velocity in flow rate was not recognized. Weirs were used to regulate bulk flow through aqueducts, and pipe area was used to meter supplies to individual buildings.

Following a number of experiments conducted using olive oil and water, Leonardo da Vinci (1452–1519) first proposed the continuity principle: that area, velocity and flow rate were related. However, most of his writings were lost until centuries later and Benedetto Castelli (ca. 1577–1644), a student of Galileo, has been credited in some texts with developing the same steady, incompressible continuity concepts in his day. Isaac Newton (1642–1727), Daniel Bernoulli (1700–1782), and Leonhard Euler (1707–1783) built the mathematical and physical bases on which modern flow meters would later be developed. By the 19th century, the concepts of continuity, energy, and momentum were sufficiently understood for practical exploitation. Relations between flow rate and pressure losses were developed that would permit the tabulation of the hydraulic coefficients necessary for the quantitative engineering design of many modern practical working flow meters.

9.3   FLOW RATE CONCEPTS

The flow rate through a pipeline, duct, or other flow system can be described by use of a control volume, a judiciously selected volume in space through which a fluid flows. The amount of fluid that passes through this volume in a given period of time will determine the flow rate. A geometrical boundary of a control volume is called a control surface. Such a control volume is shown in Figure 9.1, which consists here of a defined volume within a pipe.

The velocity of a fluid at a point can be described by the use of a three-dimensional velocity vector given here in cylindrical coordinates by

$$\mathbf{U} = \mathbf{U}(x, r, \theta) = u\hat{e}_x + v\hat{e}_r + w\hat{e}_\theta$$

FIGURE 9.1   Control volume concept applied to a pipe flow.

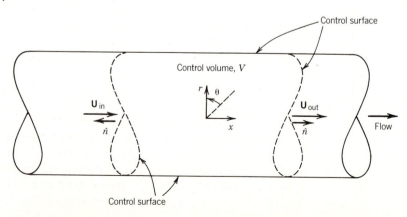

where u, v, and w are the scalar velocity magnitudes and $\hat{e}_x$, $\hat{e}_r$ and $\hat{e}_\theta$ are unit vectors in each of the component directions, x, r, and θ, respectively.

The amount of fluid of density ρ that passes through the control volume of volume V at any instant in time depends on the amount of fluid that crosses the control volume surfaces. This can be expressed by examination of mass flow into, out of, and remaining within the control volume (CV) at any instant. Conservation of mass demands that the net flow of mass through a control volume that physically crosses any of its control surfaces (CS) plus the rate at which mass accumulates within the control volume be zero. This is expressed by

$$\frac{\partial}{\partial t} \iiint_{CV} \rho \, dV + \iint_{CS} \rho \mathbf{U} \cdot \hat{n} \, dA = 0 \tag{9.1}$$

where $\hat{n}$ is the outward normal from a control surface of area A.

A steady flow situation exists when the sum of the mass flow across all control surfaces *into* the control volume equals the sum of the mass flow across all the control surfaces *out of* the control volume. For steady flows, equation 9.1 can be simplified to

$$\dot{m}_{in} = \dot{m}_{out} \tag{9.2}$$

where $\dot{m}$ is defined as the mass flow rate

$$\dot{m} = \iint_{CS} \rho \mathbf{U} \cdot \hat{n} \, dA$$

If the average mass flux over the control surface, $\overline{\rho U}$, is known then (9.2) becomes

$$\dot{m} = (\overline{\rho U}) A \tag{9.3}$$

Mass flow rate has the dimensions of mass per unit of time [e.g., units of lb_m/s, kg/h, etc.].

As a general rule, isothermal flows of liquids can be considered incompressible (i.e., constant density). This can also be assumed for isothermal gas flows which move at speeds of less than 0.3 times the speed of sound in that fluid. In these flows, equation 9.3 can be further reduced to

$$Q_{in} = Q_{out} \tag{9.4}$$

where Q is defined as the volume flow rate

$$Q = \iint_{CS} \mathbf{U} \cdot \hat{n} \, dA$$

For example, the velocity at a control surface located at axial position x along a pipe will be described by the single component $U = u(r, \theta)$. In practical terms, knowledge of $u(r, \theta)$ at x in a steady, incompressible flow would be sufficient to estimate the average velocity and yield the volume flow rate through the control surface. In a pipe of circular cross section, the volume flow rate at position x is found by

$$Q = \int_0^{r_1} \int_0^{2\pi} u(r, \theta) r \, d\theta \, dr \qquad (9.5)$$

where r_1 is the pipe radius.

If the average velocity, $\overline{U}$, over a control surface is known then the volume flow rate can be found by

$$Q = \overline{U} A \qquad (9.6)$$

The preceding analysis indicates that methods for determining steady flow rate must depend on the use of techniques that are sensitive either to the average mass flux, $\overline{\rho U}$, to estimate mass flow rate or to the average velocity, $\overline{U}$, to estimate volume flow rate.

The flow through a pipe or duct can be characterized as being laminar, turbulent, or something in between called transitional. In flow measurements this flow character can be established through the nondimensional Reynolds number, defined by

$$\text{Re}_{d_1} = \frac{\overline{U} d_1}{\nu} = \frac{4Q}{\pi d_1 \nu} \qquad (9.7)$$

where ν is the fluid kinematic viscosity and d_1 is the diameter for circular pipes or the hydraulic diameter for noncircular pipes, $4r_\text{H}$, where r_H is defined as the wetted area divided by the wetted perimeter. In pipes, turbulent flows are encountered when $\text{Re}_{d_1} > 4000$ and laminar flows are encountered when $\text{Re}_{d_1} < 2000$. In flow system design, operation in the transitional flow regime should be avoided.

9.4   VOLUME FLOW RATE THROUGH VELOCITY DETERMINATION

The practical adaptation of equation 9.5 for the determination of volume flow rate through a duct requires measurement of the velocity at points along several cross sections of a flow control surface. Methods for determining the velocity at a point include any of those previously discussed in Chapter 8. This procedure is most often used for the one-time verification and/or calibration of system flow rates, but in some situations it is used for continuous monitoring. For example, the procedure has been used in ventilation system setup and problem diagnosis where the existence or installation of an in-line flow meter is uncommon since it is not needed in regular operation.

In using this technique in circular pipes, a number of discrete measuring positions are chosen along m flow cross sections (radii) spaced at $360/m$ degrees apart, such as shown in Figure 9.2. A velocity probe is traversed along each flow cross section with readings taken at each measurement position. There are

FIGURE 9.2   Location of n measurement locations along m radials in a pipe.

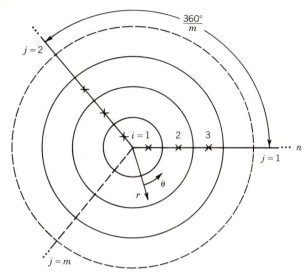

several options as to the selection of measuring positions [2] but a simple method is to divide the flow area into smaller equal areas with measurements made at the centroid of each of the smaller areas. This latter method is also useful for similar measurements in rectangular ducts. Regardless of the option selected, the average flow rate is estimated along each cross section traversed using (9.5) and the pooled mean of the flow rates for the m cross sections determined to yield the best estimate of the duct flow rate. It is important that the flow rate remain fixed during each traverse.

EXAMPLE 9.1

A steady flow of air at 70 °F passes through a 10-in.-i.d. circular pipe. A velocity-measuring probe is traversed along three cross sectional lines ($j = 1$, 2, 3) of the pipe and measurements are made at four positions ($i = 1, 2, 3, 4$) along each traverse line. The locations for each measurement are selected at the centroids of equally spaced areal increments as indicated below [2]. Determine the volume flow rate in the pipe.

Radial Location, i	$\dfrac{r}{r_1}$	$U_j\left(\dfrac{r}{r_1}\right)$ [ft/s]		
		Line 1	Line 2	Line 3
1	0.3536	8.71	8.62	8.78
2	0.6124	6.26	6.31	6.20
3	0.7906	3.69	3.74	3.79
4	0.9354	1.24	1.20	1.28

KNOWN

$$U_{ij}\left(\frac{r}{r_1}\right); i = 1, 2, 3; j = 1, 2, 3, 4$$

$d_1 = 10$ in. ($A = \pi d_1^2/4 = 0.54$ ft^2)

ASSUMPTIONS

Constant and steady pipe flow during all measurements
Incompressible flow

FIND

$\langle Q \rangle$

SOLUTION

The flow rate is found by integration of the velocity profile across the duct along each line and subsequent averaging of the three values. For discrete velocity data, equation 9.5 is written along each line, $j = 1, 2, 3$, as

$$Q_j = 2\pi \int_0^{R_1} U\left(\frac{r}{r_1}\right) r\, dr \approx 2\pi \sum_{i=1}^{4} U_{ij}\left(\frac{r}{r_1}\right) r\, \Delta r$$

where Δr is the radial distance separating each position of measurement. This can be further simplified since the velocities are located at positions that make up centroids of equal areas:

$$Q_j = \frac{A}{4}\sum_{i=1}^{4} U_{ij}\left(\frac{r}{r_1}\right) \qquad j = 1, 2, 3$$

Then, the mean flow rate along each line of traverse is

$$Q_1 = 2.71 \text{ ft}^3/\text{s} \qquad Q_2 = 2.71 \text{ ft}^3/\text{s} \qquad Q_3 = 2.73 \text{ ft}^3/\text{s}$$

The estimate of the average pipe flow rate is found from the pooled mean of the individual flow rates

$$\langle \overline{Q} \rangle = \frac{1}{3}\sum_{j=1}^{3} Q_j$$

so,

$$\langle \overline{Q} \rangle = 2.72 \text{ ft}^3/\text{s}.$$

9.5   PRESSURE DIFFERENTIAL METERS

The operating principle of a pressure differential meter is based on the relationship between volume flow rate and the pressure drop $\Delta p = p_1 - p_2$, along the flow path,

$$Q \approx (p_1 - p_2)^n$$

where the value of n equals one for laminar flow occurring between the pressure measurement locations and n equals one-half in fully turbulent flow.

The intentional alteration of the flow path by means of a local reduction in flow area will cause a measurable local pressure drop across the altered flow length. This drop is in part due to the so-called Bernoulli effect, the inverse relationship between local velocity and pressure, as well as flow energy losses. The class of pressure differential meters that incorporate a local area reduction to augment pressure drop is commonly called *obstruction meters*. A second class of pressure differential meters alter the flow in such a manner as to force the flow to become laminar ($\text{Re}_{d_1} < 2000$) between two locations of pressure measurement. This class is known as *laminar flow meters*.

Obstruction Meters

Three common obstruction meters are the *orifice plate*, the *venturi*, and the *flow nozzle*. Flow area profiles of each are shown in Figure 9.3. These meters are usually inserted in-line with a pipe. This class of meters operates using similar physical reasonings to relate volume flow rate to pressure drop. Referring to Figure 9.4, consider the energy equation written between two control surfaces for an incompressible fluid flow through the arbitrary control volume shown. It can be assumed that (1) no external energy in the form of heat is added to the flow, (2) there is no shaft work done within the control volume, and that the flow is (3) steady and (4) one-dimensional. This yields

$$\frac{p_1}{\gamma} + \frac{\overline{U}_1^2}{2g} = \frac{p_2}{\gamma} + \frac{\overline{U}_2^2}{2g} + h_{L_{1-2}} \tag{9.8}$$

FIGURE 9.3   Flow area profiles of common obstruction meters.

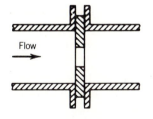

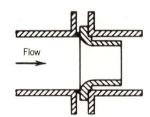

(*a*)   Square edged orifice plate meter.   (*b*)   ASME long radius nozzle.

(*c*)   ASME Herschel venturi meter.

FIGURE 9.4   Control volume concept applied between
two streamlines through an obstruction meter.

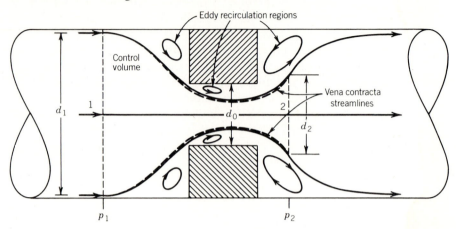

where $h_{L_{1-2}}$ denotes the head losses occurring due to frictional effects between
control surfaces 1 and 2. From conservation of mass (equations 9.4 and 9.6),

$$\overline{U}_1 = \frac{\overline{U}_2 A_2}{A_1}$$

Substituting $\overline{U}_1$ into (9.8) and rearranging yields the incompressible volume flow
rate,

$$Q_i = \overline{U}_2 A_2 = \frac{A_2}{[1 - (A_2/A_1)^2]^{1/2}} \sqrt{\frac{2(p_1 - p_2)}{\rho} + 2gh_{L_{1-2}}} \qquad (9.9)$$

where the subscript i is used only to emphasize that (9.9) yields an incompressible
flow rate. Later we drop the subscript.

When the flow area changes abruptly, the effective flow area immediately
downstream of the alteration will not necessarily be the same as the pipe flow
area. This is due to the vena contracta effect, originally investigated by Jean
Borda (1733–1799) and illustrated in Figure 9.4. This effect is brought about by
an inability of a fluid to expand immediately upon encountering an area expansion due to the inertia of each fluid particle. This forms a central core flow
bounded by regions of slower moving recirculating eddies. As a consequence,
the pressure sensed with pipe wall taps located within the vena contracta region
will correspond to the higher moving velocity within the vena contracta of unknown flow area, A_2. The unknown vena contracta area will be accounted for
by introducing the contraction coefficient, C_c, where $C_c = A_0/A_2$, into (9.9).
This yields

$$Q_i = \frac{C_c A_0}{[1 - (C_c A_0/A_1)^2]^{1/2}} \sqrt{\frac{2\Delta p}{\rho} + 2gh_{L_{1-2}}} \qquad (9.10)$$

The frictional head losses can be incorporated into a friction coefficient, C_f, such that (9.10) becomes

$$Q_i = \frac{C_f C_c A_0}{[1 - (C_c A_0/A_1)^2]^{1/2}} \sqrt{\frac{2\Delta p}{\rho}} \tag{9.11}$$

For convenience, the coefficients are replaced by a single coefficient known as the discharge coefficient, C, and (9.11) becomes

$$Q_i = CEA \sqrt{\frac{2\Delta p}{\rho}} \tag{9.12}$$

where E, known as the velocity of approach factor, is defined by

$$E \equiv \frac{1}{[1 - (A_0/A_1)^2]^{1/2}} = \frac{1}{(1 - \beta^4)^{1/2}}$$

with $\beta \equiv d_0/d_1$. In engineering handbooks, the product CE is often represented by the flow coefficient, K_0.

As shown, the discharge coefficient is derived from frictional effects and vena contracta effects. Both effects alter the flow rate from the ideal flow rate value. The discharge coefficient can be defined as the ratio of the actual flow rate through a meter to the ideal flow rate possible for the pressure drop measured. Due to its nature, C will depend on the flow Reynolds number and the β ratio, d_0/d_1, for each particular obstruction flow meter design. Because the magnitude of the vena contracta and head loss effects must vary along the length of a meter, the value of discharge coefficient is very sensitive to pressure tap location, and consistency in tap placement is imperative for correct operation [2].

In compressible gas flows, compressibility effects change the discharge coefficient. Rather than modify C, it is customary to introduce the compressible adiabatic expansion factor, Y, defined as the ratio of the actual compressible volume flow rate, Q, divided by the incompressible flow rate. Combining with (9.12) yields

$$Q = YQ_i = CEAY \sqrt{\frac{2\Delta p}{\rho_1}} \tag{9.13}$$

Equation 9.13 represents a general form of the working equation for obstruction meter volume flow rate determination.

The value for the expansion factor, Y, depends on several values: the β ratio, the particular gas specific heat ratio, k, and the relative pressure ratio across the meter, $(p_1 - p_2)/p_1$, for a particular meter type. As a general rule, compressibility effects become important when $(p_1 - p_2)/p_1 \geq 0.1$.

The flow behavior of the orifice plate, venturi, and flow nozzle has been studied to such an extent that these meters are used extensively without calibration. Values for the flow coefficients and expansion factors of these common obstruction meters have been tabulated and are available in standard flow handbooks along with standardized construction, installation, and operation techniques [2–4,12].

Orifice Meter

An orifice meter consists of a circular plate, containing a hole (orifice), which is inserted into a pipe such that the orifice is concentric with the pipe inside diameter (i.d.). Several variations in the orifice design exist but the square-edged orifice, shown in Figure 9.5, is common. Installation is simplified by housing the orifice plate between two pipe flanges. With this technique an orifice plate is

FIGURE 9.5   Square-edged orifice meter with relative pressure drop along pipe axis. 1D, 1/2D, and flange pressure taps are shown.

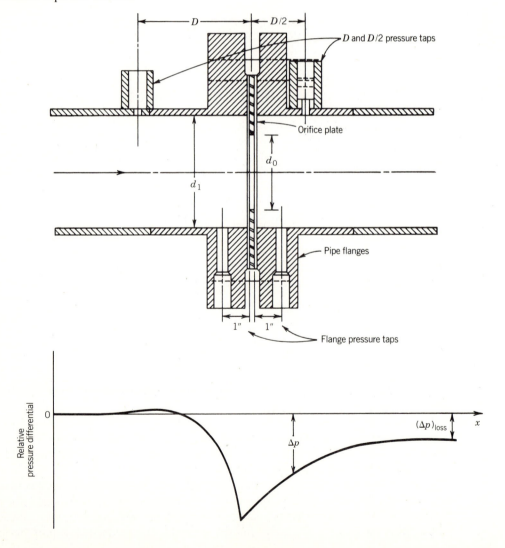

interchangeable with others of different β value. The simplicity of the installation and orifice design allows for a range of β values to be maintained on hand at modest expense. Rudimentary versions of the orifice plate have existed for several centuries. Both Torricelli and Newton used orifice plates to study the relation between pressure head and efflux from reservoirs, although neither ever got the discharge coefficients quite right [5].

For an orifice plate, equation 9.13 is used with A and β being based on the orifice hole diameter. The exact location of the pressure taps is crucial when tabulated values for flow coefficient and expansion factor are used. Standard pressure tap locations include (1) flange taps where pressure tap centers are located 25.4 mm (1 in.) upstream and 25.4 mm (1 in.) downstream of the nearest orifice face, and (2) taps located one pipe diameter upstream and one-half diameter downstream of the upstream orifice face. Nonstandard tap locations require on-site meter calibration.

Values for the flow coefficient as functions of β and Re_{d_1} and for the expansion factor as functions of k and $(p_1 - p_2)/p_1$ for a square-edged orifice plate are given in Figures 9.6 and 9.7 based on the use of flange taps. Uncertainty bias limits for discharge coefficient [12] are about 0.6% of C for $0.2 \leq \beta \leq 0.6$ and $\beta\%$ of C for all $\beta > 0.6$. Bias limits for the expansion factor are about

FIGURE 9.6   Flow coefficients for square-edged orifice meter with flange taps. (Compiled from data in [2].)

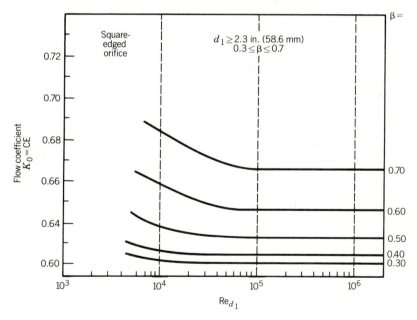

$$K_0 = \frac{1}{(1-\beta^4)^{1/2}}\,(0.5959 + 0.0312\beta^{2.1} - 0.184\beta^8$$
$$+\,B_1 d_1^{-1}\beta^4(1-\beta^4)^{-1} - B_2 d_1^{-1}\beta^3 + 91.71\beta^{2.5}\mathrm{Re}_{d_1}^{-0.75})$$

$$\text{where } B_1 = 0.09 \text{ (US)} \qquad B_2 = 0.0337 \text{ (US)}$$
$$= 2.286 \text{ (SI)} \qquad\qquad = 0.8560 \text{ (SI)}$$

FIGURE 9.7   Expansion factors for common obstruction
meters with $k = c_p/c_v = 1.4$. (Courtesy of American
Society of Mechanical Engineers, New York, NY;
compiled and reprinted from [2].)

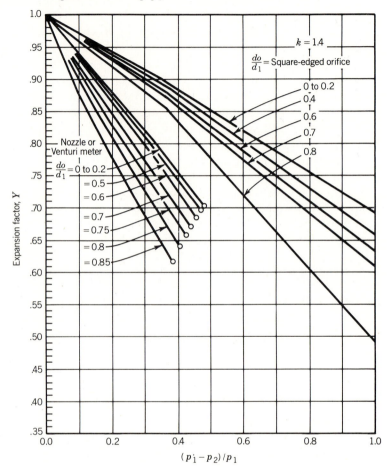

$(4(p_1 - p_2)/p_1)\%$ of Y. While the orifice plate represents a relatively inexpensive
flow meter choice that provides an easily measurable pressure drop, it introduces
a large permanent pressure loss, $(\Delta p)_{\text{loss}}$ into the flow system. The magnitude
of this drop is illustrated in Figure 9.5 with the loss estimated from Figure 9.8.

Venturi Meter

A venturi meter consists of a smooth converging contraction to a narrow
throat followed by a shallow diverging section, as shown in Figure 9.9. The
standard venturi can utilize either a 15 or 7° divergent section. The meter is
installed between two flanges intended for this purpose. Pressure is sensed between a location upstream of the throat and a location at the throat such that
equation 9.13 is used with both A and β based on the throat diameter.

FIGURE 9.8   Permanent pressure loss associated with
flow through common obstruction meters. (Courtesy of
American Society of Mechanical Engineers, New York,
NY; compiled and reprinted from [2].)

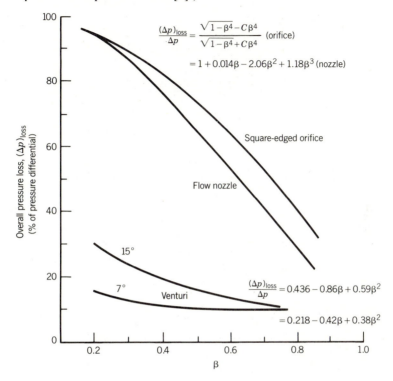

The quality of a venturi meter ranges from cast to precision machined units. The discharge coefficient varies little for pipe diameters above 11 cm (2 in.). For $2 \times 10^5 \leq Re_{d_1} \leq 2 \times 10^6$ and $0.4 \leq \beta \leq 0.75$, values for discharge coefficient of $0.984 \pm 0.7\%$ bias (95%) for cast and $0.995 \pm 1\%$ bias (95%) for machined units should be used [2,12]. Values for expansion factor are given in Figure 9.7 and have a bias limit of about $[(4 + 100\beta^2)(p_1 - p_2)/p_1]\%$ of Y [12]. Although representing a relatively expensive initial cost, as much as fivefold that of an orifice plate, Figure 9.8 demonstrates that a venturi meter shows a much smaller permanent pressure loss for a given installation.

The modern venturi meter was first proposed by Clemens Herschel (1842–1930). Herschel's design was based on his understanding of the principles developed by several men, most notably those of Daniel Bernoulli. However, he cited the studies of contraction/expansion angles and their corresponding resistance losses by Giovanni Venturi (1746–1822) and later those by James Francis (1815–1892) as instrumental to his design of a practical flow meter.

Flow Nozzles

A flow nozzle consists of a gradual contraction to a narrow throat. It needs less installation space than a venturi meter and has about 80% of the initial cost.

FIGURE 9.9   Herschel venturi meter with relative
pressure drop along axis.

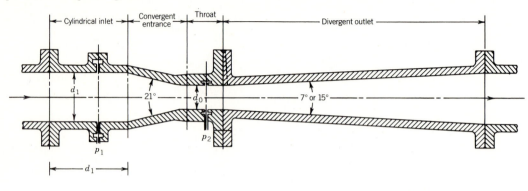

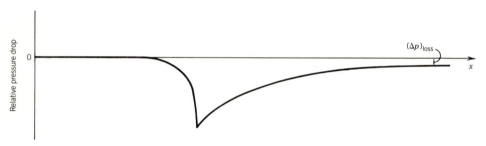

A common form for the nozzle is the ASME long radius nozzle in which the
nozzle contraction is that of the quadrant of an ellipse with major axis aligned
with the flow axis, as shown in Figure 9.10. The nozzle is typically installed
inline, but can also be used at the inlet to and the outlet from a plenum or
reservoir or at the outlet of a pipe. Pressure taps are usually located at one pipe
diameter upstream of the nozzle inlet and at the nozzle throat using either wall
or throat taps. The flow rate is determined from equation 9.13 with A and β
based on the throat diameter. Typical values for the flow coefficient and expansion factor are given in Figures 9.11 and 9.7. Bias limits for discharge coefficient
are about 2% of C and for expansion factor are about $[2(p_1 - p_2)/p_1]\%$ of Y
[12]. Because it lacks the gradual divergent section of a venturi, the permanent
loss associated with the nozzle is larger (see Figure 9.8) for the same pressure
drop.

The idea of using a nozzle as a flow meter was first proposed in 1891 by John
Ripley Freeman (1855–1932), an inspector and engineer employed by a factory
fire insurance firm. His work required tedious tests to quantify pressure losses
in pipes, hoses, and fittings. He noted a consistent relationship between pressure
drop and flow rate through fire nozzles.

EXAMPLE 9.2

A U-tube manometer filled with manometer fluid (of specific gravity S_m) is
used to measure the pressure drop across an obstruction meter. A fluid of specific

FIGURE 9.10   ASME long-radius nozzle with relative
pressure drop along axis.

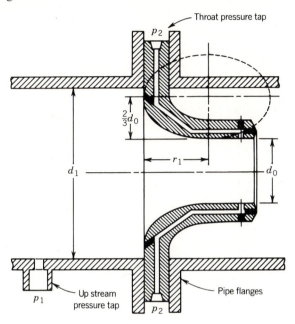

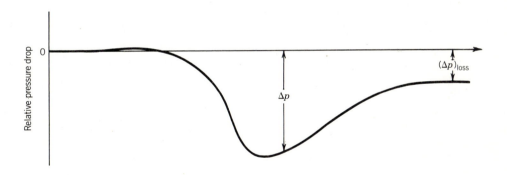

gravity, S, flows through the meter. Determine a relationship between the meter
flow rate and the measured manometer deflection, H.

KNOWN

   Fluid (of S)
   Manometer fluid (of S_m)

ASSUMPTIONS

   Density of fluids remain constant

FIND

   $Q = f(H)$

FIGURE 9.11   Flow coefficients for ASME long-radius
nozzle with throat taps. (Compiled from [2].)

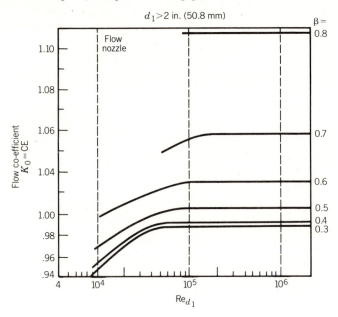

SOLUTION

Equation 9.13 provides a relationship between flow rate and a flow pressure
drop. From Figure 9.12 and hydrostatic principles, the pressure drop measured
by the manometer is

$$p_1 - p_2 = \gamma_m H - \gamma H = H(\gamma_m - \gamma) = \gamma H \left(\frac{S_m}{S} - 1 \right)$$

FIGURE 9.12   Manometer for Example 9.2.

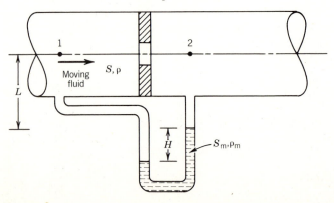

Combining this equation with equation 9.13 yields the working equation,

$$Q = CEYA \sqrt{2gH \left(\frac{S_m}{S} - 1\right)}$$

EXAMPLE 9.3

A 4.0-in.-diameter square-edged orifice plate is used to meter the steady flow of 60 °F water through an 8-in. (schedule 40) pipe. Flange taps are used and the pressure drop measured is 20 in. Hg. Determine the pipe flow rate. The specific gravity of mercury is 13.5.

KNOWN

$d_1 = 7.98$ in.
$H = 20$ in. Hg
$d_0 = 4.0$ in.

Properties

$\mu = 2.3 \times 10^{-5}$ lb-s/ft^2 (Properties are from Appendix C)
$\rho = 62.4$ lb$_m$/ft^3

ASSUMPTIONS

Steady flow
Incompressible flow ($Y = 1$)

FIND

Q

SOLUTION

Use of an orifice plate is governed by equation 9.13 which requires knowledge of E and C. Since the beta ratio for this installation is

$$\beta = \frac{d_0}{d_1} = 0.5$$

the velocity of approach factor can be calculated:

$$E = \frac{1}{(1 - \beta^4)^{1/2}} = 1.0328$$

Because $C = f(\text{Re}_{d_1}, \beta)$ the discharge coefficient, C, requires information about the flow Reynolds number. But from (9.7) with $\nu = \mu/\rho$,

$$\text{Re}_{d_1} = \frac{U d_1}{\nu} = \frac{4Q}{\pi d \nu}$$

Without information concerning either Q or U, C cannot be determined explicitly and a trial and error approach must be undertaken: Guess a value for C and iterate. A good guess is to choose a value for C at a high value of Re_{d_1}. This is the flat region of Figure 9.6. Most real flow systems operate in this region, so fewer iterations may be required if we start there. Using this as a guide, guess from Figure 9.6 a value of $K_0 \approx 0.62$ or $C = K_0/E = 0.60$.

Based on a manometer deflection (see Example 9.2),

$$Q = CEYA \sqrt{2gH \left(\frac{S_m}{S} - 1 \right)}$$

$$= 1.99 \text{ ft}^3/\text{s}$$

But this value resulted from a guessed value for C, so we must test to determine if that guessed value was correct. For this Q,

$$\mathrm{Re}_{d_1} = 4Q/\pi dv = 3.8 \times 10^5$$

From Figure 9.6, at this Reynolds number, $K_0 \approx 0.62$ or $C \approx 0.60$, so we conclude that $Q = 1.99 \text{ ft}^3/\text{s}$.

EXAMPLE 9.4

Air flows at 70 °F through a 2 in. schedule 40 pipe. A square-edged orifice plate with $\beta = 0.4$ is chosen to meter the flow rate. A pressure drop of 100 in. H_2O is measured at flange taps with an upstream pressure of 13.6 lb/in.² abs (psia). Find the flow rate.

KNOWN

$d_1 = 2.067$ in. $p_1 = 13.6$ psia
$\beta = 0.4$ $T_1 = 70 \text{ °F} = 530 \text{ °R}$
$h = 100$ in. H_2O

Properties

$v = 1.1 \times 10^{-5} \text{ ft}^2/\text{s}$ (Appendix C)
$\rho = 62.4 \text{ lb}_m/\text{ft}^3$

ASSUMPTIONS

Steady air flow
Ideal gas ($p = \rho RT$)

FIND

Q

SOLUTION

An orifice is governed by equation 9.13 which will require information about E, C, and Y. From the given information, the orifice area, $A_{d_0} = 0.0037 \text{ ft}^2$, and the velocity of approach factor, $E = 1.013$, are found. The air density is found from the ideal gas equation of state:

$$\rho_1 = \frac{p_1}{RT_1} = \frac{(13.6 \text{ lb/in.}^2)(144 \text{ in.}^2/\text{ft}^2)}{(53.3 \text{ ft-lb/lb}_m\text{-}^\circ\text{R})(530 \text{ }^\circ\text{R})}$$

$$= 0.07 \text{ lb}_m/\text{ft}^3$$

The pressure drop is found from (8.3) to be $p_1 - p_2 = \rho_{H_2O}gh/g_c = 520$ lb/ft^2.

Since the pressure ratio for this gas flow, $(p_1 - p_2)/p_1$, is 0.27, the air cannot be treated as incompressible. Using $k = 1.4$ (air) and a pressure ratio of 0.27, Figure 9.7 indicates that $Y \approx 0.92$. As in the previous example, the discharge coefficient cannot be found explicitly unless the flow rate is known, since $C = f(\text{Re}_{d_1}, \beta)$. So a trial and error approach is used.

From Figure 9.6, guess a value for $K_0 = CE \approx 0.62$ or $C = 0.60$. Then,

$$Q = CEYA \sqrt{\frac{2\Delta p}{\rho_1}} = 0.60 \times 1.013 \times 0.92 \times 0.0037 \text{ ft}^2$$

$$\times \sqrt{\frac{(2)(520 \text{ lb/ft}^2)(32.2 \text{ lb}_m\text{-ft/lb-s}^2)}{0.07 \text{ lb}_m/\text{ft}^3}}$$

$$= 1.43 \text{ ft}^3/\text{s}$$

Now check to determine whether the guessed value for C is correct. For this flow rate, $\text{Re}_{d_1} = 4Q/\pi d_1 \nu = 9.6 \times 10^5$ and from Figure 9.6, $K_0 \approx 0.62$ or $C \approx 0.60$, as assumed. The flow rate through the orifice is taken to be 1.43 ft^3/s.

Sonic Nozzles

Sonic nozzles are used to meter and to control the flow rate of compressible gases [6]. They may take the form of any of the previously described obstruction meters. If the gas flow rate through an obstruction meter becomes sufficiently high, the sonic condition will be achieved at the meter throat. At the sonic condition, the gas velocity will equal the acoustic wave velocity (speed of sound) of the gas. At that point the throat is considered to be choked and the mass flow rate through the throat will be at a maximum for the given inlet conditions regardless of any further increase in pressure drop across the meter. The theoretical basis for such a meter stems from the early work of Bernoulli, Venturi, and St. Venant (1797–1886). In 1866, Julius Weisbach (1806–1871) developed a direct relation between pressure drop and a maximum mass flow rate.

For a perfect gas, the pressure drop corresponding to the onset of the choked flow condition at the meter minimum area, the meter throat, is given by the *critical pressure ratio* (assuming an isentropic process):

$$\frac{p_2}{p_1} = \left(\frac{2}{k + 1}\right)^{k/(k-1)} \tag{9.14}$$

where p_2 is the throat pressure. A pressure ratio in excess of critical results in subsonic flow at the meter throat.

The steady-state energy equation written for a perfect gas is given by

$$c_p T_1 + \frac{\overline{U}_1^2}{2g_c} = c_p T_2 + \frac{\overline{U}_2^2}{2g_c} \tag{9.15}$$

where c_p is the constant pressure specific heat which is assumed constant. Combining equations 9.6, 9.14, and 9.15 with the ideal gas equation of state yields the mass flow at and below the critical pressure ratio:

$$\dot{m}_{\text{critical}} = \rho_1 A \sqrt{2RT_1} \sqrt{\frac{k}{k+1} \left(\frac{2}{k+1}\right)^{2/(k-1)}} \tag{9.16}$$

Equation 9.16 provides a measure of the ideal mass flow rate for a perfect gas. As with all obstruction meters, this ideal rate must be modified using a discharge coefficient to account for losses. However, the ideal and actual flow rates tend to differ by no more than 3%. When calibrations cannot be run, a $C = 0.99 \pm 2\%$ bias (95%) should be assumed [2].

The sonic nozzle provides a very convenient method to meter and to regulate a gas flow. The judicious selection of throat diameter can establish any desired fluid flow rate provided that the flow is sonic at the throat. This capability makes the sonic nozzle attractive as a local calibration standard for gases. Special orifice plate designs exist for metering at very low flow rates [2]. Both large pressure drops and system pressure losses must be tolerated with the technique, although low loss venturi designs will minimize losses [2].

EXAMPLE 9.5

A flow nozzle is to be used at choked conditions to regulate the flow of N_2 gas at 0.6 lb_m/s through a 2-in.-i.d. pipe. The pipe is pressurized at 100 lb/in.^{2} abs (psia) and gas flows at 80 °F. Determine the maximum β ratio nozzle that can be used. $R_{N_2} = 55.2$ ft-lb/lb_m-°R.

KNOWN

N_2 $(k = 1.4)$ $p_1 = 100$ psia $\dot{m} = 0.6$ lb_m/s
$d_1 = 2$ in. $T_1 = 540$ °R

ASSUMPTIONS

Perfect gas
Steady, choked flow
$C = 0.99$

FIND

Find $\beta_{\max} = d_{0,\max}/d_1$

SOLUTION

Equation 9.16 can be rewritten in terms of nozzle throat area, A:

$$A_{max} = \frac{\dot{m}}{\rho_1 (2RT)^{1/2} \{k/(k+1)[2/(k+1)]^{2/(k-1)}\}^{1/2}}$$

$$= 0.0039 \text{ ft}^2$$

Since $A = \pi d_0^2/4$, this yields a maximum nozzle throat diameter, d_0:

$$d_{0,max} = 0.84 \text{ in.}$$

Hence, $\beta = d_0/d_1 \leq 0.42$ is required to maintain the specified mass flow rate.

Obstruction Meter Selection

Selection between obstruction meter types depends on a number of factors that require some engineering compromises. Primary considerations include:

Meter placement

Overall pressure loss

Accuracy

Overall costs

Of course, the immediate availability of a particular flow meter may override other concerns in many instances.

It is very important to provide sufficient upstream and downstream pipe lengths on each side of a flow meter to provide for proper flow development. The flow development dissipates swirl, promotes a symmetric velocity distribution, and allows for proper pressure recovery downstream of the meter. For most situations, these installation effects can be minimized by placement of the flow meter in adherence to the considerations outlined in Figure 9.13. However, the engineer needs to be wary that installation effects are difficult to predict, particularly downstream of elbows (out-of-plane double elbow turns are notoriously difficult), and Figure 9.13 is to be used as a guide only [11,12]. Even with recommended lengths, an additional 0.5% uncertainty should be added to the uncertainty in the discharge coefficient to account for swirl effects [12]. Unorthodox installations will require meter calibration to determine flow coefficients or use of special correction procedures [11,12]. The physical size of a meter may become an important consideration in itself, since in large-diameter pipelines, venturi, and nozzle meters can take up a considerable length of pipe.

The unrecoverable overall pressure loss, $(\Delta p)_{loss}$, associated with a flow meter increases with β ratio and flow rate. These losses must be overcome by the system prime mover (e.g., a pump driven by a motor) in addition to any other pipe system losses and must be considered during prime mover sizing. In established systems, flow meters must be chosen on the basis of pressure loss that

FIGURE 9.13   Recommended placements for flow
meters. (Courtesy of American Society of Mechanical
Engineers, New York, NY; reprinted from [2].)

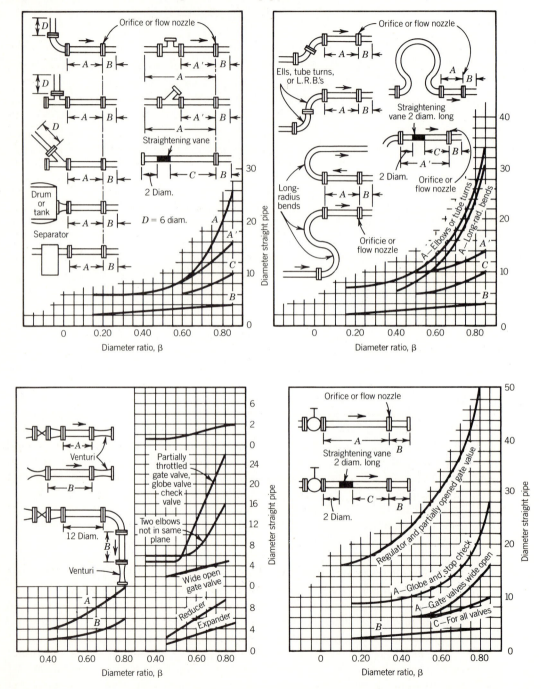

the prime mover can accommodate and still maintain a desired system flow rate. The power, $\dot{W}$, required to overcome any loss in a system is given by

$$\dot{W} = \rho \, gh_{L_{1-2}} \frac{Q}{\eta g_c} = Q \frac{(\Delta p)_{loss}}{\eta} \qquad (9.17)$$

where η is the prime mover overall efficiency.

The actual cost of any meter depends on the initial capital cost of the meter, the costs of installing the meter (including system down time), calibration costs, and the added capital and operating costs associated with flow meter pressure losses to the system. Indirect costs may include product loss due to the meter flow rate uncertainty.

The ability of an obstruction flow meter to accurately estimate the volume flow rate in a pipeline depends on both the method used to calculate flow rate and factors inherent to the meter. If standard tabulated values for meter coefficients are used, factors that contribute to the overall uncertainty of the measurement enter through the following data acquisition elemental errors:

1. Actual β ratio precision and pipe eccentricity.
2. Pressure measurement position precision.
3. Temperature effects leading to relative thermal expansion of components.
4. Upstream flow profile.

These errors are in addition to the errors inherent in the coefficients themselves and errors associated with the estimation of the upstream fluid density [2]. Direct in situ calibration of any flow meter installation can reduce these contributions to overall uncertainty. For example, the actual β ratio value, pressure tap locations, and other installation effects will be included in a system calibration and accommodated in the actual computed flow coefficient.

The selection of any meter should consider whether the system into which the meter is to be installed will be used at more than one flow rate. If so, considerations as to meter performance and its effect on system performance over the entire anticipated flow rate range should be included. The range over which any one meter can be used is known as the meter *turndown*.

EXAMPLE 9.6

For the orifice meter in Example 9.3 having a $\beta = 0.5$ and a pressure drop of 20 in. Hg, calculate the permanent pressure loss due to the meter.

KNOWN

$H = 20$ in. Hg or $p_1 - p_2 = 1411$ lb/ft^2
$\beta = 0.5$

ASSUMPTIONS

Steady flow

FIND

$(\Delta p)_{loss}$

SOLUTION

For a properly designed and installed orifice meter, Figure 9.8 indicates that the permanent pressure loss of the meter will be about 75% of the pressure drop across the meter. Hence,

$$(\Delta p)_{loss} = 0.75 \times 1411 \ \text{lb/ft}^2$$

$$= 1058 \ \text{lb/ft}^2 \ \text{or 15 in. Hg}$$

COMMENT

In comparison, for $Q = 1.99 \ \text{ft}^3/\text{s}$ a typical venturi (say 15° outlet) with a $\beta = 0.5$ would provide a pressure drop equivalent to 7.7 in. Hg with a permanent loss of only 1.2 in. Hg. A prime mover (such as the pump here) would need to be able to add additional energy to the flow to make up for the meter permanent pressure loss while providing enough power to sustain the desired flow rate. A smaller pump would be needed if using the venturi.

EXAMPLE 9.7

For the orifice of Example 9.6, estimate the operating costs required to overcome these permanent losses if electricity is available at $0.08/kW-h, the pump is used 6000 h/year and the pump-motor efficiency is 60%.

KNOWN

Water at 60 °F $(\Delta p)_{loss} = 1058 \ \text{lb/ft}^2$
$Q = 1.99 \ \text{ft}^3/\text{s}$ $\eta = 0.60$

ASSUMPTIONS

Steady flow
Meter installed according to standards [2]

FIND

Annual cost due to $(\Delta p)_{loss}$ alone

SOLUTION

We can determine the pump power required to overcome the permanent pressure losses in the system resulting from the use of the orifice meter by (9.17),

$$\dot{W} = Q \frac{\Delta p_{loss}}{\eta}$$

$$= \frac{(1.99 \ \text{ft}^3/\text{s})(1058 \ \text{lb/ft}^2)(3600 \ \text{s/h})}{0.60} = 12.6 \times 10^6 \ \text{ft-lb/h}$$

$$= 4.76 \ \text{kW}$$

The additional annual pump operating cost required to overcome this is

$$\text{Cost} = (4.76 \text{ kW})(6000 \text{ h/year})(\$0.08/\text{kW-h}) = \$2285/\text{year}$$

COMMENT

In comparison, the venturi discussed in the Comment to Example 9.6 would cost about $183/year to operate.

EXAMPLE 9.8

An ASME long radius nozzle ($\beta = 0.5$) is to be installed into a horizontal section of 12-in. (schedule 40) pipe. A long straight length of pipe exists just downstream of a fully open gate valve but upstream of an in-plane 90° elbow. Determine the minimum lengths of straight unobstructed piping that should exist upstream and downstream of the meter.

KNOWN

$d_1 = 11.94$ in.
$\beta = 0.5$
Installation layout

ASSUMPTIONS

Steady flow

FIND

Meter placement

SOLUTION

From Figure 9.13, the meter should be placed a minimum of 8 pipe diameters ($A = 8$) downstream of the valve and a minimum of 3 pipe diameters ($B = 3$) upstream of any flow fitting. Note that since a typical flow nozzle has a length of about 1.5 pipe diameters, this installation will require a straight pipe section of at least 12.5 pipe diameters ($8 + 3 + 1.5$).

COMMENT

A general rule for initial planning is to use a 10 diameter upstream length before and a 4 diameter downstream length after the meter pressure taps.

Laminar Flow Elements

The relationship between flow rate and pressure drop across a length of pipe of diameter, d_1, and containing a laminar flow follows a linear relationship. The energy equation for such a case is written as

$$\frac{p_1}{\gamma} + \frac{\overline{U}_1^2}{2g} = \frac{p_2}{\gamma} + \frac{\overline{U}_2^2}{2g} + h_{L_{1-2}}$$

The head loss term, $h_{L_{1-2}}$, can be estimated by the Darcy–Weisbach relation:

$$h_{L_{1-2}} = f \frac{L}{d_1} \frac{\overline{U}_1^2}{2g}$$

where L is the distance between pressure taps and f is the friction factor, which for a laminar flow is simply

$$f = \frac{64}{\text{Re}_{d_1}}$$

From conservation of mass (9.6),

$$Q = \overline{U}_1 \frac{\pi d_1^2}{4} = \overline{U}_2 \frac{\pi d_1^2}{4}$$

which for a constant diameter conduit requires $\overline{U}_1 = \overline{U}_2$. This yields

$$p_1 - p_2 = \gamma h_{L_{1-2}}$$

Accordingly, by substitution into the Darcy-Weisbach relation using $\overline{U} = 4Q/\pi d_1^2$ yields

$$Q = \frac{\pi d_1^4}{128\mu} \frac{p_1 - p_2}{L} \qquad \text{Re}_{d_1} < 2000 \tag{9.18}$$

Equation 9.18 reveals that volume flow rate is linear with pressure drop in a laminar pipe flow. This relation was first demonstrated by Jean Poiseulle (1799–1869) who conducted meticulous tests documenting the resistance of flow through capillary tubes.

The simplest type of laminar flow meter consists of two pressure taps separated by a length of piping.[1] However, the Reynolds number constraint for laminar flow restricts the size of pipe diameter that can be used or, alternatively, the flow rate to which (9.18) can be applied for a given fluid. This limitation is overcome in commercial units through the use of laminar flow elements (Figure 9.14) which consist of a bundle of small-diameter tubes, or some proprietary design of geometric passages, placed in parallel. The strategy of a laminar flow element is to divide up the flow by passing it through the tube bundle so as to reduce the flow rate per tube such that the individual Reynolds number in each tube remains below 2000. Pressure drop is measured between the entrance and the exit of the laminar flow element. Because of the additional entrance and exit losses associated with the laminar flow element, a flow coefficient is used to modify equation 9.18 where the coefficient must be determined by calibration. No standard tables for these coefficients exist due to a wide possible variation in flow meter design, but commercial units come supplied with individual cali-

[1]Alternately, if flow rate can be measured or is known, (9.18) provides the basis for a capillary tube viscometer.

FIGURE 9.14   Laminar flow element.

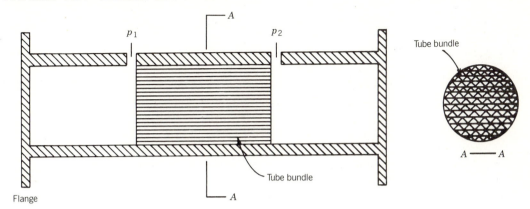

bration charts. The coefficient will be essentially constant over the useful meter range.

For any given meter there will exist a flow rate above which laminar flow will no longer exist in the laminar flow elements. Consequently, the application of any particular meter will be limited to some maximum flow rate. Various meter sizes and designs are available to accommodate user needs. Turndowns up to 100 to 1 are available.

Laminar flow elements offer some distinct advantages over other pressure differential meters. These include:

1.   A high sensitivity even at extremely low flow rates
2.   The ability to measure pipe system flows in either meter direction
3.   A wide usable flow range
4.   The ability to indicate an average flow rate in pulsating flows

The overall uncertainty bias limit in flow rate determination can be as low as 0.25% of the flow rate with these meters. However, these meters are very susceptible to clogs, restricting their use to clean fluids. All of the pressure drop measured remains a system pressure loss.

9.6   INSERTION VOLUME FLOW METERS

Myriads of volume flow meter types based on a number of different principles have been proposed, developed, and sold commercially. The largest group of meters are based on some phenomenon that is actually sensitive to the average velocity across a control surface of known area. Several of these designs are included in the discussion below.

Electromagnetic Flow Meters

The operating principle of an *electromagnetic flow meter* [7] is based on the fundamental principle that an electromotive force (emf) of electric potential, E, is induced in a conductor of length, **L**, which moves with a velocity, **U**, through a magnetic field of magnetic flux, **B**. This physical behavior was first recorded by Michael Faraday (1791–1867). In principle, we write

$$E = \mathbf{U} \times \mathbf{B} \cdot \mathbf{L} \tag{9.19}$$

A practical utilization of (9.19) is shown in Figure 9.15. The magnitude of E is determined through:

$$E = |\mathbf{U}| \cdot |\mathbf{B}| \cdot |\mathbf{L}| \cdot \sin \alpha = UBL \sin \alpha$$

where α is the angle between the mean velocity vector and the magnetic flux vector, usually at 90°. In general, electrodes are embedded in the pipe wall in a diametrical plane that is normal to the known magnetic field. As the flow moves through the magnetic field, the induced electric potential is detected and measured by the electrodes, which are separated by the length, L. The average magnitude of the velocity, $\overline{U}$, across the pipe is thus inferred through the measured emf. The flow rate is found by

$$Q = \frac{\overline{U} \pi d_i^2}{4} = \frac{E}{BL} \frac{\pi d_i^2}{4} = K_1 E \tag{9.20}$$

FIGURE 9.15   Electromagnetic principle applied to a flow meter.

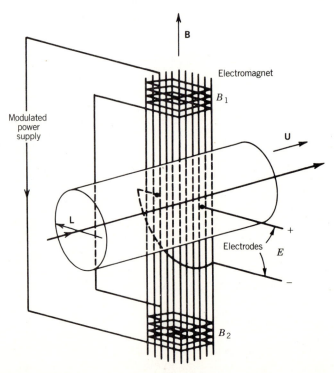

The value of L is on the order of the pipe diameter, the exact value depending on the meter construction and magnetic flux lines. The static sensitivity K_1 is a meter constant found by calibration and supplied by a manufacturer. The relationship between flow rate and measured potential is linear.

The electromagnetic flow meter comes commercially as a packaged flow device, which is installed directly inline and connected to an external electronic output unit. Units are available using either permanent magnets, called dc units, or variable flux strength electromagnets, called ac units. The magnetic flux strength of an ac unit can be increased on site for a strong signal at low flow rates of low conductivity fluids such as water. Special designs include a flow sensor unit which actually can clamp over (not in-line with) a nonmagnetic pipe, a design favored to monitor blood flow rate through major arteries during surgery.

The electromagnetic flow meter has a very low pressure loss associated with its use due to its open tube, no obstruction design and is suitable for installations that can tolerate only a small pressure drop. This absence of internal parts is very attractive for metering corrosive and "dirty" fluids. The operating principle is independent of fluid density and viscosity, responding only to average velocity, and there is no difficulty with measurements in either laminar or turbulent flows, provided that the velocity profile is reasonably symmetrical. It can be used in either steady or pulsatile flows, providing either time-averaged or instantaneous data in the latter. Data acquisition errors down to $\pm 0.25\%$ of the measured flow rate can be attained, although values between ± 1 and 5% are more common for these meters. The use of any meter is limited to fluids having a threshold value of electrical conductivity, the actual value of which depends on a particular meter's design, but fluids with values as low as 0.1 μsieman/cm have been metered. The addition of salts to a fluid will increase its conductivity.

Vortex Shedding Meters

Nearly everyone has observed an oscillating street sign or heard the "singing" of power lines on a windy day. These are but a couple of the effects induced by vortex shedding from bluff-shaped bodies, a natural phenomenon in which alternating vortices are shed in the wake of the body at a frequency that will depend on the velocity of the flow past the body. The vortices formed on opposite sides of the body are carried downstream in the body's wake forming a "vortex street," each vortex having an opposite sign of rotation. This behavior is seen in Figure 9.16, a photograph that captures the vortex shedding downstream of a section of an aircraft wing. The existence of the vortex street was deduced theoretically by the aerodynamicist Theodore von Karman (1881–1963), although da Vinci appears to have been the first to actually record the phenomenon [1]. The alternating vortices are manifestations of an oscillating pressure field about the body, a field that exerts an equal but opposite force on the body.

The vortex shedding phenomenon is used to sense average velocity in pipe flows in a *vortex flow meter*. The basic relationship between shedding frequency, ω, and average velocity, $\overline{U}$, for a given shape is given by the Strouhal number,

$$\text{St} = \frac{\omega d}{\overline{U}} \tag{9.21}$$

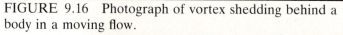

FIGURE 9.16   Photograph of vortex shedding behind a
body in a moving flow.

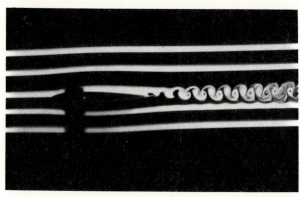

where d is a characteristic length for the body, such as diameter for a cylinder.
In general, the Strouhal number is a function of Reynolds number but various
geometrical shapes, known as shedders, can be used to produce a stable vortex
flow that has a *constant* Strouhal number over a broad range of flow Reynolds
number. Examples are given in Table 9.1. The quality and strength of the
shedding can be improved by manipulation of the design of the tail end of the
body and by providing a slightly concave upstream body face that traps the
stagnation streamline at a point giving way to a stable oscillation. Accordingly,
there are a number of proprietary designs in existence. In general, abrupt edges
on the shedder restrict the dependence on Reynolds number by fixing the flow
separation points. A typical design is shown in Figure 9.17. The shedder spans
the pipe, $d/l \approx 0.3$, to provide an average pipe velocity with $d/d_1 \approx 0.3$ for
strong, stable vortex strength.

For a constant Strouhal number, the flow rate can be deduced from continuity
based on measured shedding frequency.

$$Q = K_1 \omega$$

where K_1 is a constant value of static sensitivity provided by the manufacturer
and depends on the shedder design. Shedding frequency can be measured in
many ways. The shedder strut can be instrumented to detect the force oscillation
by using strut-mounted strain gages, for example, or to detect the pressure
oscillations by using a piezoelectric crystal.

The lower flow rate limit on vortex meters appears to be at Reynolds numbers
near 10 000, below which the Strouhal number varies nonlinearly with flow rate
and shedding becomes unstable regardless of shedder design. This can be a
problem in high-viscosity pipe flows ($\mu > 20$ cp). The upper flow bound is
limited only by the onset of cavitation in liquids and by the onset of compressibility effects in gases at Mach numbers exceeding 0.2. Property variations affect
meter performance only indirectly. Density variations affect the strength of the
shed vortex and this places a lower limit on fluid density which is based on the
sensitivity of the vortex shedding detection equipment. Viscosity affects the

TABLE 9.1 Shedder Shape and
Strouhal Number

Cross Section	Strouhal Number[a]
	0.16
	0.19
	0.16
	0.15
	0.12

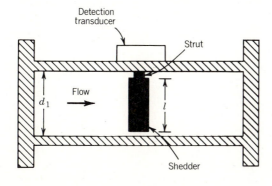

[a]For Reynolds numbers $Re_\Delta \geq 10^4$.

FIGURE 9.17 Vortex shedding flow meter.

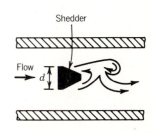

Side view   Top view

operating Reynolds number. Otherwise, within bounds, the meter is insensitive to property variations.

The meter has no moving parts and relatively low pressure losses compared to obstruction meters. A single meter can operate over a flow range of up to 20 : 1 above its minimum with a linearity in K factor of $\pm 0.5\%$. On the other hand, the shedding frequency is inversely related to the pipe diameter cubed, which suggests that meter resolution can drop quickly in large diameter pipes, placing a limit on the maximum meter size that can be used effectively.

Rotameters

The rotameter remains a widely used insertion meter for flow rate indication. As depicted in Figure 9.18, the meter consists of a float within a vertical tube, tapered to an increasing cross-sectional area at its outlet. Flow entering through the bottom passes over the float, which is free to move. The basic principle of the device is based on the simple balance between the drag force, F_D, and the weight, W, and buoyancy forces, F_B, acting on the float in the moving fluid. It is the drag force on the float that varies with the average velocity over the float.

The force balance in the vertical direction y yields

$$\Sigma F_y = 0 = -F_D + W - F_B$$

or

$$C_D \frac{1}{2} \rho A_x \overline{U}^2 = g(\rho_b - \rho) V_b$$

FIGURE 9.18   Rotameter.

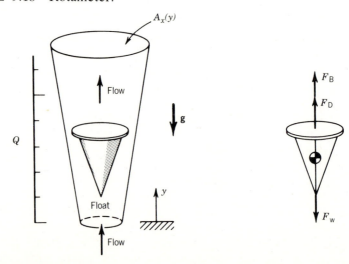

where

$$\rho = \text{density of fluid or of float, } \rho_b$$
$$C_D = \text{drag coefficient of the float; } C_D = f(\text{Re})$$
$$A_x = \text{tube cross-sectional area}$$
$$\overline{U} = \text{average velocity past the float}$$
$$V_b = \text{volume of float}$$

Over the range of the meter, the float will rise to some position within the tube at which such a force balance exists. The height of this position increases with flow velocity and, hence, flow rate. This flow rate is found by

$$Q = \overline{U}A_a(y) = |C_D K_1|^{1/2} A_a(y)$$

where $A_a(y)$ is the annular area between the float and the tube, which depends on the height of the float in the tube, and K_1 is a constant depending on the meter design and fluid in use. Since annular area is a function of float position within the vertical tube, the float's vertical position gives a direct measure of flow rate which can be read from a graduated scale, electronically sensed with an optical cell, or detected magnetically. Floats with sharp edges are less sensitive to fluid viscosity changes with temperature. Rotameters are used in noncritical applications where accuracy is not of prime concern. Uncertainty bias limits of ±2% of flow rate are common as is a turndown of 10 : 1.

Turbine Meters

Turbine meters make use of angular momentum principles to meter flow rate. In a typical design (Figure 9.19), a rotor is encased within a bored housing through which the fluid to be metered is passed. The housing contains flanges for direct insertion. The exchange of momentum between the flow and the rotor turns the rotor at a rotational speed that is proportional to the flow rate. Rotor rotation can be measured in a number of ways. For example, a reluctance pickup coil can sense the passage of magnetic rotor blades producing a pulse train signal at a frequency that is directly related to rotational speed.

The rotor angular velocity, η, will depend on the average flow velocity and fluid viscosity, ν, through the meter bore of diameter, d_1. Dimensionless analysis of these parameters [8] suggests

$$\frac{Q}{\eta D^3} = f\left(\frac{\eta D^2}{\nu}\right)$$

In practice, there will exist a region in which the rotor angular velocity will vary linearly with flow rate.

Turbine meters offer a low-pressure drop and very good accuracy. Uncertainty bias limits in flow rate down to ±0.25% are typical, with turndown of 20 to 1. They are exceptionally repeatable making them good candidates for local flow rate standards. However, their use must be restricted to clean fluids because of

FIGURE 9.19   Cutaway view of a turbine meter.
(Courtesy Schlumberger Industries, Measurements
Division, Greenwood, SC.)

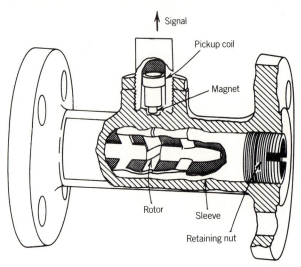

possible fouling of their rotating parts. Temperature changes affect fluid viscosity,
a property to which a turbine meter rotational speed is sensitive. Some compensation for viscosity variations can be made electronically [9]. The turbine
meter is very susceptible to installation errors caused by pipe flow swirl [11] and
a careful selection of installation position is suggested.

Positive Displacement Meters

Positive displacement meters contain mechanical elements that define a known
volume. The free-moving elements are displaced by the action of the moving
fluid. A counting mechanism counts the number of element displacements to
provide a direct reading of volume of fluid passed through the meter. Usually
used as volume meters, volume per unit time can be discerned in conjunction
with a timer. These units are typically installed where applications demand
ruggedness and high accuracy in steady flow metering. This metering method is
common to water, gasoline, and natural gas meters.

Wobble meters contain a disk that is seated in a chamber. The flow of fluid
through the chamber causes the disk to oscillate so that a known volume of fluid
moves through the chamber with each oscillation. The disk is connected to a
counter that records each oscillation as volume displaced. Wobble meters are
used for domestic water applications where they must have an uncertainty of
no greater than 1% of actual delivery. Frequently found on oil trucks and at
the gasoline pump, *rotating vane meters* use rotating cups or vanes that move
about an annular opening displacing a known volume with each rotation. Uncertainty can be as low as ±0.3% of actual delivery. Because of the good accuracy
of these meters, they are often used as local standards to calibrate volume flow
meters.

9.7 MASS FLOW METERS

There are many situations in which the mass flow rate is the quantity of interest. Direct conversion from measured volume flow rate to actual mass flow rate requires knowledge of the density of the metered fluid under the exact conditions of measurement. For many applications one can assume that both the errors in the tabulated values for fluid density and the fluid density variations within a volume meter are small enough to ignore. But not all fluids have readily established equations of state (e.g., petroleum products, polymers, and cocoa butter). When volume flow methods are used, uncertainty in the density increases the overall uncertainty in mass flow rate above the uncertainty in volume flow rate. As such, it is not uncommon to use density measuring devices just upstream of the volume flow measurement to improve the uncertainty in the mass flow determination in critical applications. The direct measurement of mass flow rate is desirable since it eliminates the uncertainties associated with density. But the difference between a meter that is sensitive to Q as opposed to $\dot{m}$ is not trivial. Prior to the 1970s reliable commercial mass flow meters with sufficient accuracy to circumvent volume flow rate corrections were generally not available, even though the basic principles and implementation schemes for such meters had been understood in theory for several decades. U.S. patents dating back to the 1940s record schemes for using heat transfer, Coriolis forces, and momentum methods to infer mass flow rate directly.

Thermal Flow Meter

The rate at which energy, $\dot{E}$, must be input to a flowing fluid to raise the temperature of the fluid some desired amount between two control surfaces is directly related to the mass flow rate by

$$\dot{E} = \dot{m}c_p \, \Delta T \qquad (9.22)$$

where c_p is the fluid specific heat. Methods to utilize this effect to directly measure mass flow rate incorporate an inline meter having some means to input energy to the fluid over the meter length. The passing of a current through an immersed filament is a common method. Fluid temperatures are measured at the upstream and the downstream locations of the meter. This type of meter is quite easy to use and appears reliable. In fact, in the 1980s, the technique was adapted for use in at least one automobile fuel injection system to provide an exact air–fuel mixture to the engine cylinders despite short-term altitude, barometric, and seasonal environmental temperature changes.

A major consideration involved with the use of this type of meter concerns the assumption that c_p is known for the fluid and remains constant over the length of the meter. For common gases, such as air, this assumption is quite valid at reasonable temperatures and pressures. Flow rate turndown of up to 100 to 1 is possible with uncertainties down to $\pm 0.5\%$ of flow rate with very little pressure drop. But the assumptions become restrictive for liquids and for many gases in which c_p either may be a strong function of temperature or may simply not be well established.

A second type of thermal mass flow meter is actually a velocity-sensing meter and thermal sensor together in one direct insertion unit. The meter uses both hot-film anemometry methods to sense fluid velocity through a conduit of known diameter and an adjacent RTD sensor for temperature measurement. For sensor and fluid temperatures, T_s and T_f, respectively, mass flow rate is inferred from the correlation

$$\dot{E} = [C + B(\rho \overline{U})^{1/n}](T_s - T_f)$$

where C, B, and n, which can be shown to depend on fluid properties [10], are determined through calibration. In a scheme to reduce the fluid property sensitivity of the meter, the RTD may be used as an adjacent resistor leg of the anemometer Wheatstone bridge circuit to provide a temperature-compensated velocity output over a wide range of fluid temperatures, with excellent repeatability ($\sigma \approx 0.25\%$). Gas velocities of up to 12 000 ft/m and flow rate turndown of 50 to 1 are possible with uncertainties down to $\pm 2\%$ of the flow rate and very little pressure drop. A series of these mass flow probes can be combined to span across large ducts or chimneys to provide better averaging information.

Coriolis Flow Meter

The term *Coriolis flow meter* refers to the family of insertion meters that meter mass flow rate by inducing a Coriolis acceleration on the flowing fluid and measuring the resulting developed force. The developed force is directly related to the mass flow rate independent of the fluid properties. The Coriolis effect was proposed by Gaspard de Coriolis (1792–1843) following his studies of accelerations in rotating systems. A large number of methods to utilize this effect have been proposed since the first U.S. patent for a Coriolis effect meter was issued in 1947.

Commercially available units for metering of liquids are based on a scheme in which the pipe flow is diverted from the main pipe and divided between two bent, parallel, adjacent tubes of equal diameter, such as shown in Figure 9.20. The tubes themselves are mechanically driven in a relative out-of-phase sinusoidal oscillation by an electromagnetic driver. In general, a fluid particle passing through the meter tube that is rotating (due to the oscillating tube) relative to the fixed pipe experiences an acceleration at any arbitrary position S. The total acceleration at S, $\ddot{\mathbf{r}}$, is composed of several components (see Figure 9.21),

$$\ddot{\mathbf{r}} = \ddot{\mathbf{R}}_{O'} + \dot{\boldsymbol{\omega}} \times \mathbf{r}_{S/O'} + \boldsymbol{\omega} \times \boldsymbol{\omega} \times \mathbf{r}_{S/O'} + \ddot{\mathbf{r}}_{S/O'} + 2\boldsymbol{\omega} \times \dot{\mathbf{r}}_{S/O'}$$

$\ddot{\mathbf{R}}_{O'}$ = translation acceleration of rotating origin O' relative to fixed origin O

$\boldsymbol{\omega} \times \boldsymbol{\omega} \times \mathbf{r}_{S/O'}$ = centripetal acceleration of S relative to O'

$\boldsymbol{\omega} \times \mathbf{r}_{S/O'}$ = tangential acceleration of S relative to O'

$\ddot{\mathbf{r}}_{S/O'}$ = translational acceleration of S relative to O'

$2\boldsymbol{\omega} \times \ddot{\mathbf{r}}_{S/O'}$ = Coriolis acceleration at S relative to O'

FIGURE 9.20   Cutaway view of Coriolis flow meter.
(Courtesy Schlumberger Industries, Measurements
Division, Greenwood, SC.)

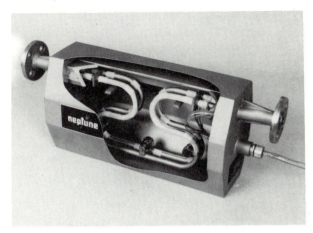

with ω the angular velocity of point S relative to O'. In these meters, the tubes are rotated but not translated, so that the translational accelerations are zero. A fluid particle will experience forces due to the remaining accelerations that cause equal and opposite reactions on the meter tube walls. The Coriolis acceleration acts in a plane perpendicular to the tube axes and develops a force gradient that creates a twisting motion or oscillating rotation about the tube plane.

The utilization of the Coriolis force depends on the shape of the meter. However, the basic principle is illustrated in Figure 9.22. Rather than rotating the tubes a complete 360° about the pipe axis, the meter tubes are vibrated continuously at a drive frequency, ω, with amplitude displacement, z, about the pipe axis. This eliminates rotational seal problems. The driving frequency is

FIGURE 9.21   Fixed and rotating reference frames.

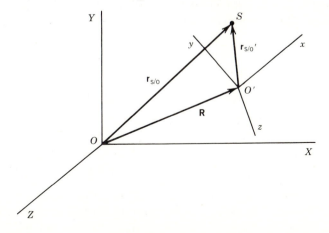

FIGURE 9.22   Concept of operating principle of Coriolis
flow meter.

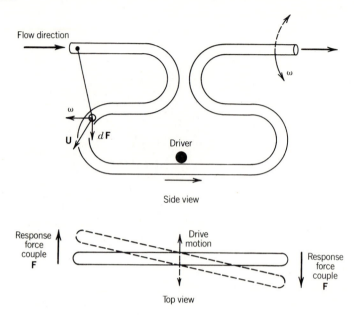

selected at the tube resonance frequency, which creates a self-sustaining tuning
fork effect since the meter will naturally respond to any disturbance at this
frequency. As a fluid particle of elemental mass, dm, flows through a section of
the flow meter, it experiences the Coriolis acceleration, $2\boldsymbol{\omega}_c \times \dot{\mathbf{r}}_{S/O'}$, and an
inertial Coriolis force

$$dF = (2\boldsymbol{\omega}_c \times \dot{\mathbf{r}}_{S/O'})\,dm$$

acting in the z direction. For the flow meter design of Figure 9.22, as the particle
travels along the meter the direction of the velocity vector changes. This results,
using the right-hand rule, in a change in direction of vector force, $d\mathbf{F}$. The
resultant forces experienced by the tube are of equal magnitude but opposite
sign of those experienced by the particle. Each tube segment senses a corresponding differential torque, $d\mathbf{T}$, a rotation about the y axis at a frequency ω_c,

$$d\mathbf{T} = \mathbf{x}' \times d\mathbf{F} = \mathbf{x}' \times dm \cdot (2\boldsymbol{\omega}_c \times \dot{\mathbf{r}}_{S/O'})$$

where x' refers to the x distance between the elemental mass and the y axis,
$\mathbf{x}' = x'\hat{e}_x$. The total torque experienced by each tube is found by integration
along the total path length of the tube, L,

$$\mathbf{T} = \int_0^L d\mathbf{T}$$

A differential element of fluid will have the mass, $dm = \rho A\,dl$, for an elemental
cross section of fluid, A, of differential length, dl, and of density, ρ. If this mass

moves with an average velocity, $\overline{U}$, then the differential mass can be written as

$$dm = \rho A \; dl = \dot{m} \frac{dl}{\overline{U}}$$

The Coriolis cross-product can be expressed as

$$2\boldsymbol{\omega}_c \times \dot{\mathbf{r}}_{S/O'} = (2 \, \omega_c \, \dot{r}_{S/O'} \sin \theta)\hat{e}_z = (2 \, \omega_c \, \overline{U} \sin \theta)\hat{e}_z$$

where θ is the angle between the Coriolis rotation and the velocity vector. Then,

$$\dot{m} = \frac{T}{2 \int_0^L (mx' \, \omega_c \sin \theta)dl} \, \hat{e}_y$$

Since the velocity direction changes by 180°, the Coriolis developed torque will act in opposite directions on each side of the tube.[2] The meter tubes will twist about the y axis of the tube at an angle δ. For small angles of rotation the twist angle is related to torque by

$$\delta = k_S T = \text{constant} \times \dot{m}$$

where k_S is related to the stiffness of the tube. The objective becomes to measure the twist angle, which can be accomplished in many ways. For example, by driving the two meter tubes 180° out of phase, the relative phase at any time between the two tubes will be directly related to the mass flow rate. The exact relationship is linear over a wide flow range and determined by calibration with any fluid.

The tangential and the centripetal accelerations remain of minor consequence due to the tube stiffness in the directions in which they act. However, at high mass flow rates they can excite modes of vibration in addition to the driving mode, ω, and response mode, ω_c. In doing so, they affect the meter linearity and zero error (drift), which affects mass flow uncertainty and meter turndown. Essentially, the magnitude of these undesirable effects are inherent to the particular meter shape and are controlled, if necessary, through tube-stiffening members. The meter principle is unaffected by changing fluid properties but temperature changes will affect the overall meter stiffness, an effect that can be compensated for electronically. A very desirable feature is an apparent insensitivity to installation position. Commercially available Coriolis flow meters can measure flow rate with uncertainty bias limits as low as ±0.25% of mass flow rate, with turndown of about 20 to 1. However, uncertainties as low as ±0.10% are conceivably achievable but must await improved certification methods. The meter is also used as an effective densitometer.

---

[2]It is not difficult to envision a design in which the velocity does not reverse direction but ω does.

9.8   FLOW METER CALIBRATION AND STANDARDS

A fundamental primary standard for flow rate does not exist. But a number of calibration procedures do exist in lieu of this. The general procedure for the calibration of most in-line flow meters requires the establishment of steady flow in a calibration flow loop and the determination of the volume or mass of flowing fluid passing through the flow meter in an accurately determined time interval. Flow loop systems in which the flow rate is established by some means so as to calibrate a flow meter are known as *provers*. Several methods to establish the flow rate are discussed.

In liquids, variations of the "catch-and-weigh" technique are often employed in flow provers. One variation of the technique consists of a calibration loop with dual catch tanks. A flow gate diverts the flow between the two tanks, A and B. Tank A is a large tank from which fluid is pumped back to a constant head reservoir which supplies the loop with a steady flow. Tank B is the catch-and-weigh tank into which liquid can be diverted for an accurately determined period of time. The liquid volume is measured, either directly, or indirectly through its weight, and flow rate deduced through time. The ability to determine the volume and the uncertainty in the initial and final time of diversion of the liquid are fundamental limitations to the accuracy of this technique. Neglecting installation effects, the ultimate limits of uncertainty (at 95%) in flow rate of liquids are on the order of 0.03%, a number based on $u_m \approx 0.02\%$, $u_\rho \approx 0.02\%$, and $u_t \approx 0.01\%$.

Flow meter calibration by velocity profile determination is particularly effective for in situ calibration in both liquids and gases, provided that the gas velocity does not exceed about 70% of the sonic velocity. Velocity traverses at any cross-sectional location some 20 to 40 pipe diameters downstream of any pipe fitting in a long section of straight pipe are preferred.

Comparison calibration by insertion of a local standard flow meter into a calibration loop is another common means of establishing the flow rate through a system. Other flow meters can be installed in tandem with the standard and directly calibrated against it. Such a method has the advantage of being useful with either liquids or gases. Turbine and vortex meters and Coriolis mass flow meters have consistent, highly accurate calibration curves and are often used as local standards. Other provers use an accurate positive displacement meter to determine flow volume over time. Of course, these standards must be periodically recalibrated. In the United States, NIST maintains flow meter calibration facilities for this purpose. However, installation effects in the end-user facility will not be accounted for in an NIST calibration. This last point accounts for much of the uncertainty in a flow meter calibration. Lastly, a sonic nozzle can also be used as a standard to establish the flow rate of a gas in a comparison calibration. In any of these methods, the calibration uncertainty is limited by the standard used, installation effects, and the inherent limitations of the flow meter calibrated.

9.9   SUMMARY

Flow quantification has been an important engineering task for well over two millennia. In this chapter, methods to determine the volume rate of flow and the mass rate of flow were presented. The engineering decision involving the selection of a particular meter was found to depend on a number of constraining factors. In general, flow rate can be determined to within about ±0.25% of actual flow rate with the best of present technology. However, new methods may push these limits even further, provided that calibration standards can be developed to document method accuracies and the effects of installation.

REFERENCES

1. da Vinci, L., *Del Moto e Misura Dell'Aqua*, E. Carusi and A. Favaro, eds., Bologna, 1924.
2. *Flow Meters, Theory and Applications*, 6th ed., American Society of Mechanical Engineers, New York, 1971.
3. *Flow of Fluids through Valves and Fittings*, Technical Paper No. 410, Crane Co., Chicago, 1969.
4. *ASHRAE Fundamentals*, American Society of Heating, Refrigeration and Air Conditioning Engineers, 1977.
5. Rouse, H., and S. Ince, *History of Hydraulics*, Dover, New York, 1957.
6. Arnberg, B. T., A review of critical flowmeters for gas flow measurements, *Transactions of the ASME*, 84:447, 1962.
7. Shercliff, J. A., *Theory of Electromagnetic Flow Measurement*, Cambridge University Press, New York, 1962.
8. Hochreiter, H. M., Dimensionless correlation of coefficients of turbine-type flowmeters, *Transactions of the ASME*, 1363, October 1958.
9. Lee, W. F., and H. Karlby, A study of viscosity effects and its compensation on turbine flow meters, *Journal of Basic Engineering*, 717–728, September 1960.
10. Hinze, J. O., *Turbulence*, McGraw-Hill, New York, 1953.
11. Mattingly, G., Fluid measurements: Standards, calibrations and traceabilities, Proc. ASME/AIChE National Heat Transfer Conference, Philadelphia, PA, 1989.
12. *Measurement of Fluid Flow in Pipes Using Orifice, Nozzle and Venturi*, ASME Standard MFC-3M-1985, American Society of Mechanical Engineers, New York, 1985.

NOMENCLATURE

d	diameter $[l]$	$h_{L_{1-2}}$	energy head loss between points 1 and 2 $[l]$
d_0	flow meter throat or minimum diameter $[l]$		
d_1	upstream pipe diameter $[l]$	k	ratio of specific heats
		$\dot{m}$	mass flow rate $[m\text{-}t^{-1}]$
d_2	diameter (see Figure 9.4) $[l^2]$	p	pressure $[m\text{-}l^{-1}\text{-}t^{-2}]$
f	friction factor		
g	gravitational acceleration constant $[l\text{-}t^{-2}]$	$p_1 - p_2, \Delta p$	pressure drop $[m\text{-}l^{-1}\text{-}t^{-2}]$

$(\Delta p)_{loss}$ permanent pressure loss $[m\text{-}l^{-1}\text{-}t^{-2}]$

r radial coordinate $[l]$

r_1 pipe radius $[l]$

r_H hydraulic radius $[l]$

A area $[l^2]$

A_0 area based on d_0 $[l^2]$

A_1 area baed on d_1 $[l^2]$

A_2 area based on d_2 $[l^2]$

B   magnetic field flux vector

C discharge coefficient

C_D drag coefficient

E velocity of approach factor; voltage

Gr   Grashof number

H manometer deflection $[l]$

K_0 flow coefficient

K_1 flow meter constant

L length $[l]$

Q volume flow rate $[l^3\text{-}t^{-1}]$

R gas constant $[l^2\text{-}t^{-2}\text{-}°]$

Re_{d_1} Reynolds number, $\mathrm{Re} = \dfrac{\overline{U}d_1}{\nu}$

S specific gravity

T temperature $[°]$; torque $[m\text{-}l^2\text{-}t^{-2}]$

U velocity $[l\text{-}t^{-2}]$

V volume $[l^3]$

Y expansion factor

β diameter ratio

δ twist angle

ν kinematic viscosity $[l^2\text{-}t^{-1}]$

η rotational speed $[t^{-1}]$

ρ density $[m\text{-}l^{-3}]$

ω circular frequency $[t^{-1}]$; angular velocity $[t^{-1}]$

PROBLEMS

9.1 Determine the average mass flow rate of 5 °C air at 14.7 psia through a 5-cm-i.d. pipe whose velocity profile is found to be symmetric and described by

$$U(r) = 25\left[1 - \left(\frac{r}{r_1}\right)^2\right] \text{ cm/s}$$

9.2 A 10-cm-i.d. pipe of flowing 10 °C air is traversed along three radial lines with measurements taken at five equidistant stations along each radial. Determine the best estimate of the pipe flow rate.

Location	Radial r (cm)	U(r) [cm/s] Line 1	U(r) [cm/s] Line 2	U(r) [cm/s] Line 3
1	1.0	25.31	24.75	25.10
2	3.0	22.48	22.20	22.68
3	5.0	21.66	21.53	21.79
4	7.0	15.24	13.20	14.28
5	9.0	5.12	6.72	5.35

9.3 A manometer is used in a 2-in. water line to measure the pressure drop across a flow meter. Mercury is used as the manometric fluid ($S = 13.57$). If the manometer deflection is 4 in. Hg, determine the pressure drop in lb/in.^{2}.

9.4 A bourdon tube pressure gauge indicates a 10-lb/in.^{2} drop across an orifice meter located in a 10-in. pipe containing flowing air at 90 °F. Find the equivalent pressure head in inches H_2O.

9.5 For a square-edged orifice plate using flange taps to meter the flow of O_2 at 60 °F through a 3-in.-i.d. pipe, the upstream pressure is 100 lb/in.^{2} and the downstream pressure is 76 lb/in.^{2}. If the orifice diameter is 1.5 in., can this flow be treated as incompressible for the purposes of estimating the flow rate? Demonstrate your reasoning. $R_{O_2} = 48.3$ ft-lb/lb$_m$-°R

9.6 Find the discharge coefficient of a 2-in.-diameter, square-edged orifice plate using flange taps and located in a 6-in. pipe if the Reynolds number is 250 000.

9.7 At what flow rate of 20 °C water through a 10-cm-i.d. pipe would the discharge coefficient of a square-edged orifice plate ($\beta = 0.4$) become essentially independent of Reynolds number?

9.8 Estimate the flange tap pressure drop across a square-edged orifice plate ($\beta = 0.5$) within a 12-cm-i.d. pipe if 25 °C water flows at 50 L/s. Compute the permanent pressure loss associated with the plate.

9.9 Determine the flow rate of 100 °F air through a 2-in. pipe if a flow nozzle with a 1-in. throat is used, the pressure drop measured is 30 in. H_2O, and the upstream pressure is 13.7 psia.

9.10 A square-edged orifice ($\beta = 0.5$) is used to meter N_2 at 520 °R through a 4-in. pipe. If the upstream pressure is 20 psia and the downstream pressure is 15 psia, determine the flow rate through the pipe. Flange taps are used. $R_{N_2} = 55.13$ ft-lb/lb$_m$-°R

9.11 Size a suitable orifice plate to meter water at 20 °C flowing through a 38-cm pipe if the nominal flow rate expected is 200 kg/s and the pressure drop must not exceed 15 cm Hg.

9.12 A venturi meter is to be used to meter the flow of 60 °F water through a 4-in. pipe. For a maximum differential pressure of 30 in. H_2O and a nominal 17-ft^3/min flow rate select a suitable throat size.

9.13 For 120 ft^3/m of 60 °F water flowing through a 6-in. pipe, size a suitable orifice, venturi, and nozzle flow meter if the maximum pressure drop cannot exceed 20 in. Hg. Estimate the permanent losses associated with each meter. Compare the annual operating cost associated with each if power cost is $0.10/kW-h, a 60% efficient pump motor is used, and the meter is operated 6000 h/year.

9.14 Estimate the flow rate of water through a 6-in.-i.d. pipe that contains an ASME long radius nozzle ($\beta = 0.6$) if the pressure drop across the nozzle is 10 in. Hg. Water temperature is 80 °F.

9.15 Select a location for a 2-in. square-edged orifice along a 10-ft straight run of 4-in. pipe if the orifice must be situated downstream of a 90° elbow. The orifice is to be operated uncalibrated under ASME codes.

9.16 An orifice ($\beta = 0.4$) is used as a sonic nozzle to meter a 100 °F air flow in a 2-in. pipe. If upstream pressure is 13.4 psia and downstream pressure is 6.4 psia, estimate the mass flow rate through the meter. $R = 53.3$ (ft-lb)/(lb$_m$-°R), $k = 1.4$.

9.17 Compute the flow rate of 20 °C air through a 0.5-m square-edged orifice plate that is situated in a 1.0-m-i.d. pipe. The pressure drop across the plate is 90 mm H$_2$O and upstream pressure is 2 atm.

9.18 An ASME long radius nozzle ($\beta = 0.5$) is to be used to meter the flow of 70 °F water through a 3-in.-i.d. pipe. If flow rates will range from 10 to 25 gal/min, select the range required of a single pressure transducer if that transducer will be used to measure pressure drop over the flow rate range. If the type of transducer to be used has a typical error of 0.25% of full scale, estimate the uncertainty in flow rate at the design stage.

9.19 A square-edged orifice plate is selected to meter the flow of water through a 2-in.-i.d. pipe over the range of 10 to 50 gal/min at 60 °F. It is desired to operate the orifice in the range where C is independent of the Reynolds number for all flow rates. A pressure transducer having an accuracy to within 0.5% of reading is to be used. Select a suitable orifice plate and estimate the design-stage uncertainty in measured flow rate at 10, 25, and 50 gal/min. Assume flange pressure taps and reasonable values for the bias limits on pertinent parameters.

9.20 Estimate the error contribution to the uncertainty in flow rate due to the effect of the relative humidity of air on air density. Consider the case of using an orifice plate to meter air flow if the relative humidity of the air can vary from 10 to 80% but a density equivalent to 45% relative humidity is used in the computations.

9.21 For Problem 9.20, suppose the air flow rate is 10 ft^3/m at 70 °F through a 2-in.-i.d. pipe, a square-edged orifice ($\beta = 0.4$) is used with flange taps, and the pressure drop can be measured to within ±0.5% of reading for all pressures above 2 in. H$_2$O using a manometer. If basic dimensions are maintained to within 0.005 in., estimate a design-stage uncertainty in the flow rate. Use $p_1 = 14$ psia, $R = 53.3$ ft-lb/lb$_m$-°R.

9.22 It is desired to regulate the flow of air through an air sampling device using a sonic nozzle. A flow rate of 45 ft^3/m relative to 70°F is to be maintained. For an ASME long radius nozzle, determine the downstream pressure required to choke the nozzle if air is supplied at 14.1 psia. Select a suitable nozzle throat diameter. $k = 1.4$, $R = 53.3$ (ft-lb)/(lb$_m$-°R).

9.23 A vortex flow meter uses a shedder having a Strouhal number of 0.20. Estimate the mean duct velocity if the shedding frequency indicated is 77 Hz and the shedder characteristic length is 0.5 in.

9.24 A thermal mass flow meter is used to meter 80 °F air flow through a 1-in.-i.d. pipe. If 25 W of power are required to maintain a 2 °F temperature rise across the meter, estimate the mass flow rate through the meter. Clearly state any assumptions. $c_P = 0.24$ BTU/(lb$_m$-°R).

9.25 When used with air (or any perfect gas), a sonic nozzle can be used to regulate volume flow rate. However, uncertainty arises due to variations in density brought on by local atmospheric pressure and temperature changes. For a range of 14.7 ± 1.0 psia and 50 ± 25 °F, estimate the error in regulating a 50-ft^3/min flow rate at critical pressure ratio due to these variations alone.

9.26 Estimate an uncertainty in the determined flow rate in Example 9.4 assuming that dimensions are known to within 0.001 in., that pressure is known to within 0.1 in. H$_2$O, and that the pressure drop shows a variance of 0.2 in. H$_2$O in 20 readings. Upstream pressure is constant. All assumptions should be justified and reasonable.

CHAPTER 10
METROLOGY, DISPLACEMENT, AND MOTION MEASUREMENTS

10.1   INTRODUCTION

In this chapter, a brief introduction to common techniques for dimensional measurements, displacement measurements, and mass, vibration, and acceleration measurements will be presented. These measurements each involve only the fundamental units of mass, length, and time.

The measurement of displacement, velocity, and acceleration for certain specific applications are referred to as vibration or shock measurements. The measurement of these quantities will be discussed in the context of a specific instrument design based on a second-order dynamic system, the seismic instrument.

10.2   DIMENSIONAL MEASUREMENTS: METROLOGY

The term *metrology* implies the science of weights and measures, referring primarily to the measurements of lengths, angles, and weight, but including other measurements essential to engineering practice such as the establishment of a flat, plane reference surface. Length is one of the most common specifications and measurements. Measurements of length span an amazing range, for example extending from the distance to the Great Nebula in Andromeda, which is 2×10^{22} m, to the radius of a hydrogen atom, 5×10^{-11} m. The tremendously difficult task of making accurate measurements on these scales requires highly specialized measurement techniques and systems that are not applicable to most engineering measurements. We wish to limit the scope of the present discussion to common dimensional measurements, such as would be necessary in design and manufacturing.

Several concepts based on accurate dimensional measurement that are now generally accepted as commonplace are actually quite recent. We certainly expect to be able to purchase standard threaded bolts and nuts that are completely interchangeable, and even to be able to purchase precision parts for automobiles, such as pistons and camshafts, which are finished to such exacting standards that they are completely interchangeable. In fact, mass production requires the ability to accurately measure specified dimensions. The techniques and instruments for accurate dimensional measurements have not always existed in manufacturing, and could only be achieved through the adoption of common standards for length, and the development of sufficiently accurate measurement techniques and instruments.

The need for accurately reproducible weights and measure has been recognized since ancient times. For example, the ancient Egyptians maintained a primary standard measuring stick that defined the unit known as the cubit. More modern development of accurate measuring instruments and machine tools in American industry can be traced to the early nineteenth century. The Brown and Sharpe company was founded in 1833, and by 1851 was manufacturing steel rules, scales, and vernier calipers [1]. Industrial practice in this period was based upon tailoring individual parts to fit each item produced. The firearms industry was the first to attempt to produce interchangeable parts, demonstrating completely inter-changeable parts for rifles around 1850. The desire to manufacture parts to specified tolerances, and thus eliminate manually fitting parts to a specific item, prompted the development of standards for dimensional measurements. These standards were first established within a particular company, but it soon became apparent that universal standards were needed.

The gauge block is the major innovation that allowed a dimensional standard to exist in a practical sense on every shop floor. Gauge blocks were developed in the 1890s in Sweden by Carl Edvard Johansson (1864–1934). He developed the concept of using precisely dimensioned blocks in combination to produce a wide range of dimensions. His original concept entailed 102 blocks that would allow measurement of 20 000 distinct dimensions through combinations of these gauge blocks.

Principles of Linear Measurement

Accurate measurements of length can be accomplished only through comparison with a standard, preferably one that is traceable to a primary standard. Local standards for the measurement of length, such as line standards, end standards, and gauge blocks, are essential to manufacturing. Two marks on a dimensionally stable material define length on a *line standard*. The lines are produced by a diamond tool to a width as small as 0.002 mm. The original SI[1] length standard was such a line standard. The distance between marks is estab-lished using interferometric methods. The length of an end standard is the distance between its flat, parallel end faces.

---

[1]The acronym SI is derived from the French *Le Systéme International d'Unités*.

FIGURE 10.1   Gauge block set. (Courtesy of The L. S. Starrett Company.)

Gauge blocks are the most often used length standard for machining processes, and are a necessary tool for dimensional quality control in the manufacture of interchangeable parts. Figure 10.1 shows a set of gauge blocks. The process whereby a length standard is established involves a unique property of highly polished, extremely flat surfaces on which there is a thin lubricant film. When such surfaces are slid together with a slight contact pressure the surfaces adhere with a significant force; this procedure for combining gauge blocks is called wringing. By combining gauge blocks, a range of lengths can be produced; for example, a set of 81 blocks can yield a range of lengths from 0.100 to 12.000 in. in increments of 0.001 in. Thus, a set of gauge blocks forms a useful and accurate local standard for length measurement. Table 10.1 lists the federally established accuracy standards and designations for gauge blocks.

Hand measuring tools form the basis for sufficiently accurate measurements of length for most engineering applications. The most common instruments for the measurement of length are undoubtedly the ruler and measuring tape. A measuring tape can be used to measure distances on the order of 100 feet with total uncertainties as low as 0.05%.

FIGURE 10.1   Continued.

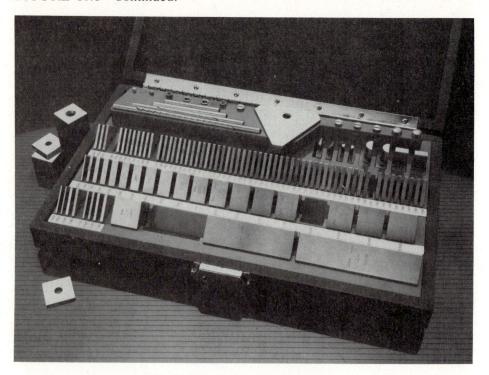

The sources of error in a ruler or tape include resolution errors and bias errors due to thermal expansion; in addition, error in a measurement using a tape may arise from improper tension in the tape. Typically, bias errors in the scale markings are negligible compared to the resolution errors in reading the scale.

EXAMPLE 10.1

A steel ruler 1 foot long is used to measure a length of 8 inches. The ruler is graduated in 1/16 inch increments and was manufactured in a controlled temperature environment of 77 °F. At this reference temperature, the ruler has

TABLE 10.1   Federal Accuracy Grades for Gauge Blocks

Federal Accuracy Grade	Accuracy [in.]	Accuracy [mm]
0.5	0.000001	0.00003
1	0.000002	0.00005
2	+0.000004	+0.0001
	−0.000002	−0.00005
3	+0.000006	+0.00015
	−0.000002	−0.00005

negligible bias error. Determine the design-stage uncertainty in a reading of 8 inches if the measurement is taken in a 100 °F environment, but not corrected for thermal expansion.

KNOWN

A ruler referenced to 77 °F is used in a 100 °F environment. The coefficient of thermal expansion of the steel is 9.9×10^{-6} in./in. °F.

FIND

The uncertainty in the measured length.

SOLUTION

The uncertainty analysis will include two contributions: a contribution due to thermal expansion, and a contribution due to interpolation errors. The interpolation uncertainty, u_o, is estimated in the usual manner as $\pm\frac{1}{2}$ the least division, which in this case is $\frac{1}{32}$ of an inch. The instrument uncertainty, u_c, contains only one elemental error and is found from the equation

$$\Delta L = (L)(C_\alpha)(\Delta T) \tag{10.1}$$

where

$$
\begin{aligned}
\Delta L &= \text{change in length due to thermal expansion} \\
L &= \text{length} \\
C_\alpha &= \text{coefficient of thermal expansion} \\
\Delta T &= \text{change in temperature}
\end{aligned}
$$

This relationship yields

$$\Delta L = (8 \text{ in.})(9.9 \times 10^{-6} \text{ in./in. °F})(100 - 77 \text{ °F}) = 1.82 \times 10^{-3} \text{ in.}$$

The uncertainty contributions may be combined as

$$u_d = \sqrt{(u_o)^2 + (u_c)^2} = \sqrt{(\tfrac{1}{32})^2 + (1.82 \times 10^{-3})^2} = 0.0313 \text{ in.}$$

which yields a design-stage uncertainty of 0.031 in.

COMMENTS

The uncertainty of 0.031 in. results entirely from the interpolation error of the ruler. Thermal expansion associated with a temperature change of 23 °F is negligible for this application. However, the error is approximately 0.002 in. over the 8-in. measured distance. In some applications, such an error would be totally unacceptable. Environmental controls for temperature are a common practice in machine tool manufacturing and precision machining facilities.

FIGURE 10.2   Vernier caliper. (Courtesy of The L. S. Starrett Company.)

Other measurement tools are designed for specific kinds of measurements, depending on the required accuracy and the specific geometry being measured. Vernier calipers, as shown in Figure 10.2, allow precise measurement of both inside and outside dimensions. Consider the construction of a Vernier scale, named for the French mathematician Pierre Vernier (1580–1673) who invented this method for increased resolution in measurements in the early seventeenth century. The basic principle of a Vernier scale is illustrated in Figure 10.3. The upper stationary scale is divided into equal lengths of one unit each. Suppose it is desired to further resolve the measured distance to the nearest $\frac{1}{4}$ unit. The movable Vernier scale would be graduated in four equal divisions of length $\frac{3}{4}$ of a unit. As the Vernier scale moves to the right, say between 2 and 3, at $2\frac{1}{4}$ the $\frac{1}{4}$ mark would exactly line up with the number 3 above it. Reading the Vernier scale is then accomplished by locating the line on the Vernier that is most closely aligned with a line on the main scale. In this case the reading is $2\frac{3}{4}$. Figure 10.4 shows a Vernier caliper and its associated reading.

A micrometer is commonly used for accurate measurements of a part that is small enough to be located between the frame and the spindle of the micrometer, as shown in Figure 10.5. The micrometer spindle moves by virtue of a precise screw thread, typically 40 threads per inch for U.S. customary units system

FIGURE 10.3   Operating principle for a Vernier scale.

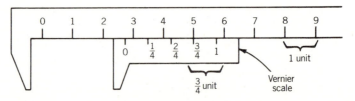

Length indicated is $2\frac{3}{4}$ units

FIGURE 10.4   Reading a Vernier scale. The metric reading is 27.42 mm. (Courtesy of The L. S. Starrett Company.)

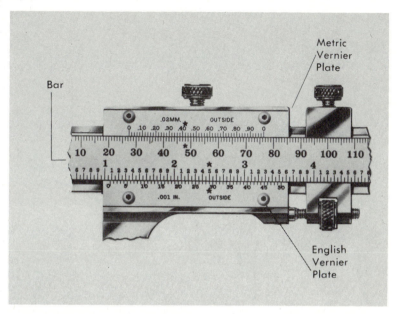

micrometers. Equipped with a Vernier scale, such a micrometer has a resolution of 0.0001 in. Further information on errors in micrometer measurements and calibration techniques may be found in [2].

Dial indicators measure distance of travel of a spindle; the construction of a typical dial indicator is shown in Figure 10.6. Dial indicators are most useful in setup work for machining, such as alignment of machine components, and in inspection of uniformly dimensioned parts. Dial indicators can be used as depth gauges and hole gauges, and may be employed in a variety of caliper gauges. The principle employed in a dial indicator provides the means of output for a variety of specialized measurement devices.

Optical Methods

The primary standard for length is currently based on the standard for time, such that one meter is the length of the path traveled by light in a vacuum in 1/299 792 458 of a second. However, until 1982 the standard for length was based on a specific number of wavelengths of light having a particular wavelength, specifically one meter equaled 1 650 763.73 wavelengths of the orange-red light of a krypton-86 lamp.

One of the most accurate means of measuring length currently is through the use of an interferometer [3]. The history of the interferometer includes the progress of the standard for length measurement from a single prototype line standard to a universal light wave standard. The meter was originally defined by a platinum–iridium bar, called the International Prototype Meter, from late in the eighteenth century to well into the twentieth century. The physicist Albert

FIGURE 10.5   Micrometer construction. (Courtesy of The L. S. Starrett Company.)

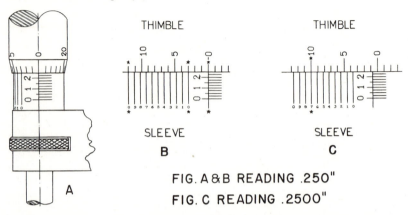

FIG. A & B READING .250"

FIG. C READING .2500"

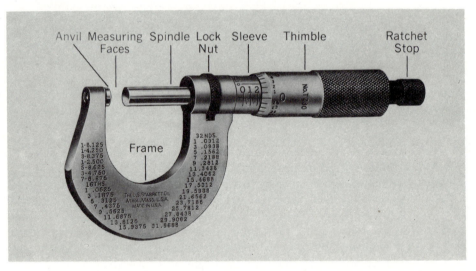

A. Michelson (1852–1931) designed an interferometer for length comparisons between the primary standard meter bar and copies, which were to be used as laboratory standards at locations remote from the primary standard. The discovery that these metal bars changed in length over time led to the adoption of a light wave standard for length in 1960. In France in 1799 the definition of the meter based on terrestrial distances was accurate to one part in 100 000. The next standard, the prototype platinum–iridium bar, was accurate to one part in one million, and current standards are accurate to approximately one part in 10^8.

Figure 10.7 shows a schematic diagram of an interferometer as first employed by Michelson. A light ray from the source is divided into two beams, one of which is incident upon mirror M_A and the other upon mirror M_B. These waves form an interference pattern that is visible at the location of the observer. As the mirror M_B is moved, interference fringes will pass the field of view; the number of interference fringes passing the field of view is a measure of the

FIGURE 10.6   Construction of a dial indicator. (Courtesy of The L. S. Starrett Company.)

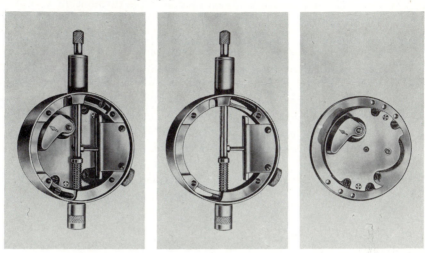

distance that M_B has moved. Thus, it is important to note that the interferometer measures the change in distance between the two mirrors, not the absolute distance to the movable mirror. Length measurements made in this manner are extremely accurate, since a large number of fringes are counted. Each fringe corresponds to a movement of the mirror of one-half of a wavelength. The advent of the laser as a light source has made it possible to use a laser interferometer to measure significant distances with resolution into the microinch range [4].

FIGURE 10.7   Schematic diagram of an interferometer.

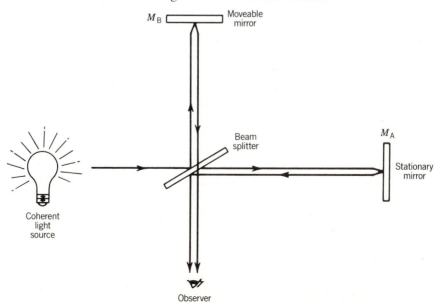

Optical means for length measurement also include measuring microscopes, which allow the measurement of small dimensions, typically less than 2 mm, by direct comparison with a scale that is visible through the reticle. A length measurement may also be accomplished by moving a telescope with a very precise, fine-thread screw that is calibrated to yield length measurements. A hairline in the telescope optics allows greater resolution in length measurements than can be accomplished using some other techniques, and allows for remote measurement. Such a device is called a cathetometer. Cathetometers consist of a vertical post mounted on a tripod that may be adjusted so that the post is oriented vertically. A telescope is mounted on a rack and pinion, such that the telescope can be moved vertically along the post. Using a scale on the post, and a Vernier scale that moves with the telescope, the vertical position of the telescope may be determined. Accuracies to ± 0.05 mm or ± 0.001 in. are possible. One application of a cathetometer is to remotely read a liquid level in a manometer.

10.3   DISPLACEMENT MEASUREMENTS

The transducers and physical principles used to measure displacement are highly dependent upon the particular application; as such, only the most common methods will be discussed. The measurement of displacement involves the determination of the relative motion of two points, of which one is usually fixed. The special subject of the measurement of the small displacements that occur in engineering materials under load will be reserved for treatment in Chapter 11.

Potentiometers

A potentiometer[2] or variable electrical resistance transducer is depicted in Figure 10.8. The transducer is composed of a sliding contact and a winding. The winding is made of many turns of wire, wrapped around a nonconducting substrate. Output signals from such a device can be realized by imposing a known voltage across the total resistance of the winding, and measuring the output voltage, which is proportional to the fraction of the distance the contact point has moved along the winding. Potentiometers can also be configured in a rotary form, with numerous total revolutions of the contact possible in a helical arrangement. The output from the sliding contact as it moves along the winding is actually discrete, as illustrated in Figure 10.8; the resolution is limited by the number of turns per unit distance. The loading errors associated with voltage-dividing circuits, discussed in Chapter 6, should be considered in choosing a measuring device for the output voltage.

---

[2]The potentiometer-transducer should not be confused with the potentiometer-instrument. Although both are based on voltage-divider principles, the latter measures emf as is described in Chapter 6.

FIGURE 10.8   Potentiometer construction.

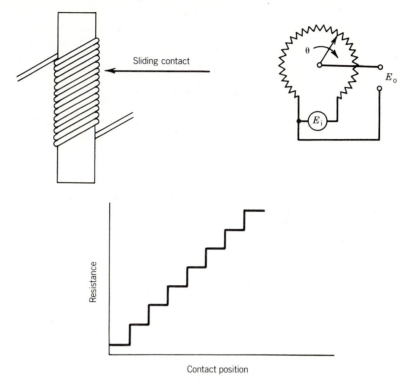

Contact position

Linear Variable Differential Transformers

The linear variable differential transformer (LVDT), as shown in Figure 10.9, produces an electrical output that is proportional to the displacement of a movable core. The movement of the core causes a mutual inductance in the secondary coils, for an ac voltage applied to the primary coil. In 1831 the English physicist Michael Faraday (1791–1867) demonstrated that a current could be induced in a conductor by a changing magnetic field. An interesting account of the development of the transformer may be found in [5].

Recall that for two coils in close proximity, a change in the current in one coil will induce an emf in the second coil according to Faraday's law. The application of this inductance principle to the measurement of distance begins by applying an ac voltage to the primary coil of the LVDT. The two secondary coils are connected in a series circuit, such that when the iron core is centered between the two secondary coils the output voltage is zero (see Figure 10.9). Motion of the magnetic core changes the mutual inductance of the coils, which causes a different emf to be induced in each of the two secondary coils. Over the range of operation, the output voltage is essentially linear with core displacement, as first noted in a U.S. patent by G. B. Hoadley in 1940 [6].

The output of a differential transformer is illustrated in Figure 10.10. Over a specific range of core motion the output is essentially linear. Beyond this linear range, the output voltage will rise in a nonlinear manner to a maximum, and eventually fall to zero. The output voltages on either side of the zero displace-

FIGURE 10.9   Construction of an LVDT. (Courtesy of
Schaevitz Engineering; from [6].)

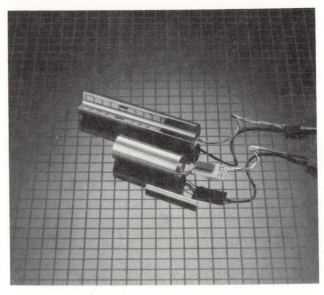

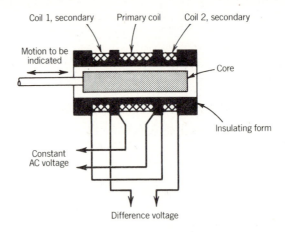

ment position are 180° out of phase. With appropriate circuitry it is possible to
determine positive or negative displacement of the core. However, note that
due to harmonic distortion in the supply voltage and the fact that the two
secondary coils are not identical, the output voltage with the coil centered is
not zero but instead reaches a minimum. Resolution of an LVDT strongly depends on the resolution of the measurement system used to determine its output.
Resolutions in the microinch range can be accomplished.

The differential voltage output of an LVDT, as shown in Figure 10.10, may
be analyzed by assuming that the magnetic field strengths are uniform along the
axis of the coils, neglecting end effects, and limiting the analysis to the case
where the core does not move beyond the ends of the coils [6]. Under these
conditions the differential voltage may be expressed in terms of the core displacement. The sensitivity of the LVDT in the linear range is a function of the

FIGURE 10.10   LVDT output as a function of core position.

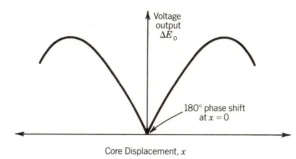

Core Displacement, x

number of turns in the primary and secondary coils, the root-mean-square (rms) current in the primary coil and the physical size of the LVDT. The output of an LVDT is not linear for all positions of the core, but in practical applications a range of core motion over which the nonlinearity is negligible is specified as the operating range.

Excitation Voltage and Frequency

The dynamic response of an LVDT is directly related to the frequency of the applied ac voltage, since the output voltage of the secondary coil is induced by the variation of the magnetic field induced by the primary coil. For this reason the excitation voltage should have a frequency at least 10 times the maximum frequency in the measured input. An LVDT can be designed to operate with input frequencies ranging from 60 Hz up to 25 kHz (for specialized applications, frequencies in the megahertz range can be used).

The maximum allowable applied voltage for an LVDT is determined by the current-carrying capacity of the primary coil, typically in the 1- to 10-V range. A constant current source is preferable for an LVDT, to limit temperature effects. For other than a sine wave input voltage form, harmonics in the input signal will increase the voltage output at the null position of the core.

The measurement of distance using an LVDT is accomplished using an assembly known as an LVDT gauge head. Such devices are widely used in machine tools and various types of gauging equipment. Control applications will have similar transducer designs. The basic construction is shown in Figure 10.11, which can yield instrument errors as low as 0.05% and repeatability of 0.0001 mm.

Angular displacement can also be measured using inductance techniques employing a rotary variable differential transformer (RVDT). The output curve of an RVDT and a typical construction are shown in Figure 10.12, where the linear output range is approximately ±40°.

10.4   MEASUREMENT OF MASS

Newton's second law states that the fundamental quantity force is proportional to the product of the mass and acceleration,

$$F = \frac{ma}{g_c} \tag{10.2}$$

FIGURE 10.11   LVDT gauge head. (Courtesy of Schaevitz Engineering; from [6].)

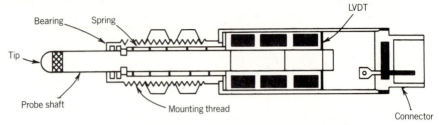

(*a*)   Cross section of typical LVDT gauge head.

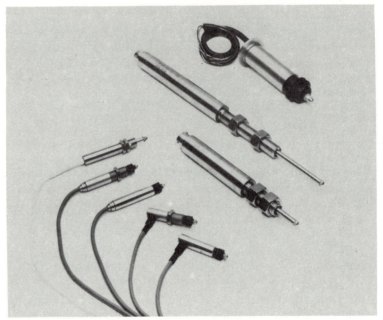

(*b*)   Several variations of LVDT gauge heads.

where g_c is the proportionality constant. Units systems vary in the choice of defined and derived units. In the SI system, the defined quantities are

Mass	kilogram
Length	meter
Time	second

and g_c is defined as 1.0 kg-m/s²-N. The resulting expression for Newton's second law does not explicitly require the inclusion of g_c. However, in the U.S. customary system, the defined quantities are

Mass	lb$_m$
Length	foot

FIGURE 10.12   Rotary variable differential transformer.
(Courtesy of Schaevitz Engineering; from [6].)

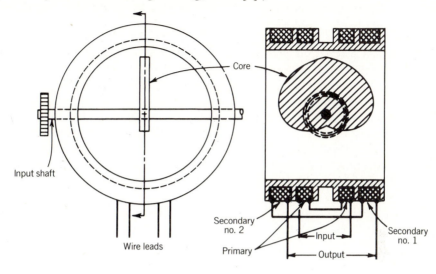

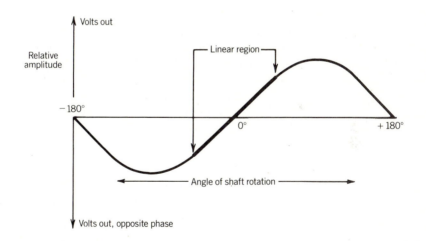

| Time | second |
| Force | lb (or lb_f) |

The units of force and mass are related by a definition: One lb_m exerts a force of 1 lb in a standard gravitational field. Thus Newton's second law, based on this definition, is

$$1 \text{ lb} = \frac{(1 \text{ lb}_m)(32.174 \text{ ft/s}^2)}{g_c}$$

Thus, g_c is seen to have a value of 32.174 ft-lb_m/lb-s^2. However, in the SI system g_c is defined as unity, and force becomes a derived unit, whereas in the U.S.

customary system the unit of force is the defined quantity, and g_c must be derived from Newton's second law.

In the mid-seventeenth century, Newton postulated the law of universal gravitation. This statement of the proportionality between mass and gravitational force allows the measurement of mass by comparison with the known gravitational force on a standard mass (or through the direct measurement of gravitational force under a known gravitational acceleration). The instrument for determining an unknown mass by comparison with a standard mass is known as an analytical balance. This null balance device compares the moment that results from the force exerted by a known mass in a gravitational field with the moment created by an unknown mass. Such a balance is shown in Figure 10.13.

To achieve accurate comparisons, such a balance must be properly designed and utilized. The sensitivity of the balance increases with increasing length of the balance arms, but decreases with increasing weight of the balance arm. The sensitivity is also approximately inversely proportional to the vertical distance from the point P to the center of gravity of the arm/pointer assembly, which must be below the point P for a stable system.

EXAMPLE 10.2

The mass of a Teflon® object is to be determined through direct comparison with a standard weight on an analytical balance. The density of the Teflon is 2200 kg/m³, and the density of the material from which the standard weights are constructed is 7900 kg/m³. Determine the error that would occur if the mass of the Teflon was measured without accounting for buoyancy forces. The comparison is made in air at 25 °C and 1 atm, and a standard mass of 0.4 kg is required to balance the unknown mass of Teflon.

KNOWN

An unknown mass of Teflon is in equilibrium with a standard mass, made of a material of known density, on an analytical balance.

FIGURE 10.13   Construction of an analytical balance.

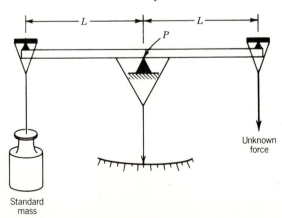

Standard
mass

Unknown
force

FIND

The effect of buoyancy forces on the measurement, expressed as the error resulting from ignoring buoyancy.

SOLUTION

A free-body diagram of the standard mass and the Teflon object results in three forces acting on each mass: the weight, W, the supporting force from the balance, F_{bal}, and the buoyancy force, F_{air}, as shown in Figure 10.14. The volume of the standard mass can be determined from

$$V_s = \frac{M_s}{\rho_s} = \frac{0.4 \text{ kg}}{7900 \text{ kg/m}^3} = 5.063 \times 10^{-5} \text{ m}^3$$

From the ideal gas equation of state, the density of air at 25 °C and 1 atm is calculated as 1.181 kg/m³. The buoyancy force on the standard mass is then

$$F_{air} = V_s \rho_{air} g = (5.063 \times 10^{-5} \text{ m}^3)(1.81 \text{ kg/m}^3)(9.8 \text{ m/s}^2)$$

$$= 5.8602 \times 10^{-4} \text{ N}$$

which yields the force that the standard mass exerts on the balance as 3.919 N (which corresponds to an effective mass of 0.39994 kg).

For the Teflon, the value of F_{bal} must be the same, if the balance is in static equilibrium, which requires

$$(F_{bal})_{Teflon} = (F_{bal})_{standard} = 3.919 \text{ N}$$

and by examining a free-body diagram of the Teflon or standard mass,

$$F_{bal} = W - F_{air}$$

FIGURE 10.14   Free-body diagram for Example 10.2.

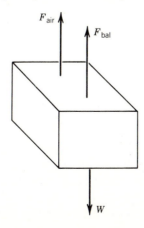

which implies

$$\rho_{\text{Teflon}} V_{\text{Teflon}} g - \rho_{\text{air}} V_{\text{Teflon}} g = F_{\text{bal}}$$

$$3.919 \text{ N} = V_{\text{Teflon}} g (\rho_{\text{Teflon}} - \rho_{\text{air}})$$

and yields for the volume of the Teflon

$$V_{\text{Teflon}} = \frac{3.919 \text{ N}}{(9.8 \text{ m/s}^2)(2200 - 1.181) \text{ kg/m}^3} = 1.819 \times 10^{-4} \text{ m}^3$$

Thus, the mass of the Teflon is 0.400155, or a difference of 0.039% between the corrected value of the mass and 0.4 kg. The analysis of the forces on the two masses can be combined to yield the general expression to correct for buoyancy forces as

$$m_u = \left(\frac{\rho_u}{\rho_u - \rho_{\text{air}}}\right)\left(1 - \frac{\rho_{\text{air}}}{\rho_s}\right) \tag{10.3}$$

where

$$m_u = \text{the unknown mass}$$
$$\rho_u = \text{density of the unknown mass}$$
$$\rho_{\text{air}} = \text{density of air at the local conditions}$$
$$\rho_s = \text{density of the standard mass}$$

Various adaptations of the analytical balance to eliminate the need for standard masses have been devised. Figure 10.15 shows an unequal arm balance and a schematic diagram of a pendulum scale. As previously stated, a balance compares the moment created by a standard mass to the moment created by an unknown mass. To achieve an equilibrium condition, any change in the system that changes the moment can be utilized. The moment about the pivot point for a balance may be expressed as

$$\mathbf{M} = \mathbf{r} \times \mathbf{F} \tag{10.4}$$

where

$$\mathbf{r} = \text{radius to the pivot (vector)}$$
$$\mathbf{F} = \text{force created by the mass}$$
$$\mathbf{M} = \text{moment}$$

The unequal arm balance changes the moment applied at the pivot by changing the position of the mass along the beam; the output is determined by a scale on the beam. For the pendulum scale, the vector cross product is affected by the change in the angle between the radius and the line of action of the force. The output of the pendulum scale is the angle, α. The cam changes the inherently

FIGURE 10.15 Unequal arm balance and pendulum scale.

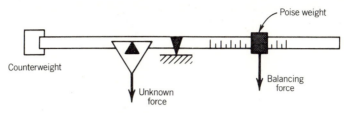

(a) Unequal arm balance.

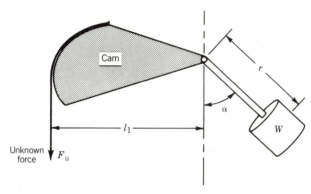

(b) Principle of operation of a pendulum scale.

nonlinear relationship between the unknown force and the angle into a linear one. For the pendulum scale, summing moments about the pivot yields

$$F_u l_1 = Wr \sin \alpha \qquad (10.5)$$

which may be directly solved for F_u.

10.5 MEASUREMENT OF ACCELERATION AND VIBRATION

The measurement of acceleration is required for a variety of purposes, ranging from machine design to guidance systems. Because of the range of applications for acceleration and vibration measurements, there exists a wide variety of transducers and measurement techniques, each associated with a particular application. In this chapter we address some fundamental aspects of these measurements along with some common applications.

Displacement, velocity, or acceleration measurements are also referred to as shock or vibration measurements, depending on the waveform of the forcing function that causes the acceleration. A forcing function that is periodic in nature generally results in accelerations that are analyzed as vibrations. On the other hand, a force input having a short duration and a large amplitude would be classified a shock load.

The fundamental aspects of acceleration, velocity, and displacement measurements can be discerned through examination of the most basic device for measuring acceleration and velocity, a seismic transducer.

Seismic Transducer

A seismic transducer consists of three basic elements, as shown in Figure 10.16: a spring–mass–damper system, a protective housing, and an appropriate output transducer. Through the appropriate design of the characteristics of this spring–mass–damper system, the output is a direct indication of either displacement or acceleration. To accomplish a specific measurement, this basic seismic transducer is rigidly attached to the object experiencing the motion that is to be measured.

Consider the case where the output transducer senses the position of the seismic mass; a variety of transducers could serve this function. Under some conditions, the displacement of the seismic mass serves as a direct measure of the acceleration of the housing, and the object to which it is attached. To illustrate the relation between the relative displacement of the seismic mass and acceleration, consider the case where the input to the seismic instrument is a constant acceleration. The response of the instrument is illustrated in Figure 10.17. At steady-state conditions, under this constant acceleration, the mass will be at rest with respect to the housing. The force required to accelerate the seismic mass is transmitted from the housing through the spring. The spring deflects an amount proportional to the force required to accelerate the seismic mass, and since the mass is known, Newton's second law yields the corresponding acceleration. The relationship between a constant acceleration and the displacement of the seismic mass is linear for a linear spring (where $F = kx$).

FIGURE 10.16   Seismic transducer.

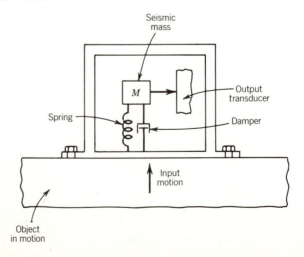

FIGURE 10.17   Response of a seismic transducer to a constant acceleration.

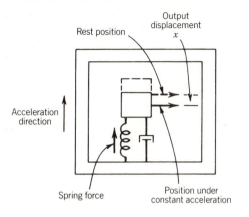
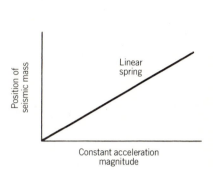

We wish to measure not only constant accelerations, but also complex acceleration waveforms. Recall from Chapter 2 that a complex waveform can be represented as a series of sine or cosine functions, and that by analyzing a measuring system response to a periodic waveform input we can discern the allowable range of frequency inputs. Consider an input to our seismic instrument such that the displacement of the housing is a sine wave, $y_h = A \sin \omega t$, and the absolute value of the resulting acceleration of the housing is $A\omega^2 \sin \omega t$. If we consider a free-body diagram of the seismic mass, the spring force and the damping force must balance the inertial force for the mass. (Notice that gravitational effects could play an important role in the analysis of this instrument, for instance, if the instrument were installed in an aircraft.) The spring force and damping force are proportional to the relative displacement and velocity between the housing and the mass, whereas the inertial force is dependent only on the absolute acceleration of the seismic mass. Since,

$$y_m = y_h + y_r \tag{10.6}$$

Newton's second law may be expressed

$$m \frac{d^2 y_m}{dt^2} + c \frac{dy_r}{dt} + k y_r = 0 \tag{10.7}$$

Substituting (10.6) in (10.7) yields

$$m \left(\frac{d^2 y_h}{dt^2} + \frac{d^2 y_r}{dt^2} \right) + c \frac{dy_r}{dt} + k y_r = 0 \tag{10.8}$$

But we know that $y_h = A \sin \omega t$ and

$$\frac{d^2 y_h}{dt^2} = -A \omega^2 \sin \omega t \tag{10.9}$$

Thus,

$$m \frac{d^2 y_r}{dt^2} + c \frac{dy_r}{dt} + k y_r = m A \omega^2 \sin \omega t \qquad (10.10)$$

This equation is identical in form to equation 3.11. As in the development for a second-order system response, we will examine the steady-state solution to this governing equation. The transducer will sense the relative motion between the seismic mass and the instrument housing. Thus, for the instrument to be effective, the value of y_r must provide indication of the desired output.

For the input function $y_h = A \sin \omega t$, the steady-state solution for y_r is

$$(y_r)_{\text{steady}} = \frac{(1/\omega_n^2) A \omega^2 \cos (\omega t - \phi)}{\{[1 - (\omega/\omega_n)^2]^2 + [2\zeta(\omega/\omega_n)]^2\}^{1/2}} \qquad (10.11)$$

where

$$\omega_n = \sqrt{\frac{k}{m}} \qquad \zeta = \frac{c}{2(km)^{1/2}} \qquad \phi = \tan^{-1} \frac{2\zeta(\omega/\omega_n)}{1 - (\omega/\omega_n)^2} \qquad (10.12)$$

The characteristics of this seismic instrument can now be discerned by examining equations 10.11 and 10.12. The natural frequency and damping will be fixed for a particular design. We wish to examine the motion of the seismic mass and the resulting output for a range of input frequencies.

For vibration measurements, it is desired to measure the amplitude of the displacements associated with the vibrations; thus, the desired behavior of the seismic instrument would be to have an output that gave a direct indication of y_h. For this to occur, the seismic mass should remain essentially stationary in an absolute frame of reference, and the housing and output transducer should move with the vibrating object. To determine the conditions under which this behavior would occur, the amplitude of y_r at steady state can be examined. The ratio of the maximum amplitude of the output divided by the equivalent static output in the present case would be $(y_r)_{\text{max}}/A$. For vibration measurements, this ratio should have a value of 1. Figure 10.18 shows $(y_r)_{\text{max}}/A$ as a function of the ratio of the input frequency to the natural frequency. Clearly, as the input frequency increases, the output amplitude, y_r, approaches the input amplitude, A, as desired. Thus, a seismic instrument that is to be used as a vibrometer should have a natural frequency smaller than the expected input frequency. Damping ratios near 0.7 are common for such an instrument. The seismic instrument designed for this application is called a vibrometer.

EXAMPLE 10.3

A seismic instrument like the one shown in Figure 10.16 is to be used to measure a periodic vibration having an amplitude of 0.05 in. and a frequency of 15 Hz.

FIGURE 10.18 Displacement amplitude at steady state
as a function of input frequency for a seismic transducer.

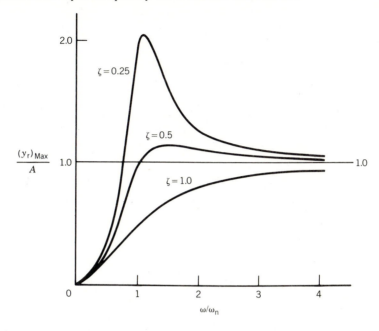

(a) Specify an appropriate combination of natural frequency and damping
ratio such that the amplitude error in the output is less than 5%.
(b) What spring constant and damping coefficient would yield these values
of natural frequency and damping ratio?
(c) Determine the phase lag for the output signal. Would the phase lag
change if the input frequency were changed?

KNOWN

Input function $y_h = 0.05 \sin 30\pi t$

FIND

Values of ω_n, ζ, k, m, and c to yield a measurement with less than 5%
magnitude error. Examine the phase response of the system.

SOLUTION

Numerous combinations of the mass, spring constant, and damping coefficient
would yield a workable design. Let's choose $m = 0.05$ lb$_m$ and $\zeta = 0.7$. We
know that

$$\omega_n = \sqrt{\frac{k}{m}} \qquad c_c = 2\sqrt{km}$$

for a spring–mass–damper system. With $\zeta = c/c_c = 0.7$, the damping coefficient, c, is found as

$$c = 2(0.7)\sqrt{km}$$

We can now examine the values of $(y_r)_{max}/A$ for $\zeta = 0.7$. The results are shown below, in tabular form:

$\dfrac{\omega}{\omega_n}$	$\dfrac{(y_r)_{max}}{A}$
10	1.000
8	1.000
6	1.000
4	0.999
3	0.996
2	0.975
1.7	0.951

Acceptable behavior is achieved for values of $\omega/\omega_n \geq 1.7$ for $\zeta = 0.7$, and an acceptable maximum value of the natural frequency is

$$\omega_n = \frac{\omega}{1.7} = \frac{30\pi}{1.7} = 55.4 \text{ rad/s}$$

Since[3]

$$\omega_n = \sqrt{\frac{k}{m}}$$

we find that with $m = 0.05$ lb$_m$, the value of k is 4.8 lb/ft. Then the value of the damping coefficient is found as

$$c = 2(0.7)\sqrt{km} = 0.12$$

The phase behavior for this system is given by equation 10.12, and results in the phase response tabulated as

$\dfrac{\omega}{\omega_n}$	$\phi \, [°]$
10	172
8	169.9
6	166.5
4	159.5
3	152.3
2	137.0
1.7	128.5

---

[3]Note that if k has units of lb/ft, and mass is in units of lb$_m$, g_c is required in the expression for ω_n.

COMMENTS

The design of such an instrument would have other constraints that would have to be considered in the choice of design parameters. The design would be influenced by such factors as size, cost, and operating environment. The phase behavior determined for this combination of design parameters does not result in a linear relationship between phase shift and frequency. As such, the possibility of distortion of complex waveform inputs does exist.

If it is desired to measure acceleration, the behavior of the seismic mass must be quite different. The amplitude of the acceleration input signal is $A\omega^2$. To have the output value y_r represent the acceleration, it is clear from equation 10.11 that the value of $y_r/A(\omega/\omega_n)^2$ must be a constant over the design range of input frequencies. If this is true, the output will be proportional to the acceleration. The amplitude of $y_r/A(\omega/\omega_n)^2$ can be expressed

$$\frac{(y_r)_{\text{steady}}}{A(\omega/\omega_n)^2} = \frac{\cos(\omega t - \phi)}{\{[1 - (\omega/\omega_n)^2]^2 + [2\zeta(\omega/\omega_n)]^2\}^{1/2}} \tag{10.13}$$

and

$$M(\omega) = \frac{1}{\{[1 - (\omega/\omega_n)^2]^2 + [2\zeta(\omega/\omega_n)]^2\}^{1/2}} \tag{10.14}$$

where $M(\omega)$ is the magnitude ratio as defined by equation 3.18. The magnitude ratio is plotted as a function of input frequency and damping ratio in Figure 3.15.

To achieve the desired behavior it is clear that the magnitude ratio should be unity. Over a range of input frequency ratios from 0 to 0.4, as determined from Figure 3.15, the magnitude ratio is approximately 1.0. Typically, in an accelerometer the damping ratio is designed to be near 0.7, so that the phase shift is linear with frequency, and distortion is minimized.

In summary, the seismic instrument can be designed so that the output can be interpreted in terms of either the input displacement or the input acceleration. Acceleration measurements may be integrated to yield velocity information; the differentiation of displacement data to determine velocity or acceleration introduces significantly more difficulties than the integration process.

Transducers for Shock and Vibration Measurement

In general, the destructive forces generated by vibration and shock are best quantified through the measurement of acceleration. Although a variety of accelerometers are available, strain gauge and piezoelectric transducers are widely employed for the measurement of shock and vibration [7].

A piezoelectric accelerometer employs the principles of a seismic transducer through the use of a piezoelectric element to provide a portion of the spring

FIGURE 10.19   Basic piezoelectric accelerometer.

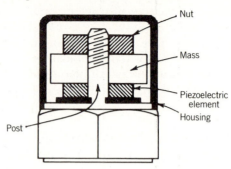

force. Figure 10.19 illustrates one basic construction of a piezoelectric acceler-ometer. A preload is applied to the piezoelectric element simply by tightening the nut that holds the mass and piezoelectric element in place. Upward or downward motion of the housing will change the compressive forces in the piezoelectric element, resulting in an appropriate output signal. Instruments are available with a range of frequency response from 0.03 to 10 000 Hz. Depending on the piezoelectric material used in the transducer construction, the static sensitivity can range from 1 to 100 mV/g. Steady accelerations cannot be effec-tively measured with such a piezoelectric transducer.

Strain gauge accelerometers are generally constructed using a mass supported by a flexure member, with the strain gauge sensing the deflection that results from an acceleration of the mass. The frequency response and range of accel-eration of these instruments are related, such that instruments designed for higher accelerations have a wider bandwidth, but significantly lower static sen-sitivity. Table 10.2 provides typical performance characteristics for two strain gauge accelerometers that employ semiconductor strain gauges.

Many other accelerometer designs are available, including potentiometric, reluctive, or accelerometers that use closed loop servo systems to provide a high output level. Piezoelectric transducers have the highest frequency response and range of acceleration, but have relatively lower sensitivity. Amplification can overcome this drawback to some degree. Semiconductor strain gauge transducers have a lower frequency response limit than piezoelectric, but can be used to measure steady accelerations. Other transducers generally have lower frequency response behaviors compared to piezoelectric and strain gauge units.

TABLE 10.2   Representative Performance Characteristics for Piezoresistive Accelerometers[a]

Characteristic	25-g Range	2500-g Range
Sensitivity [mV/g]	50	0.1
Resonance frequency [Hz]	2700	30 000
Damping ratio	0.4–0.7	0.03
Resistance [Ω]	1500	500

[a]Adapted from reference [7].

10.6 SUMMARY

The design and fabrication of any finished product requires the ability to ensure that the product meets design specifications; in part, this requires accurate dimensional measurements. Engineering measurements of size are accomplished using a variety of methods and instruments, with the most basic tools, such as calipers and micrometers, providing sufficient accuracy for most applications. Gauge blocks serve as a local standard for length, and are widely used in quality control for machining.

The measurement of displacement is used in construction of sensors for a variety of measured quantities. The most common means of displacement measurement include LVDTs and potentiometric methods. The seismic instrument provides a means for measuring acceleration and vibration, through appropriately chosen parameters for the spring–mass–damper system, which forms the sensor stage for the seismic instrument.

This chapter has provided a brief introduction to basic measurements associated with length and displacement. However, many applications and measurement techniques exist for the establishment of the flatness of surfaces, surface finish, curvature, and so on, all of which are related to dimensional measurement. Further information is found in [1,2].

REFERENCES

1. Moore, W. R., *Foundations of Mechanical Accuracy*, The Moore Special Tool Company, Bridgeport, CT, 1970.
2. Sharp, K. W. B., *Practical Engineering Metrology*, Pitman, London, 1970.
3. Pontius, P. E., *Measurement Assurance Program—A Case Study: Length Measurements*, Part 1, *Long Gage Blocks (5 to 20 in.)*, National Bureau of Standards Monograph 149, 1975.
4. Dukes, J. N., and G. B. Gordon, A two-hundred-foot yardstick with graduations every microinch, *Hewlett Packard Journal*, 1970; reprinted in *Inventions of Opportunity: Matching Technology with Market Needs*, Hewlett Packard Co., Palo Alto, CA, 1983.
5. Coltman, J. W., The transformer, *Scientific American*, 86, January 1988.
6. Herceg, E. E., *Schaevitz Handbook of Measurement and Control*, Schaevitz Engineering, Pennsauken, NJ, 1976.
7. Bredin, H., Measuring shock and vibration, *Mechanical Engineering*, 30, February 1983.

NOMENCLATURE

a	acceleration $[l\text{-}t^2]$	g_c	proportionality constant in Newton's second law
c	damping coefficient $[m\text{-}t]$		
c_c	critical damping coefficient $[m\text{-}t]$	k	spring constant $[m\text{-}t^2]$
		m	mass $[m]$
g	acceleration of gravity $[l\text{-}t^2]$	m_s	standard mass $[m]$

m_u unknown mass, equal arm balance $[m]$

$\mathbf{r}$ radius, vector $[l]$

t time $[t]$

u uncertainty

y_h displacement of housing, seismic instrument $[l]$

y_m displacement of seismic mass $[l]$

y_r relative displacement between seismic mass and housing in a seismic instrument $[l]$

A amplitude

C_α coefficient of thermal expansion $[°]$

E_i input voltage $[V]$

E_o output voltage $[V]$

$\mathbf{F}$ force, vector $[m\text{-}l\text{-}t^2]$

F_{air} buoyancy force of air $[m\text{-}l\text{-}t^2]$

F_{bal} force exerted by equal arm balance $[m\text{-}l\text{-}t^2]$

F_u unknown force $[m\text{-}l\text{-}t^2]$

L length $[l]$

ΔL change in length $[l]$

$\mathbf{M}$ moment, vector $[m\text{-}l^2\text{-}t^2]$

T temperature $[°]$

ΔT change in temperature $[°]$

V volume $[l^3]$

V_s volume of standard mass $[l^3]$

W weight $[m\text{-}l\text{-}t^2]$

α angle

ζ damping ratio

θ angle

ρ_s density of standard mass $[m\text{-}l^3]$

ρ_{air} density of air $[m\text{-}l^3]$

ϕ phase angle

ω_n natural frequency $[t^{-1}]$

PROBLEMS

10.1 Identify techniques for the measurement of lengths in the following size ranges, and specify the expected uncertainties:
a.   0.1 to 1 m
b.   20 to 100 Å
c. 10^6 miles
d.   1 km
e.   100 light-years

10.2 A steel tape that is graduated in $\frac{1}{16}$-in. increments is used to measure 2×6-in. planking for constructing a deck. Identify sources of error in the measurement. If the planks could be cut to the measured size with negligible error, within what range of values would you expect the length of a single plank to fall with 95% confidence? What errors are introduced in the cutting process?

10.3 Provide a list of possible error sources for use in uncertainty analysis for the following instruments having the specified range:
a.   Ruler (12 in.)
b.   Micrometer (2 in.)
c.   Interferometer (1 in.)
d.   Calipers (4 in.)

10.4 A micrometer has a least division of 0.0001 in. and is calibrated using two gauge blocks (wrung together) which form a value of 0.1501 in. with a stated accuracy for each block of $+4$ and -2 μin. If the micrometer is subsequently used to measure 0.1501 in., estimate the design-stage uncertainty in the measured length assuming only these elemental errors.

FIGURE 10.20   Circuit for Problem 10.5.

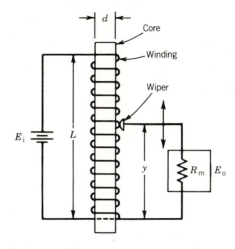

10.5 Consider a linear potentiometer as shown in Figure 10.8. The potentiometer consists of 0.1-mm copper wire ($\rho_e = 1.7 \times 10^{-8}$ Ω-m) wrapped around a core to form a total resistance of 1 kΩ. The sliding contact surface area is very small.

a.   Estimate the range of displacement that could be measured with this potentiometer for a 1.5-cm core.

b.   The circuit shown in Figure 10.20 is used to record position. On a single plot, show the loading error in an indicated displacement as a function of displacement, over the range found in (a), for values of the meter resistance, R_m, of 1, 10, and 100 kΩ. For practical meters, would the loading error be significant?

10.6 Determine the design-stage uncertainty in the measured force, F_u, for a pendulum scale having the following parameters and associated uncertainties:

$$l_1 = 25 \text{ cm} \pm 0.5 \text{ mm}$$
$$r = 12 \text{ cm} \pm 0.5 \text{ mm}$$
$$W = 350 \pm 0.1 \text{ g}$$
$$F_u = 100 \text{ g (nominally)}$$

All uncertainty values represent total uncertainties for the indicated variables. The uncertainty in the angle α may be assumed to be ± 0.02 rad.

10.7 A seismic instrument, as shown in Figure 10.16, is used to measure a vibration given by

$$y = 0.2 \cos 10t + 0.3 \cos 20t$$

where

$$y = \text{displacement in inches}$$
$$t = \text{time in seconds}$$

The seismic instrument is to have a damping ratio of 0.7, and a spring constant of 1.2 lb/ft.

a.   Select a combination of seismic mass and damping coefficient to yield less than a 10% amplitude error in measuring the input signal.

b.   Describe the phase response of the system, either in a plot or tabular form. Under what conditions would the output signal experience significant distortion?

10.8   A seismic instrument has a natural frequency of 20 Hz and a damping ratio of 0.65. Determine the maximum input frequency for a vibration such that the amplitude error in the indicated displacement is less than 5%.

10.9   Determine the bandwidth for an accelerometer having a seismic mass of 0.2 g and a spring constant of 20 000 N/m, with very low damping. Discuss the advantages of a high natural frequency and a low damping ratio. Piezoelectric sensors are well suited for the construction of accelerometers, since they possess these characteristics.

CHAPTER 11
STRAIN
MEASUREMENT

11.1 INTRODUCTION

The design of load-carrying components for machines and structures requires information concerning the distribution of forces within the particular component. Proper design of load-carrying devices, such as shafts, pressure vessels, and a variety of support structures must consider load-carrying capacity and allowable deflections. Mechanics of materials provides the basis for predicting these essential characteristics of a mechanical design, and provides the fundamental understanding of the behavior of load-carrying parts. However, for complex geometries and loadings, available theoretical analysis may not be sufficient, and experimental analysis is required to achieve a final design.

Our interest in this chapter is the measurement of physical displacements in engineering components, which impacts directly upon the mechanical design of load-carrying parts in machines or structures. Engineering designs are based on a safe level of stress within a material. In an object that is subject to loads, forces within the material act to balance the external load.

As a simple example, consider a slender rod that is placed in axial tension, as shown in Figure 11.1. If the rod is sectioned at $B - B$, a force within the material at $B - B$ is necessary to maintain static equilibrium for the sectioned rod. Such a force within the rod, per unit area, is called *stress*. Design criteria are based on stress levels within a part. In most cases stress cannot be measured directly. But the length of the rod in Figure 11.1 will change when the load is applied, and such changes in length or shape of a material can be measured. The stress is calculated from these measured deflections. Before we can proceed to develop techniques for these measurements, we will briefly review the relationship between deflections and stress.

11.2 STRESS AND STRAIN

The experimental analysis of stress is accomplished by measuring the deformation of a part under load, and inferring the existing state of stress from the measured deflections. Consider the rod in Figure 11.1. If the rod has a cross-

FIGURE 11.1   Free-body diagram illustrating internal
forces for a rod in uniaxial tension.

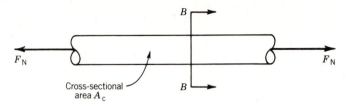

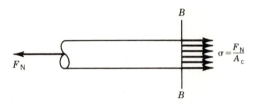

sectional area of A_c, and the load is applied only along the axis of the rod, the
normal stress is defined as

$$\sigma_a = \frac{F_N}{A_c} \tag{11.1}$$

where A_c is the cross-sectional area and F_N is the tension force applied to the
rod, normal to the area A_c. The ratio of the change in length of the rod (which
results from applying the load) to the original length is the axial strain, defined
as

$$\varepsilon_a = \frac{\delta L}{L} \tag{11.2}$$

where ε_a is the average strain over the length L, δL is the change in length, and
L is the original unloaded length. For most engineering materials, strain is a
small quantity; strain is usually reported in units of 10^{-6} in./in. or 10^{-6} m/m.
These units are equivalent to a dimensionless unit called a microstrain (μs).

Stress–strain diagrams are very important in understanding the behavior of a
material under load. Figure 11.2 is such a diagram for mild steel (a ductile
material). For loads less than that required to permanently deform the material,
most engineering materials display a linear relationship between stress and strain.
The range of stress over which this linear relationship holds is called the elastic
region. The relationship between uniaxial stress and strain for this elastic behavior is expressed as

$$\sigma_a = E_m \varepsilon_a \tag{11.3}$$

where E_m is the modulus of elasticity, or Young's modulus, and the relationship
is called Hooke's law. Hooke's law applies only over the range of applied stress

FIGURE 11.2   A typical stress–strain curve for mild
steel.

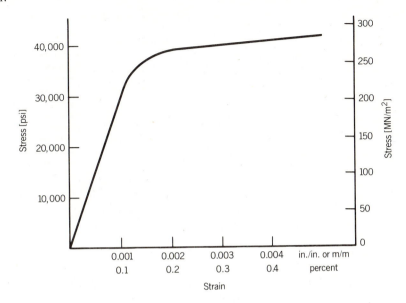

where the relationship between stress and strain is linear. Different materials
respond in a variety of ways to loads beyond the linear range, largely depending
on whether the material is ductile or brittle. For almost all engineering components, stress levels are designed to remain well below the elastic limit of the
material; thus, a direct linear relationship may be established between stress
and strain. Under this assumption, Hooke's law forms the basis for experimental
stress analysis, through the measurement of strain.

Consider the elongation of the rod shown in Figure 11.1, which occurs as a
result of the load F_N. As the rod is stretched in the axial direction, the cross-
sectional area must decrease, since the total mass (or volume for constant
density) must be conserved. Similarly, if the rod were compressed in the axial
direction, the cross-sectional area would increase. This change in cross-sectional
area is most conveniently expressed in terms of a lateral strain. For a circular
rod, the lateral strain is defined as the change in the diameter divided by the
original diameter. In the elastic range, there is a constant rate of change in the
lateral strain as the axial strain increases. In the same sense that the modulus
of elasticity is a property of a given material, the ratio of lateral strain to axial
strain is also a material property. This property is called Poisson's ratio, defined
as

$$\nu_p \equiv \frac{|\text{lateral strain}|}{|\text{axial strain}|} \qquad (11.4)$$

Engineering components are seldom subject to one-dimensional axial loading.
The relationship between stress and strain must be generalized to a multidi-
mensional case. Consider a two-dimensional geometry, as shown in Figure 11.3,

FIGURE 11.3   Biaxial state of stress.

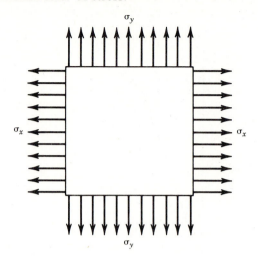

subject to tensile loads in both the x and y directions, resulting in normal stresses σ_x and σ_y. In this case, for a biaxial state of stress the stresses and strains are

$$\varepsilon_y = \frac{\sigma_y}{E_m} - v_p \frac{\sigma_x}{E_m} \qquad \varepsilon_x = \frac{\sigma_x}{E_m} - v_p \frac{\sigma_y}{E_m}$$

$$\sigma_x = \frac{E_m(\varepsilon_x + v_p\varepsilon_y)}{1 - v_p^2} \qquad \sigma_y = \frac{E_m(\varepsilon_y + v_p\varepsilon_x)}{1 - v_p^2} \tag{11.5}$$

$$\tau_{xy} = G\gamma_{xy}$$

In this case, all of the stress and strain components lie in the same plane. The state of stress in the elastic condition for a material is similarly related to the strains in a complete three-dimensional situation [1,2]. Since stress and strain are related, it is possible to determine stress from measured strains under appropriate conditions. However, strain measurements are made at the surface of an engineering component. The measurement yields information about the state of stress on the surface of the part. The analysis of measured strains requires application of the relationship between stress and strain at a surface. Such analysis of strain gauge data is described in [3]. Our emphasis in this chapter will be on techniques for the measurement of strain.

11.3   RESISTANCE STRAIN GAUGES

The measurement of the small displacements that occur in a material or object that is subject to a mechanical load determines the strain. Strain can be measured by methods as simple as observing the change in the distance between two scribe marks on the surface of a load carrying member, or as advanced as holography. In any case, the ideal sensor for the measurement of strain would:

1. Have good spatial resolution, implying that the sensor would measure strain at a point.

2. Be unaffected by changes in ambient conditions.
3. Have a high frequency response for dynamic strain measurements.

A common sensor that nearly attains these ideal characteristics is the bonded resistance strain gauge.

The most common method of measuring strain is through the change in electrical resistance that materials exhibit under load. Both metallic and semiconductor materials experience a change in electrical resistance when they are subjected to a strain. In an 1856 publication in the *Proceedings of the Royal Society* in England, Lord Kelvin (William Thomson) [4] laid the foundations for understanding the changes in electrical resistance that metals undergo when subjected to loads.

The resistance strain gauge is not only used for direct measurement of strain for the determination of stress, but also forms the basis for a variety of other transducers such as load cells, pressure transducers, and torque meters. Two individuals began the modern development of strain measurement in the late 1930s, Edward Simmons at the California Institute of Technology and Arthur Ruge at the Massachusetts Institute of Technology. Their development of the bonded metallic wire strain gauge lead to commercially available strain gauges.

Metallic Gauges

Consider a conductor having a uniform cross-sectional area, A_c, and a length, L, made of a material having a resistivity, ρ_c. For this electrical conductor, the resistance, R, is given by

$$R = \frac{\rho_c L}{A_c} \tag{11.6}$$

If the conductor is subjected to a normal stress along the axis of the wire, the cross-sectional area and the length will change, resulting in a change in the total electrical resistance, R. The total change in R is due to several effects, as illustrated in the total differential:

$$dR = \frac{A_c(\rho_c dL + L d\rho_c) - \rho_c L dA_c}{A_c^2} \tag{11.7}$$

which may be expressed in terms of Poisson's ratio as

$$\frac{dR}{R} = \frac{dL}{L}(1 + 2\nu_p) + \frac{d\rho_c}{\rho_c} \tag{11.8}$$

The changes in resistance are due to two basic effects: the change in geometry as the length and cross-sectional area change, and the change in the value of the resistivity, ρ_c. The dependence of resistivity on mechanical strain is called piezoresistance, and may be expressed in terms of a piezoresistance coefficient, π_1 defined by

$$\pi_1 = \frac{1}{E_m} \frac{d\rho_c/\rho_c}{dL/L} \tag{11.9}$$

With this definition, the change in resistance may be expressed

$$\frac{dR/R}{dL/L} = 1 + 2v_p + \pi_1 E_m \tag{11.10}$$

EXAMPLE 11.1

Determine the total resistance of a copper wire having a diameter of 1 mm and a length of 5 cm. The resistivity of copper is 1.7×10^{-8} Ω-m.

KNOWN

$D = 1$ mm
$L = 5$ cm
$\rho_c = 1.7 \times 10^{-8}$ Ω-m

FIND

The total electrical resistance.

SOLUTION

The resistance may be calculated from equation 11.6 as

$$R = \frac{\rho_c L}{A_c}$$

where

$$A_c = \frac{\pi}{4} D^2 = \frac{\pi}{4} (1 \times 10^{-3})^2 = 7.85 \times 10^{-7} \text{ m}^2$$

The resistance is then

$$R = \frac{(1.7 \times 10^{-8} \text{ } \Omega\text{-m})(5 \times 10^{-10} \text{ m}^2)}{7.85 \times 10^{-7} \text{ m}^2} = 1.08 \times 10^{-3} \text{ } \Omega$$

COMMENT

If the material were nickel instead of copper, for the same diameter and length of wire, what would the resistance be? The resistivity of nickel is 7.8×10^{-8} Ω-m, which results in a resistance of 5×10^{-3} Ω.

EXAMPLE 11.2

A very common material for the construction of strain gauges is the alloy constantan (55% copper with 45% nickel), having a resistivity of 49×10^{-8}

Ω-m. A typical strain gauge might have a resistance of 120 Ω. What length of constantan wire of diameter 0.025 mm would yield a resistance of 120 Ω?

KNOWN

The resistivity of constantan is 49×10^{-8} Ω-m.

FIND

The length of constantan wire needed to produce a total resistance of 120 Ω.

SOLUTION

From equation 11.6, we may solve for the length, which yields in this case

$$L = \frac{RA_c}{\rho_c} = \frac{(120 \; \Omega)(4.91 \times 10^{-10} \; \text{m}^2)}{49 \times 10^{-8} \; \Omega\text{-m}} = 0.12 \text{ m}$$

The wire would then be 12 cm in length to achieve a resistance of 120 Ω.

As shown in Example 11.2 single conductors cannot be used as strain gauges since the length of wire needed to achieve a reasonable total resistance would not allow sufficiently small measurement lengths for the strain gauge. Strain gauges are generally constructed by bending the conductor so that several lengths of wire are oriented along the axis of the strain gauge, as shown in Figure 11.4. For complex loadings, strain at the surface of an engineering component varies with position, and thus ideally the measurement of strain would take place at a single point on the surface. However, available strain sensors measure the average strain over a finite length called the *gauge length*.

Figure 11.5 illustrates the construction of a typical metallic-foil bonded strain gauge. Such a strain gauge consists of a metallic foil pattern that is formed in a manner similar to the process used to produce printed circuits. This photoetched

FIGURE 11.4   Detail of a basic strain gauge construction. (Courtesy of Micro-Measurements Division, Measurements Group, Inc. Raleigh, NC 27611.)

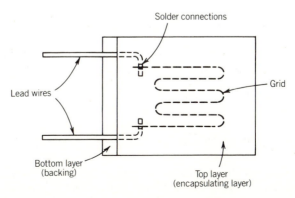

FIGURE 11.5 Construction of a typical metallic foil strain gauge. (Courtesy of Micro-Measurements Division, Measurements Group, Inc. Raleigh, NC 27611.)

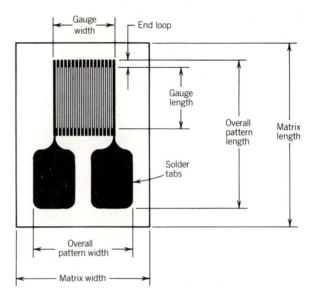

metal foil pattern is mounted on a plastic backing material. The gauge length, as illustrated in Figure 11.5, is an important aspect of selecting an appropriate strain gauge for a particular application. Strain is usually measured at the location on an engineering component where the stress is a maximum, and the stress gradients are high. The strain gauge averages the measured strain over the gauge length, and since the maximum strain is the quantity of interest, errors can result from improper choice of a gauge length [5].

The strain gauge backing serves several important functions. It electrically isolates the metallic gauge from the test specimen, and transmits the applied strain to the sensor. A bonded resistance strain gauge must be appropriately mounted to the specimen for which the strain is to be measured. The backing provides the surface used for bonding with an appropriate adhesive. Backing materials are available that are useful over temperatures that range from −270 to 290 °C.

The variety of conditions encountered in particular applications require special construction and mounting techniques, including design variations in the backing material, the grid configuration, bonding techniques, and total gauge electrical resistance. The adhesives used in the bonding process and the mounting techniques for a particular gauge and manufacturer will vary according to the specific application. However, there are some fundamental aspects that are common to all bonded resistance gauges.

The adhesive bond serves as a mechanical and thermal coupling between the metallic gauge and the test specimen. As such the strength of the adhesive should be sufficient to accurately transmit the strain experienced by the test specimen, and should have thermal conduction and expansion characteristics suitable for the application. If the adhesive shrinks or expands during the curing process, apparent strain can be created in the gauge. A wide array of adhesives are available for bonding strain gauges to a test specimen. Among these are epoxies, cellulose nitrate cement, and ceramic-based cements. Specific procedures and substances are required for proper bonding of strain gauges; however, with proper surface preparation and choice of adhesive, strain gauges can be successfully bonded to almost any surface.

If a strain gauge is placed in a state of stress where there are both normal and transverse components of strain, the total change in resistance of the gauge will be a result of both the axial and transverse components of strain. Strictly speaking, the relationship for change in resistance expressed in equation 11.8 is true only for a single conductor in a state of uniaxial tension or compression. The goal of the design of electrical resistance strain gauges is to create a sensor that is sensitive only to strain along one axis, and therefore has zero sensitivity to transverse and shearing strains. Actual strain gauges are sensitive to transverse strains to some degree; for most gauges the sensitivity to shearing strains may be neglected.

Gauge Factor

The change in resistance of a strain gauge is normally expressed in terms of an empirically determined parameter called the gauge factor, GF. For a particular strain gauge, the gauge factor is supplied by the manufacturer. The gauge factor can be expressed as

$$\text{GF} \equiv \frac{dR/R}{dL/L} \qquad (11.11)$$

Relating this definition to equation 11.10, we see that the gauge factor is dependent upon the Poisson ratio for the gauge material and its piezoresistivity. For metallic strain gauges, the Poisson ratio is approximately 0.3 and the resulting gauge factor is about 2.

The gauge factor represents the total change in resistance for a strain gauge, under a calibration loading condition. The calibration loading condition generally creates a biaxial strain field, and the transverse sensitivity of the gauge influences the measured result. Strictly speaking then, the sensitivity to normal strain of the material used in the gauge and the gauge factor are not the same. Generally gauge factors are measured in a biaxial strain field which results from the deflection of a beam having a value of Poisson's ratio of 0.285. Thus, for any other strain field there is an error in strain indication due to the transverse sensitivity of the strain gauge. The following equation expresses the percentage error due to transverse sensitivity for a strain gauge mounted on any material, at any orientation in the strain field:

$$e_t = \frac{K_t(\varepsilon_t/\varepsilon_a + \nu_{p0})}{1 - \nu_{p0}K_t} \times 100 \qquad (11.12)$$

where

$\varepsilon_a, \varepsilon_t$ = axial and transverse strains,
respectively (with respect the axis of the gauge)

ν_{p0} = Poisson's ratio of the material on which
the manufacturer measured GF (usually 0.285 for
steel)

e_t = error as a percentage of axial strain
(with respect to the axis of the gauge)

K_t = transverse sensitivity of the strain
gauge

Typical values of the transverse sensitivity for commercial strain gauges range
from 0.05 to -0.19. Figure 11.6 shows a plot of the percentage error for a strain

FIGURE 11.6   Strain measurement error due to strain
gauge transverse sensitivity. (Courtesy of Measurements
Group, Inc. Raleigh, NC 27611.)

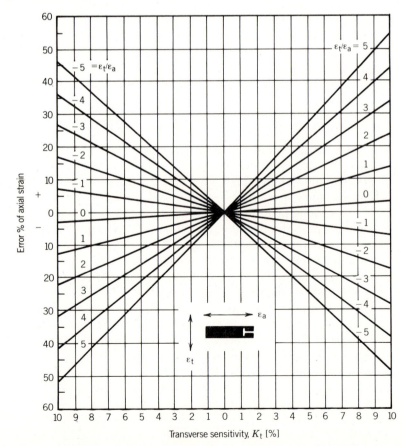

gauge as a function of the ratio of transverse loading to axial loading and the transverse sensitivity. It is possible to correct for the transverse sensitivity effects [6].

Semiconductor Strain Gauges

When subjected to a load, a semiconductor material exhibits a change in resistance, and therefore can be used for the measurement of strain. Silicon crystals are the basic material for semiconductor strain gauges; the crystals are sliced into very thin sections to form strain gauges. Mounting such gauges in a transducer, such as a pressure transducer, or on a test specimen requires backing and adhesive techniques similar to those used for metallic gauges. Because of the high piezoresistance coefficient, the semiconductor gauge exhibits a very high gauge factor, as high as 200 for some gauges. These gauges also exhibit higher resistance, longer fatigue life, and lower hysteresis under some conditions than metallic gauges. However, the output of the semiconductor strain gauge is nonlinear with strain, and the strain sensitivity or gauge factor may be markedly dependent on temperature.

In general, materials exhibit a change in resistivity with strain, characterized by the piezoresistance coefficient, π_1. For a semiconductor, this change in resistivity with strain can be very large. The effect of applied stress is to change the number and the mobility of the charge carriers within the material, thus causing large changes in the resistivity. Resistivity is a direct measure of the charge carrier density. The semiconductor crystals from which strain gauges are constructed contain a certain amount of impurities. These impurities control the number and mobility of the charge carriers within the gauge, and thus allow control of the gauge characteristics during the manufacturing process. Semiconductor materials for strain gauge applications have resistivities ranging from 0.001 to 1.0 Ω-cm. Semiconductor strain gauges may have a relatively high or low density of charge carriers [3,7]. Semiconductor strain gauges made of materials having a relatively high density of charge carriers($\approx 10^{20}$ carriers/cm^3) exhibit little variation of their gauge factor with strain or temperature. On the other hand, for the case where the crystal contains a low number of charge carriers (less than 10^{17} carriers/cm^3), the gauge factor may be approximated as

$$ \text{GF} = \frac{T_0}{T} \text{GF}_0 + C_1 \left(\frac{T_0}{T}\right)^2 \varepsilon \tag{11.13} $$

where GF_0 is the gauge factor at the reference temperature T_0, under conditions of zero strain [8], and C_1 is a constant for a particular gauge. The behavior of a high-resistivity P-type semiconductor is shown in Figure 11.7.

Because of the capability for producing small gauge lengths, silicon semiconductor strain gauge technology provides for the construction of very small transducers. These transducers may, for example, be used for the measurement of pressure and by virtue of their size would allow measurement of strain at higher frequencies than previously was possible. Some problems exist with environmental conditions for the use of silicon. For example, silicon diaphragm pressure transducers require special procedures for measuring in liquid environments. Semiconductor strain gauges are somewhat limited in the maximum strain that

FIGURE 11.7   Temperature effect on resistance for
various impurity concentrations for P-type semiconductors
(reference resistance at 81 °F). (Courtesy of Kulite
Semiconductor Products, Inc.)

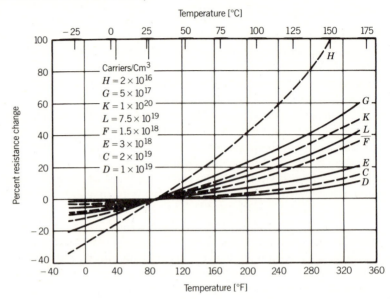

they can measure, approximately 5000 in./in. for tension, and higher in compression.

Semiconductor strain gauges find their primary application in the construction of transducers, such as load cells and pressure transducers. Because of the possibility of an inherent sensitivity to temperature, careful consideration must be made for each application to provide appropriate temperature compensation or correction. Temperature effects can result, for a particular measurement, in zero drift during the duration of a measurement.

11.4   STRAIN GAUGE ELECTRICAL CIRCUITS

The output of a bonded resistance strain gauge is a change in resistance, δR. A Wheatstone bridge is generally used to detect the small changes in resistance that form the output of a strain gauge measurement circuit. A typical strain gauge measuring installation on a steel specimen will have a sensitivity of 10^{-6} $\Omega/(kN\ m^{-2})$. As such, a high-sensitivity device is desirable for measuring resistance changes for strain gauges. Equipment is commercially available that can measure changes in gauge resistance of less than $0.0005\ \Omega$ ($0.000001\ \mu s$).

In many applications, strain measurements under both static and dynamic loading conditions are desired. A Wheatstone bridge circuit, or a similar bridge circuit, is the most often used means of measuring the resistance changes associated with static and dynamic loading of a strain gauge. The fundamental

FIGURE 11.8   Basic strain gauge Wheatstone bridge circuit.

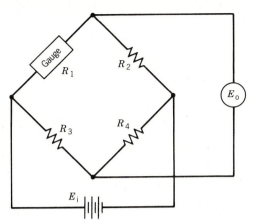

relationships for the analysis of such bridge circuits are discussed in Chapter 6. Their application to strain gauge circuits will be discussed in this section.

A simple strain gauge Wheatstone bridge circuit is shown in Figure 11.8. The bridge output under these conditions is given by equation 6.23 (repeated here):

$$E_\mathrm{o} = E_\mathrm{i} \frac{(R_1 + \delta R)R_4 - R_3 R_2}{(R_1 + \delta R + R_2)(R_3 + R_4)}$$

FIGURE 11.9   Balancing schemes for bridge circuits.

Circuit arrangement for shunt balance

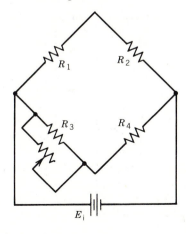

Differential shunt balance arrangement

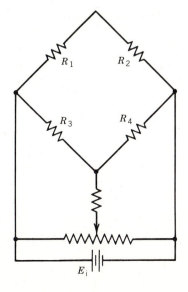

Consider the case where all the fixed resistors and the strain gauge resistance are initially equal, and the bridge is balanced. If the strain gauge is then subjected to a state of strain and experiences a change in resistance δR, the change in the output voltage will be

$$\frac{\delta E_{\mathrm{o}}}{E_{\mathrm{i}}} = \frac{\delta R/R}{4 + 2(\delta R/R)} \approx \frac{\delta R/R}{4} \tag{11.14}$$

The Wheatstone bridge has several distinct advantages for use with electrical resistance strain gauges. The bridge may be balanced by changing the resistance of one arm of the bridge. Therefore, once the gauge is mounted in place on the test specimen under a condition of zero loading, the output from the bridge may be zeroed. Two schemes for circuits to accomplish this balancing are shown in Figure 11.9. Shunt balancing provides the best arrangement for strain gauge applications, since the changes in resistance for a strain gauge are small.

EXAMPLE 11.3

A strain gauge, having a gauge factor of 2, is mounted on a rectangular steel bar ($E_{\mathrm{m}} = 200 \times 10^6$ kN/m^2), as shown in Figure 11.10. The bar is 3 cm wide and 1 cm high, and is subjected to a tensile force of 30 kN. Determine the resistance change of the strain gauge, if the resistance of the gauge was 120 Ω in the absence of the axial load.

KNOWN

GF $= 2$ $E_{\mathrm{m}} = 200 \times 10^6$ kN/m^2 $F_{\mathrm{N}} = 30$ kN

$R = 120\ \Omega$ $A_{\mathrm{c}} = 0.03$ m $\times$ 0.01 m

FIGURE 11.10   Strain gauge circuit subject to uniaxial tension.

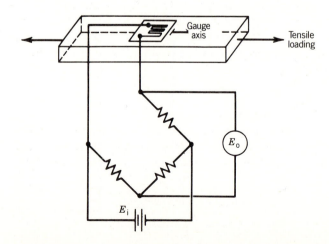

FIND

The resistance change of the strain gauge for a tensile force of 30 kN.

SOLUTION

The stress in the bar under this loading condition is

$$\sigma_a = \frac{F_N}{A_c} = \frac{30 \text{ kN}}{(0.03 \text{ m})(0.01 \text{ m})} = 1 \times 10^5 \text{ kN/m}^2 \qquad (11.15)$$

and the resulting strain is

$$\varepsilon_a = \frac{\sigma_a}{E_m} = \frac{1 \times 10^5 \text{ kN/m}^2}{200 \times 10^6 \text{ kN/m}^2} = 5 \times 10^{-4} \text{ m/m} \qquad (11.16)$$

By definition, the gauge factor relates the strain to the change in resistance of the strain gauge. For strain along the axis of the strain gauge, the change in resistance is

$$\frac{\delta R}{R} = \varepsilon GF$$

or

$$\delta R = R\varepsilon GF = (120 \text{ } \Omega)(5 \times 10^{-4})(2.0)$$
$$= 0.12 \text{ } \Omega$$

EXAMPLE 11.4

Suppose the strain gauge described in Example 11.3 is to be connected to a measurement device capable of determining a change in resistance with a stated accuracy to ± 0.05 Ω (95%). This stated accuracy includes a resolution of 0.01 Ω. What uncertainty in stress would result using this resistance measurement device?

KNOWN

A stress is to be inferred from a strain measurement using a strain gauge having a gauge factor of 2 and a zero load resistance of 120 Ω. The measurement of resistance has a stated accuracy to ± 0.05 Ω (95%).

FIND

The design-stage uncertainty in stress.

SOLUTION

The design-stage uncertainty in stress, $(u_d)_\sigma$, is given by

$$(u_d)_\sigma = \frac{\partial \sigma}{\partial(\delta R)} (u_d)_{\delta R}$$

with

$$\sigma = \varepsilon E_m = \frac{\delta R/R}{GF} E_m$$

Then with

$$\frac{\partial \sigma}{\partial(\delta R)} = \frac{E_m}{R(GF)}$$

we can express the uncertainty as

$$(u_d)_\sigma = \frac{E_m}{R(GF)} (u_d)_{\delta R} = \frac{200 \times 10^6 \text{ kN/m}^2}{120 \ \Omega(2.0)} (0.05 \ \Omega)$$

This results in a design-stage uncertainty in stress of $(u_d)_\sigma = 41\ 666.7$ kN/m^2 (95%).

11.5   PRACTICAL CONSIDERATIONS FOR STRAIN MEASUREMENT

The selection of strain gauges for specific applications involves considerations of the size, grid pattern, and gauge length of the strain gauge. In addition, the total resistance, backing material, dynamic response, and the hysteresis and zero drift of the gauge must usually be considered. This section will describe some characteristics of strain gauge applications that allow practical implementation of strain measurement.

Apparent Strain

A strain gauge will respond to inputs other than the strain experienced by the material on which it is mounted. *Apparent strain* is manifested as any change in resistance of the strain gauge that is not due to an applied stress in the material to which the gauge is bonded. The primary influence other than mechanical strain on the resistance is the temperature of the gauge. However, the gauge may be strained either by applied forces or differential thermal expansion. Differential thermal expansion between the gauge and the specimen on which it is mounted creates an apparent strain in the strain gauge. Thus, temperature sensitivity of strain gauges is a result of both the changes in resistance due to temperature changes in the gauge itself, and the strain experienced by the gauge due to differential thermal expansion between the gauge and the material on which it is mounted. The temperature sensitivity of a strain gauge is a significant obstacle to accurate mechanical strain measurements. With any resistance measurement technique, heating of the strain gauge as a result of current flow from the measuring device may be a source of significant error, since the gauge is also a temperature sensitive element.

Numerous methods have been devised to provide compensation for changes in temperature of the strain gauge during a particular measurement. Figure 11.11 shows two circuit arrangements that provide temperature compensation for a strain measurement. The strain gauge mounted on the test specimen experiences changes in resistance due to temperature changes and due to applied strain, while the compensating gauge experiences resistance changes due only to temperature changes. As long as the compensating gauge, as shown in Figure 11.11, experiences an identical thermal environment as the measuring gauge, temperature effects will be eliminated from the circuit.

Consider method 1, as shown in Figure 11.11a. The compensating gauge is assumed to experience identically the same thermal environment as the active gauge. Consider the case when all the bridge resistances are initially equal, and the bridge is therefore balanced. If the temperature of the strain gauges now changes, their resistance will change; suppose the temperature change produces

FIGURE 11.11   Bridge arrangements for temperature compensation.

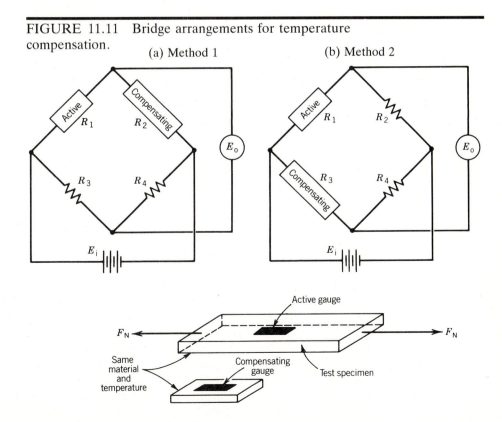

(a) Method 1   (b) Method 2

a small resistance change of a positive δR. The output of the bridge will be given by equation 11.17, for small δR:

$$E_o = E_i \frac{(R_1 + \delta R)\,R_4 - (R_2 + \delta R)R_3}{(R_1 + R_2)(R_3 + R_4)} \qquad (11.17)$$

where R_1, R_2, R_3, and R_4 are the initial, equal values of resistance. Under this condition, the output value will remain zero. Since this relationship holds in general, the compensating gauge serves to offset the effects due to temperature changes as long as the active and compensating gauges remain at the same temperature.

The sensitivity of the bridge arrangement in Figure 11.11a (method 1) is

$$K_B = \frac{E_o}{\varepsilon} = E_i \frac{R_1 R_2}{(R_1 + R_2)^2}\,GF \qquad (11.18)$$

and with $E_i = (2R_g)I_g$ and $R_g = R_1 = R_2$, the sensitivity may be expressed in terms of the current flowing through the gauge (I_g) as

$$K_B = \tfrac{1}{2}GF\sqrt{(I_g^2 R_1)R_1} \qquad (11.19)$$

Note that $(I_g)^2 R_1$ is the power dissipated in the strain gauge as a result of the bridge current. Excessive power dissipation in the gauge would cause temperature changes and introduce uncertainty into a strain measurement. These effects can be minimized by good thermal coupling between the strain gauge and the object to which it is bonded, to allow effective dissipation of thermal energy.

A second arrangement that also provides temperature compensation is shown in Figure 11.11b. With identical gauges at positions R_1 and R_3 and equal resistance changes for the two gauges, no change in bridge output would occur. However, the sensitivity for this arrangement is not the same as for method 1, but is given by

$$K_B = \frac{R_1/R_2}{1 + R_1/R_2}\,GF\sqrt{(I_g^2 R_1)R_1} \qquad (11.20)$$

Here the sensitivity is the same as for a bridge having a single active gauge and without temperature compensation. However, the sensitivity depends on the choice of the fixed resistor R_2. If $R_1 = R_2$, the resulting sensitivity will be the same as for method 1. However, the resistor R_2 can be chosen to provide the desired sensitivity for the circuit, within the limitations of measurement capability and allowable bridge current.

Although the apparent strain induced by temperature changes is the single most important error source in the measurement of strain using bonded resistance strain gauges, additional apparent strain can be created by differential thermal expansion between the strain gauge and the material to which it is bonded. Since the strain gauge is rigidly bonded to the part in which the strain is being measured, the strain gauge experiences dimensional changes due to temperature changes that are identical to the test specimen. The ideal solution to this problem is to construct a strain gauge that has a thermal expansion coefficient identical to that of the material in which the mechanical strain is being measured. However, the total temperature-induced strain is a result of the combined effects of differential thermal expansion and the temperature dependence of resistance. Self-temperature-compensated gauges can minimize the errors due to these effects [5]. This method allows the coefficient of thermal expansion of the strain gauge to be controlled through metallurgical processing. In this way the changes in resistance of the gauge due to differential thermal expansion and due to the material resistance dependence on temperature, taken together, result in acceptable magnitudes of the apparent strain, over a small range of temperature change. The desired value of the thermal expansion coefficient depends on the material to which the gauge is to be bonded in a specific application. Constantan is the material most commonly employed for self-temperature-compensated gauges. Manufacturers of strain gauges may also supply information for the correction of errors due to apparent strain.

Bridge Constant

A fundamental circuit analysis of any arrangement of strain gauges in a bridge circuit yields the relationship between the input strains and the output voltage of the bridge circuit. Consider the case when all four resistances in the bridge circuit of Figure 11.8 represent active strain gauges. In general, the bridge output is given by

$$E_o = E_i \left(\frac{R_1}{R_1 + R_2} - \frac{R_3}{R_3 + R_4} \right) \tag{11.21}$$

The strain gauges R_1, R_2, R_3, and R_4 are assumed initially to be in a state of zero strain. If these gauges are now subjected to strains such that the resistances change by δR_i, where $i = 1, 2, 3,$ and 4, then the change in the output voltage can be expressed

$$dE_o = \sum_{i=1}^{4} \frac{\partial E_o}{\partial R_i} dR_i$$

Evaluating the appropriate partial derivatives from equation 11.21 yields

$$dE_o = E_i \left[\frac{R_2 dR_1 - R_1 dR_2}{(R_1 + R_2)^2} + \frac{R_3 dR_4 - R_4 dR_3}{(R_3 + R_4)^2} \right] \tag{11.22}$$

Then from equations 11.2 and 11.11, $dR_i = R_i \epsilon_i GF_i$, and the value of dE_o can be determined. Assuming $dR_i \ll R_i$, the resulting change in output voltage may be expressed as

$$\delta E_o = E_i \left[\frac{R_1 R_2}{(R_1 + R_2)^2} (\epsilon_1 GF_1 - \epsilon_2 GF_2) \right.$$

$$\left. + \frac{R_3 R_4}{(R_3 + R_4)^2} (\epsilon_4 GF_4 - \epsilon_3 GF_3) \right] \tag{11.23}$$

If $R_1 = R_2 = R_3 = R_4$, then

$$\frac{\delta E_o}{E_i} = \frac{1}{4} \left[\varepsilon_1 GF_1 - \varepsilon_2 GF_2 + \varepsilon_4 GF_4 - \varepsilon_3 GF_3 \right] \qquad (11.24)$$

It is often possible to purchase matched sets of strain gauges for a particular application, so that $GF_1 = GF_2 = GF_3 = GF_4$, and

$$\frac{\delta E_o}{E_i} = \frac{GF}{4} (\varepsilon_1 - \varepsilon_2 + \varepsilon_4 - \varepsilon_3) \qquad (11.25)$$

(Compare this equation with 11.14).

Equation 11.25 shows that for a bridge containing four active strain gauges, equal strains on opposite bridge arms sum, whereas equal strains on adjacent arms of the bridge cancel. These characteristics can be used to increase the output of the bridge, to provide temperature compensation, or to cancel unwanted components of strain. Practical means of achieving these desirable characteristics will be explored further, after the concept of the bridge constant is developed.

Commonly used strain gauge bridge arrangements may be characterized by a *bridge constant*, κ, defined as the ratio of the actual bridge output to the output of a single gauge sensing the maximum strain, ε_{max}, (assuming the remaining bridge resistances remain fixed). The output for a single gauge experiencing the maximum strain may be expressed

$$\frac{\delta R}{R} = \varepsilon_{max} GF$$

So that, again for a single gauge,

$$\frac{\delta E_o}{E_i} = \frac{\varepsilon_{max} GF}{4 + 2\varepsilon_{max} GF} \qquad (11.26)$$

The bridge constant, κ, is found from the ratio of the actual bridge output to the output for a single gauge as given by equation 11.26.

EXAMPLE 11.5

Determine the bridge constant for two strain gauges mounted on a member in uniaxial tension, as shown in Figure 11.12. The member is subject to uniaxial tension, which produces an axial strain ε_a and a lateral strain $\varepsilon_L = -\varepsilon_a$. Assume that all the resistances in Figure 11.12 are initially equal, and therefore the bridge is initially balanced.

KNOWN

Strain gauge installation shown in Figure 11.12.

FIGURE 11.12   Bridge circuit with two arms active;
strain gauge installation for increased sensitivity.

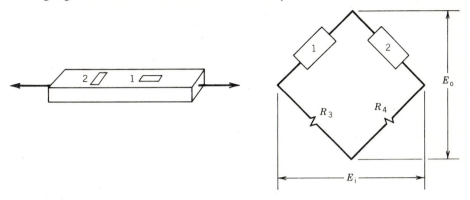

FIND

The bridge constant for this installation.

ASSUMPTIONS

The change in the strain gauge resistances are small compared to the initial resistance (see explanation below).

SOLUTION

If the gauges are mounted so that gauge 1 is aligned with the axial tension and gauge 2 is mounted transversely on the member, the output of the bridge will be greater than for gauge 1 alone. Since strain gauges generally experience small resistance changes, it is often convenient to develop approximate relationships for specific bridge circuit arrangements and strain gauge installations under this assumption.

The changes in resistance for the gauges may be expressed

$$\frac{\delta R_1}{R_1} = \varepsilon_a (GF)_1 \qquad (11.27)$$

and

$$\frac{\delta R_2}{R_2} = -\nu_p \varepsilon_a (GF)_2 = -\nu_p \frac{\delta R_1}{R_1} \qquad (11.28)$$

With only one gauge active, equation 11.14 (or equation 11.26) is applicable:

$$\frac{\delta E_o}{E_i} = \frac{\delta R_1 / R_1}{4 + 2(\delta R_1 / R_1)}$$

But with both gauges installed and active, as shown in Figure 11.12, the output of the bridge can be determined from a fundamental analysis of the bridge response, which results in

$$\frac{\delta E_o}{E_i} = \frac{(\delta R/R)(1 + \nu_p)}{4 + 2(\delta R/R)(1 - \nu_p)} \qquad (11.29)$$

Therefore, the bridge constant is

$$\kappa = \frac{(\delta R/R)(1 + \nu_p)/[4 + 2(\delta R/R)(1 - \nu_p)]}{(\delta R/R)/[4 + 2(\delta R/R)]} \qquad (11.30)$$

or

$$\kappa = \frac{(1 + \nu_p)[4 + 2(\delta R/R)]}{4 + 2(\delta R/R)(1 - \nu_p)} \qquad (11.31)$$

If $(\delta R/R) \ll 1$, then this relation may be reduced to

$$\kappa \cong 1 + \nu_p \qquad (11.32)$$

The use of two gauges oriented as described has increased the output from the bridge by a factor of $1 + \nu_p$.

Techniques for accomplishing temperature compensation, eliminating certain components of strain, and increasing the value of the bridge constant can be devised by examining more closely equation 11.25. The bridge constant is influenced by:

1. The location of strain gauges on the test specimen.
2. The connection scheme employed to construct a bridge circuit.

The combined effect of these two factors is determined by examining the existing strain field and using equation 11.25 to determine the resulting bridge output.

Consider a beam having a rectangular cross section and subject to the loading condition shown in Figure 11.13a where the beam is subject to an axial load F_N and a bending moment, M. The stress distribution in this cross section is given by

$$\sigma_x = \frac{-12My}{bh^3} + \frac{F_N}{bh} \qquad (11.33)$$

Identical strain gauges are mounted on the beam as shown in Figure 11.13b, and connected in bridge locations 1 and 4. The gauges experience equal but opposite bending strains (see equation 11.33), and both strain gauges are subject to the same axial strain caused by F_N. The bridge output under these conditions is

$$\frac{\delta E_o}{E_i} = \frac{GF}{4}(\varepsilon_1 + \varepsilon_4) \qquad (11.34)$$

FIGURE 11.13　Strain gauge installation for bending compensation.

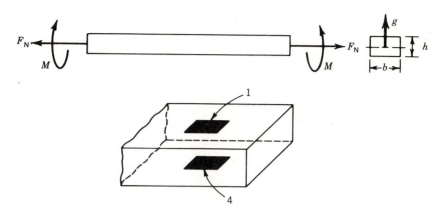

where $\varepsilon_1 = \varepsilon_{a_1} + \varepsilon_{b_1}$ and $\varepsilon_4 = \varepsilon_{a_4} - \varepsilon_{b_4}$ with subscripts a and b referring to axial and bending strain, respectively. Hence, the bending strains cancel but the axial strains will sum.

For a single gauge experiencing the maximum strain,

$$\frac{\delta E_o}{E_i} = \frac{GF}{4}\,\varepsilon_1$$

Therefore, the bridge constant is 2. Temperature compensation could be provided for this installation by having strain gauges that are the same temperature as gauges 1 and 4 and occupy arms 2 and 3 in the bridge.

Construction and Installation

Figure 11.14 shows a variety of strain gauge configurations. Practical construction of gauges is accomplished by techniques similar to those used in printed circuit board technology, and result in a thin-film metallic grid of conductors which form the strain gauge. The first metallic foil gauges were developed around 1950, and most commercial gauges are of this type. Standard gauge resistances are 120 and 350 Ω. Because of the accuracy of the process that produces the metal foil, a variety of patterns for the metal film are possible.

The operating assumption that leads to the definition of the gauge factor is that the change in resistance of the gauge is linear with applied strain, for a particular gauge. However, a strain gauge installed in a measurement environment will exhibit some nonlinearity. Also, in cycling between a loaded and unloaded condition, there will be some degree of hysteresis and a shift in the resistance for a state of zero strain. A typical cycle of loading and unloading is shown in Figure 11.15. The strain gauge will typically indicate lower values of strain during unloading than are measured as the load is increased. The extent of these behaviors is determined not only by the strain gauge characteristics, but also by the characteristics of the adhesive, and by the previous strains that the gauge has experienced. For properly installed gauges, the deviation from linearity should be on the order of 0.1% [3]. On the other hand, first-cycle

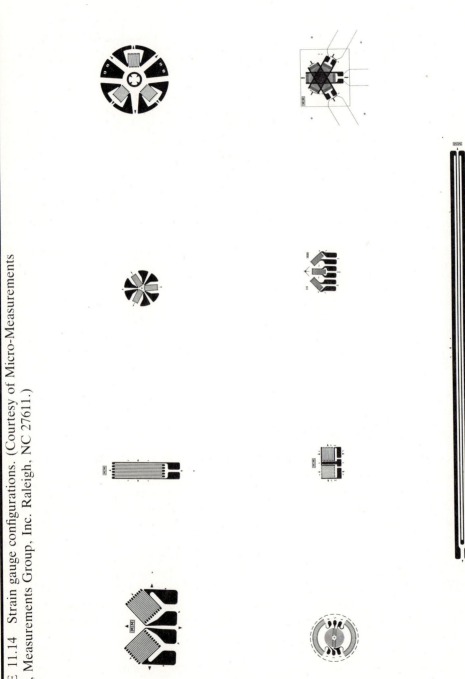

FIGURE 11.14 Strain gauge configurations. (Courtesy of Micro-Measurements Division, Measurements Group, Inc. Raleigh, NC 27611.)

FIGURE 11.15   Hysteresis in initial loading cycles for
two strain gauge materials (Courtesy of Micro-
Measurements Division, Measurements Group, Inc.
Raleigh, NC 27611.)

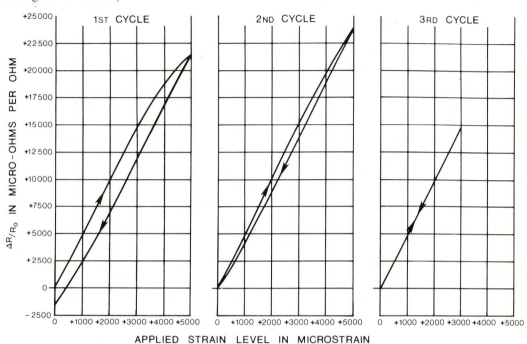

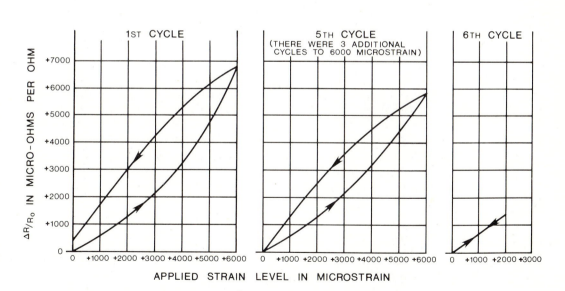

hysteresis and zero shift are difficult to predict. The effects of first-cycle hysteresis and zero shift can be minimized by cycling the strain gauge between zero strain and a value of strain above the maximum value to be measured.

In dynamic measurements of strain, the dynamic response of the strain gauge itself is generally not the limiting factor for such dynamic measurements. However, the transient response of bonded resistance strain gauges has been experimentally examined. A strain wave is made to propagate through a specimen to which a strain gauge is mounted, and the response of the gauge observed. The rise time (90%) of a bonded resistance strain gauge may be approximated as [9]

$$t_{90} \approx 0.8 \frac{L}{a} + 0.5 \ \mu s \qquad (11.35)$$

where L is the gauge length and a is the speed of sound in the material on which the gauge is mounted. Typical response times for gauges mounted on steel specimens are on the order of 1 μs.

Analysis of Strain Gauge Data

Strain gauges mounted on the surface of a test specimen respond only to the strains that occur at the surface of the test specimen. As such, the results from strain gauge measurements must be analyzed to determine the state of stress occurring at the strain gauge locations. The complete determination of the stress at a point on the surface of a particular test specimen will in general require the measurement of three strains at the point under consideration. The result of these measurements will yield the principal strains and allow determination of the maximum stress [3].

If more information is available concerning the expected state of stress, less than three strain gauges may be employed. When the directions of the principal axes are known in advance, only two strain measurements are necessary to calculate the maximum stress at the measured point. The multiple-element strain gauges used to measure more than one strain at a point are called strain gauge rosettes. An example of a two-element rosette is shown in Figure 11.16. For general measurements of strain and stress, commercially available strain rosettes can be chosen that have a pattern of multiple-direction gauges that is compatible with the specific nature of the particular application [5]. In practice, the applicability of the single-axis strain gauge is extremely limited, and improper use can result in large errors in the measured stress.

EXAMPLE 11.6

The strain developed at M different locations about a large test specimen is to be measured at each location using a setup similar to that shown in Figure 11.17. At each location similar strain gauges are appropriately mounted and connected to a Wheatstone bridge that is powered by an external supply. The bridge deflection voltage is to be amplified and measured. A similar setup is used at all M locations. The output from each amplifier is input through an M-channel multiplexer to an automated data acquisition system. If N (say 30)

FIGURE 11.16 Biaxial strain gauge rosettes. (Courtesy of Micro-Measurements Division, Measurements Group, Inc. Raleigh, NC 27611.)

(*a*)   Single-plane type.

(*b*)   Stacked type.

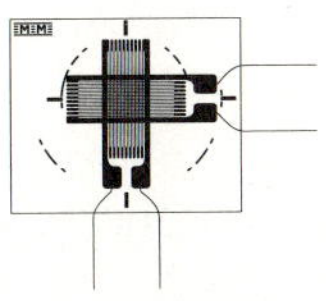

FIGURE 11.17   Data acquisition and reduction system
for Example 11.6.

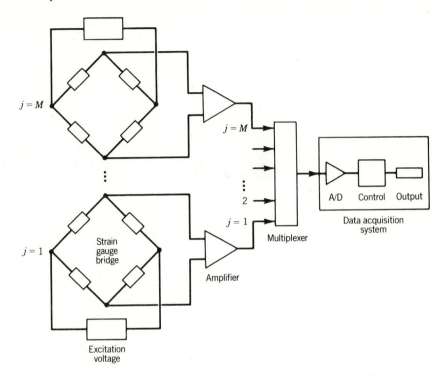

readings for each of the M setups are taken while the test specimen is maintained
in a condition of uniform zero strain, what information is obtained?

KNOWN

Strain setup of Figure 11.17
$M(j = 1, 2, \ldots, M)$ setups
$N(i = 1, 2, \ldots, N)$ readings per setup

ASSUMPTIONS

All setups are to be operated in a similar manner. Gauges are operated in
their linear regimes.

SOLUTION

Consider the data that will become available. First, each setup is exposed to
a similar strain. Hence, each setup should indicate the same strain. We can
calculate the pooled mean value of the $M \times N$ readings to obtain

$$\langle \overline{\varepsilon} \rangle = \frac{1}{MN} \sum_{j=1}^{M} \sum_{i=1}^{N} \varepsilon_{ij} = \frac{1}{M} \sum_{j=1}^{M} \overline{\varepsilon}_{j}$$

The difference between the pooled mean strain and the applied strain (zero here) is an estimate of the bias limit that can be expected from any channel during data acquisition.

Precision errors will manifest themselves through scatter in the data set. The pooled standard deviation

$$\langle S_\varepsilon \rangle = \sqrt{\frac{\sum\limits_{j=1}^{M}\sum\limits_{i=1}^{N}(\varepsilon_{ij} - \bar{\varepsilon}_j)^2}{M(N-1)}}$$

will provide a representative estimate of the precision error to be expected from any channel due to the data acquisition and reduction instrumentation.

COMMENT

This test yields an estimate of the bias and precision error due to propagation of the elemental errors amid M setups due to:

- Excitation voltage errors (differences in settings; variations)
- Amplifier error (differences in gain; noise)
- A/D converter, multiplexer, and conversion errors
- Computer errors (noise and roundoff)
- Bridge null errors
- Apparent strain error during the test
- Variations in gauge factors and gauge heating

The test data do not include temporal variation of the measurands or procedural variations under loading, instrument calibration errors, temperature variation effects and electrical noise induced by operation of the loading test of the specimen, dynamic effects on the gauges, including differences in creep and fatigue, or reduction curve fit errors.

11.6 OPTICAL STRAIN MEASURING TECHNIQUES

Optical methods for experimental stress analysis can provide fundamental information concerning directions and magnitudes of the stresses in parts under design loading conditions. Optical techniques have been developed for the measurement of stress and strain fields, either in models made of materials having appropriate optical properties, or through coating techniques for existing specimens. Photoelasticity takes advantage of the changes in optical properties of certain materials that occur when these materials are strained. For example, some plastics display a change in optical properties when strained that causes an incident beam of polarized light to be split into two polarized beams that travel with different speeds and that vibrate along the principal axes of stress. Since the two light beams are out of phase, they can be made to interfere; measuring the resulting light intensity yields information concerning applied stress. To implement this method, a model is constructed of an appropriate material, or a coating is applied to an existing part.

A second optical method of stress analysis is based on the development of a moiré pattern, which is an optical effect resulting from the transmission or reflection of light from two overlaid grid patterns. The fringes that result from relative displacement of the two grid patterns can be used to measure strain; each fringe corresponds to the locus of points of equal displacement.

Recent developments in strain measurement include the use of lasers and holography to very accurately determine whole field displacements for complex geometries.

Basic Characteristics of Light

To utilize optical strain measurement techniques, we must first examine some basic characteristics of light. Electromagnetic radiation, such as light, may be thought of as a transverse wave with sinusoidally oscillating electric and magnetic field vectors that are at right angles to the direction of propagation. In general, a light source emits a series of waves containing vibrations in all perpendicular planes, as illustrated in Figure 11.18. A light wave is said to be plane-polarized

FIGURE 11.18   Polarization of light. (Courtesy of Measurements Group, Inc. Raleigh, NC 27611.)

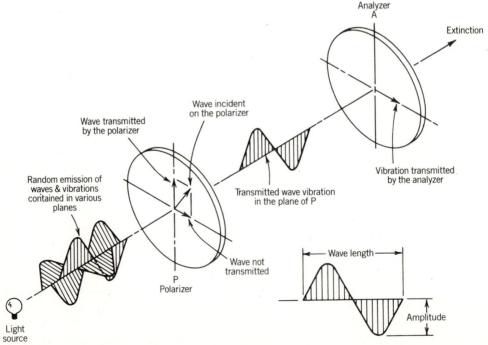

if the transverse oscillations of the electric field are parallel to each other at all points along the direction of propagation of the wave.

Figure 11.18 illustrates the effect of a polarizing filter on an incident light wave; the transmitted light will be plane polarized, with a known direction of polarization. Complete extinction of the light beam could be achieved by introduction of a second polarizing filter, with the axis of polarization at 90° to the first filter (labeled an Analyzer in Figure 11.18). These behaviors of light are employed to measure direction and magnitude of strain in photoelastic materials.

Photoelastic Measurement

Photoelastic methods of stress analysis take advantage of the anisotropic optical characteristics of some materials, notably plastics, when subject to an applied load. Stress analysis may be accomplished either by constructing a model of the part to be analyzed from a material selected for its optical properties, or by coating the actual part or prototype with a photoelastic coating. If a model is constructed from a suitable plastic, the required loads for the model are significantly less than the service loads of the actual part, which reduces effort and expense in testing.

The changes in optical properties, known as artificial birefringence, which occur in certain materials subject to a load or loads was first observed by Sir David Brewster [10] in 1815. He observed that when light passes through glass that is subject to uniaxial tension, such that the stress is perpendicular to the direction of propagation of the light, the glass becomes doubly refracting, with the axes of polarization in the glass aligned with and perpendicular to the stress. Maxwell [11] and Neumann [12] first put forward the mathematical observation that the relationship between artificial birefringence and applied stress or strain is linear. These relations are known as the stress–optic law.

The anisotropy that occurs in photoelastic materials results in two refracted beams of light and one reflected beam, produced for a single incident beam of appropriately polarized light. The two refracted beams propagate at different velocities through the material because of an anisotropy in the index of refraction. In an appropriately designed photoelastic (two-dimensional) model, these two refracted components of the incident light travel in the same direction and can be examined in a polariscope. The degree to which the two light waves are out of phase is related to the stress by the stress–optic law:

$$\delta \propto n_x - n_y \tag{11.36}$$

where

$$\delta = \text{relative retardation between the two light beams}$$
$$n_x, n_y = \text{indices of refraction in the directions of}$$
$$\text{the principal strains}$$

The index of refraction changes in direct proportion to the amount the material is strained, such that

$$n_x - n_y = K(\varepsilon_x - \varepsilon_y) \tag{11.37}$$

The strain optical coefficient, K, is generally assumed to be a material property that is independent of the wavelength of the incident light. However, if the photoelastic material is strained beyond the elastic limit, this constant may become wavelength dependent, a phenomenon known as photoelastic dispersion.

Figure 11.19 shows the use of a plane polariscope to examine the strain in a photoelastic model. Plane polarized light enters the specimen and emerges with two planes of polarization along the principal strain axes. This light beam is then passed through a polarizing filter, called the analyzer, which transmits only the component of each of the light waves that is parallel to the plane of polarization. The transmitted waves will interfere, since they are out of phase, and the resulting light intensity will be a function of the angle between the analyzer and the principal strain direction and the phase shift between the beams. The variations in strain in the specimen produce a pattern of fringes, which can be related to the strain field through the strain optic relation.

When a photoelastic model is observed in a plane polariscope, a series of fringes is observed. The complete extinction of light occurs at locations where the principal strain directions coincide with the axes of the analyzer or where either the strain is zero or $\epsilon_x - \epsilon_y = 0$. These fringes are termed isoclinics, and are used to determine the principal strain directions at all points in the photoelastic model. Figure 11.20 shows the isoclinics in a ring subject to a compression

FIGURE 11.19   Construction of a plane polariscope.
(Courtesy of Measurements Group, Inc. Raleigh, NC 27611.)

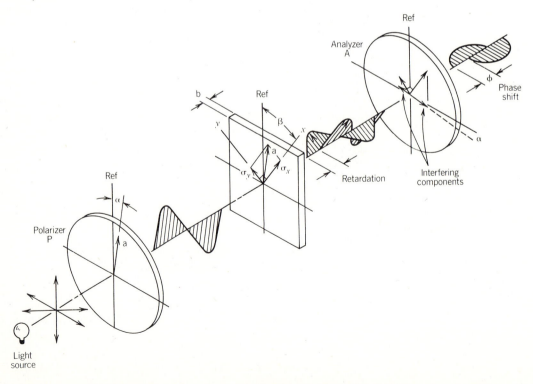

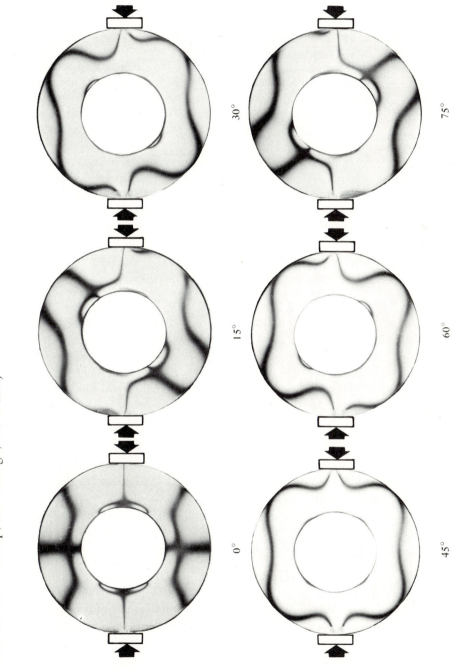

FIGURE 11.20 Isoclinic fringes in a ring loaded in compression. (Courtesy of Measurements Group, Inc. Raleigh, NC 27611.)

load (as shown in the figure). A reference direction is selected along the horizontal compression load and labeled 0°. For each measurement angle, one of the principal strains at a point on an isoclinic is parallel to the specified angle and the other is perpendicular. For the 0° isoclinics, the principal strains are oriented at 0 and 90°.

Using the fact that the direction of the principal axes is known at a free surface and the fact that the shear stress is zero on a free surface, the magnitude of the stress on the boundary can be determined. The primary applications for photoelasticity, especially in a historical sense, have been in the study of stress concentrations around holes or reentrant corners. In these cases, the maximum stress is at the boundary and corresponds to one of the principal stresses. This maximum stress can be obtained directly by the optical method, since the shear stress is zero on the boundary.

Optical methods provide information about the strain and stress at every point in the object being examined, in contrast to a strain gauge which supplies information about the strain at a single location on the object. The optical methods provide the possibility of identifying stress concentration locations and may allow for an improvement in design or guide detailed measurements with strain gauges.

Moiré Methods

A moiré pattern results from two overlaid, relatively dense patterns that are displaced relative to each other. This observable optical effect occurs, for example, in color printing where patterns of dots form an image. If the printing is slightly out of register, a moiré pattern will result. Another common example is the striking shimmering effect that occurs with some patterned clothing on television. This effect results when the size of the pattern in the fabric is essentially the same as the resolution of the television image.

In experimental mechanics, moiré patterns are used to measure surface displacements, typically in a model constructed specifically for this purpose. The technique uses two gratings, or patterns of parallel lines spaced equally apart. Figure 11.21 shows two line gratings. There are two important properties of line gratings for moiré techniques. The pitch is defined as the distance between the centers of adjacent lines in the grating, and for typical gratings has a value of from 1 to 40 lines/mm. The second characteristic of gratings is the ratio of the open, transparent area of the grating to the total area, or, for a line grating, the ratio of the distance between adjacent lines to the center-to-center distance, as illustrated in Figure 11.21. Clearly, a greater density of lines per unit width allows a greater sensitivity of strain measurement; however, as line densities increase, coherent light is required for practical measurement.

To determine strain using the moiré technique, a grating is fixed directly to the surface to be studied. This can be accomplished through photoengraving, cementing film copies of a grating to the surface, or interferometric techniques. The master or reference grating is next placed in contact with the surface, forming a reference for determining the relative displacements under loaded conditions. A series of fringes result when the gratings are displaced relative to each other; the bright fringes are the loci of points where the projected displacements of the surface are integer multiples of the pitch. The technique then is two-dimensional, providing information concerning the projection of the displacements

FIGURE 11.21 Moiré gratings.

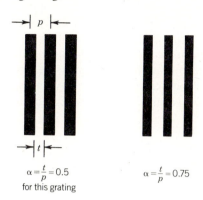

$$\alpha = \frac{t}{p} = 0.5$$
for this grating

$$\alpha = \frac{t}{p} = 0.75$$

into the plane of the master grating. Once a fringe pattern is recorded, data reduction techniques are employed to determine the stress and strain field. Graphical techniques exist that allow the strain components in two orthogonal directions to be determined. Further information on moiré techniques, and additional references may be found in the review article by Sciammarella [13].

Recently, techniques such as moiré-fringe multiplication have greatly increased the sensitivity of moiré techniques, with possible grating frequencies of 1200 lines/mm. Moiré interferometry is an extension of moiré-fringe multiplication that uses coherent light and has sensitivities on the order of 0.5 µm/fringe [14]. A reflective grating is applied to the specimen, which experiences deformation under load conditions. The technique provides whole field readings of in-plane strain, with a four-beam optical arrangement currently in use [15].

11.7 SUMMARY

Experimental stress analysis can be accomplished through several practical techniques, including electrical resistance, photoelastic, and moiré strain measurement techniques. Each of these methods yield information concerning the surface strains for a test specimen. The design and selection of an appropriate strain measurement system begins with the choice of a measurement technique.

Optical methods are useful in the initial determination of a stress field for complex geometries, and the determination of whole-field information in model studies. Such whole-field methods provide the basis for design and establish information necessary to make detailed local strain measurements.

The bonded electrical resistance strain gauge provides a versatile means of measuring strain at a specific location on a test specimen. Strain gauge selection involves the specification of strain gauge material, the backing or carrier material, and the adhesive used to bond the strain gauge to the test specimen, as well as the total electrical resistance of the gauge. Other considerations include the orientation and pattern for a strain gauge rosette, and the temperature limit and maximum allowable elongation. In addition, for electrical resistance strain gauges appropriate arrangement of the gauges in a bridge circuit can provide temperature compensation and elimination of specific components of strain.

The techniques for strain measurement described in this chapter provide the basis for determining surface strains for a test specimen. While the focus here has been the measurement techniques, the proper placement of strain gauges and the interpretation of the measured results requires further analysis. The inference of load-carrying capability and safety for a particular component is a result of the overall experimental program.

REFERENCES

1. Bauld, N. R., *Mechanics of Materials*, Brooks/Cole, Monterey, CA, 1982.
2. Timoshenko, S. P., and J. M. Goodier, *Theory of Elasticity*, Engineering Society Monographs, McGraw-Hill, New York, 1970.
3. Dally, J. W., and W. F. Riley, *Experimental Stress Analysis*, 2d ed., Wiley, New York, 1978.
4. Thomson, W. (Lord Kelvin), On the electrodynamic qualities of metals, *Philosophical Transactions of the Royal Society (London)* 146: 649–751, 1856.
5. Micro-Measurements Division, Measurements Group, Inc., *Strain Gauge Selection: Criteria, Procedures, Recommendations*, Technical Note 505-1, Raleigh, NC, 1989.
6. Micro-Measurements Division, Measurements Group, Inc., *Strain Gauge Technical Data*, Catalog 500, Part B, and TN-509: *Errors Due to Transverse Sensitivity in Strain Gauges*, Raleigh, NC, 1982.
7. Kulite Semiconductor Products, Inc, Bulletin KSG-5E: *Semiconductor Strain Gauges*, Leonia, NJ.
8. Weymouth, L. J., J. E. Starr, and J. Dorsey, Bonded resistance strain gauges, *Experimental Mechanics* 6(4):19A, 1966.
9. Oi, K. Transient response of bonded strain gauges, *Experimental Mechanics* 6(9):463, 1966.
10. Brewster, D., On the effects of simple pressure in producing that species of crystallization which forms two oppositely polarized images and exhibits the complementary colours by polarized light, *Philosophical Transactions A (GB)* 105:60, 1815.
11. Maxwell, J. C., On the equilibrium of elastic solids, *Transactions of the Royal Society* 20(Part I):87, 1853.
12. Neumann, F. E., Uber Gesetze der Doppelbrechung des Lichtes in comprimierten oder ungleichformig Erwamten unkrystallischen Korpern, *Abh. Akad. Wiss. Berlin*, Part II: 1, 1841. (In German)
13. Sciammarella, C. A., The moiré method—A review, *Experimental Mechanics* 22(11):418, 1982.
14. Post, D. Moiré interferometry at VPI & SU, *Experimental Mechanics* 23(2):203, 1983.
15. Post, D. Moiré interferometry for deformation and strain studies, *Optical Engineering* 24(4):663, 1985.

NOMENCLATURE

b width $[l]$
c speed of sound $[l\text{-}t]$
h height $[l]$
n_i index of refraction in the direction of the principal strain in the i direction
$(u_d)_x$ design-stage uncertainty in the variable x
A_c cross-sectional area $[l^2]$
D diameter $[l]$
E_m modulus of elasticity $[m\text{-}l\text{-}t^2]$
E_i input voltage $[V]$
E_o output voltage $[V]$
E_t strain gauge transverse sensitivity error as a percentage of axial strain
F_N force normal to A_c $[m\text{-}l\text{-}t^2]$
G shear modulus $[m\text{-}t^2]$
GF   gauge factor
K strain optical coefficient
K_B bridge sensitivity $[V]$
K_t strain gauge transverse sensitivity
L length $[l]$
δL change in length $[l]$
M bending moment $[m\text{-}l^2\text{-}t]$

R electrical resistance $[\Omega]$
T temperature $[°]$
T_0 reference temperature $[°]$
α Moiré grating width to spacing parameter
γ_{xy} shear strain in the xy plane
δ relative retardation between two light beams in a photoelastic material
ε_a axial strain
ε_i strain in the i coordinate direction (i.e., x direction)
ε_t transverse strain
κ bridge constant
ν_p Poisson ratio
ν_{po} Poisson ratio for gauge factor calibration test specimen
π_1 piezoresistance coefficient $[t^2\text{-}l\text{-}m]$
ρ_e electrical resistivity $[\Omega\text{-}l]$
σ stress $[m\text{-}l\text{-}t^2]$
σ_a axial stress $[m\text{-}l\text{-}t^2]$
τ_{xy} shear stress in the xy plane

PROBLEMS

11.1 Calculate the change in length of a steel rod ($E_m = 30 \times 10^6$ psi) having a circular cross section, a length of 10 in., and a diameter of $\frac{1}{4}$ in. The rod supports a weight of 40 lb_m in such a way that a state of uniaxial tension is created in the rod.

11.2 Calculate the change in length of a steel rod ($E_m = 20 \times 10^{10}$ Pa) which has a length of 0.3 m and a diameter of 5 mm. The rod supports a mass of 50 kg in a standard gravitational field in such a way that a state of uniaxial tension is created in the rod.

11.3 An electrical coil is made by winding copper wire around a core. What is the resistance of 20 000 turns of 16-gauge wire (0.051 in. diameter) at an average radius of 2.0 in.?

11.4 Compare the resistance of a volume of $\pi \times 10^{-5}$ m^3 of aluminum wire having a diameter of 2 mm, with the same volume of aluminum formed into 1-mm-diameter wire. (The resistivity of aluminum is 2.66×10^{-8} Ω-m.)

11.5   A conductor made of nickel ($\rho_c = 6.8 \times 10^{-8}$ Ω-m) has a rectangular cross section 5×2 mm and is 5 m long. Determine the total resistance of this conductor. Calculate the diameter of a 5-m-long copper wire having a circular cross-section which yields the same total resistance.

11.6   Consider a Wheatstone bridge circuit having all resistances equal to 100 Ω. The resistance R_1 is a strain gauge that cannot sustain a power dissipation of more than 0.25 W. What is the maximum applied voltage that can be used for this bridge circuit? At this level of bridge excitation, what is the bridge sensitivity?

11.7   A resistance strain gauge with $R = 120$ Ω and a gauge factor of 2 is placed in an equal-arm Wheatstone bridge in which all the resistances are equal to 120 Ω. If the maximum gauge current is to be 0.05 A, what is the maximum allowable bridge excitation voltage?

11.8   A strain gauge having a nominal resistance of 350 Ω and a gauge factor of 1.8 is mounted in an equal-arm bridge, which is balanced at a zero applied strain condition. The gauge is mounted on a 1-cm^2 aluminum rod, having $E_m = 70$ GPa. The gauge senses axial strain. The bridge output is 1 mV for a bridge input of 5 V. What is the applied load, assuming the rod is in uniaxial tension?

11.9   Consider a structural member subject to loads that produce both axial and bending stresses, as shown in Figure 11.13. Two strain gauges are to be mounted on the member and connected in a Wheatstone bridge in such a way that the bridge output indicates the axial component of strain only (the installation is bending compensated). Show that the installation of the gauges shown in Figure 11.13 will not be sensitive to bending.

11.10   A steel beam member ($\nu_p = 0.3$) is subjected to simple axial tensile loading. One strain gauge aligned with the axial load is mounted on the top and center of the beam. A second gauge is similarly mounted on the bottom of the beam. If the gauges are connected as arms 1 and 4 in a Wheatstone bridge (Figure 11.12), determine the bridge constant for this installation. Is the measurement system temperature compensated? (Clearly explain why or why not). If $\delta E_o = 10$ μV and $E_i = 10$ V, determine the axial and transverse strains. The gauge factor for each gauge is 2, and all resistances are initially equal to 120 Ω.

11.11   An axial strain gauge and a transverse strain gauge are mounted to the top surface of a steel beam that experiences a uniaxial stress of 2222.2 psi. The gauges are connected to arms 1 and 2 (Figure 11.12) of a Wheatstone bridge. With a purely axial load applied, determine the bridge constant for the measurement system. If $\delta E_o = 250$ μV and $E_i = 10$ V, estimate the average gauge factor of the strain gauges. For this material Poisson's ratio is 0.3 and the modulus of elasticity is 29.4×10^6 psi.

11.12   A strain gauge is mounted on a steel cantilever beam of rectangular cross section. The gauge is connected in a Wheatstone bridge; initially $R_{gauge} = R_2 = R_3 = R_4 = 120$ Ω. A gauge resistance change of 0.1 Ω is measured for the loading condition and gauge orientation shown in

Figure 11.22. If the gauge factor is $2.05 \pm 1\%$ (95%) estimate the strain. Suppose the uncertainty in each resistor value is 1% (95%). Estimate an uncertainty in the measured strain due to the uncertainties in the bridge resistances and gauge factor. Assume that the bridge operates in a null mode, which is detected by a galvanometer. Also assume reasonable values for other necessary uncertainties and parameters, such as input voltage or galvanometer sensitivity.

FIGURE 11.22   Loading for Problem 11.12.

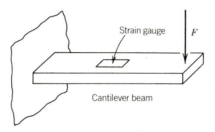

11.13   Two strain gauges are mounted so that they sense axial strain on a steel member in uniaxial tension. The 120-Ω gauges form two legs of a Wheatstone bridge, and are mounted on opposite arms. For a bridge excitation voltage of 4 V and a bridge output voltage of 120 μV under load, estimate the strain in the member. What is the resistance change experienced by each gauge? The gauge factor for each of the strain gauges is 2 and E_m for steel is 29×10^6 psi.

11.14   A galvanometer having an internal impedance of 100 $\Omega \pm 0.5\%$ and a 1-μA resolution is used to indicate strain in a strain gauge Wheatstone bridge circuit arrangement. A single gauge, R_1, is aligned on a steel member in uniaxial tension, so as to sense axial strain. The bridge is composed of two fixed resistors ($R_2 = R_3 = 120 \Omega \pm 1\%$) and a null adjust potentiometer (R_p known to $\pm 1\%$). With no applied load R_p must be set to 120.07 Ω to balance the bridge. Under load the galvanometer indicates a current flow of 2 μA. Estimate the strain and the uncertainty in this measured value. The gauge factor is 2 and the input voltage is 4 V.

FIGURE 11.23   Bridge arrangement for Problem 11.15.

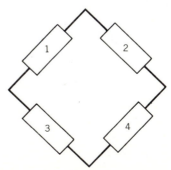

11.15 For each of the bridge configurations shown in Figure 11.23, determine the bridge constant. Assume that all of the active gauges are identical, and that all of the fixed resistances are equal.

Bridge Arrangement	Description
(a) 1: Active gauge 2,3,4: Fixed resistors	Single gauge in uniaxial tension.

(b) 1: Active gauge
    3: Poisson gauge
    3,4: Fixed resistors
	Two active gauges in uniaxial stress field. Gauge 1 aligned with maximum axial stress; gauge 2 transverse.

(c) 1: Active gauge
    3: Active gauge
    2,4: Fixed resistors
	Equal and opposite strains applied to the active gauges (bending compensation).

(d) Four active gauges	Gauges 1 and 4 aligned with uniaxial stress; gauges 2 and 3 transverse.
(e) Four active gauges	Gauges 1 and 2 subject to equal and opposite strains; gauges 3 and 4 subject to the same equal but opposite strains.

APPENDIX A
DATA ANALYSIS SOFTWARE

A.1 LEAST-SQUARES REGRESSION ANALYSIS PROGRAM

The following program can determine up to a fifth-order polynomial of the form

$$y = a_0 + a_1 x + a_2 x^2 + \cdots + a_m x^m$$

using a linear least-squares regression analysis routine. It can be easily incorporated into the utility software files of a personal computer or a mainframe and compiled using any FORTRAN-77 compiler. Dimension statements should be expanded to accommodate larger data sets. The program will request the order of polynomial to be fit, m, and the number of data pairs, N, available. The data pairs (x_i, y_i) where $i = 1, 2, \ldots, N$, are then requested. The program outputs the regression coefficients $a_0, a_1, \ldots, a_m$ and the standard error of the fit.

```
      PROGRAM CFIT
C****** LEAST SQUARES CURVE FITTING PROGRAM ******
      IMPLICIT REAL*8 (A-H,O-Z)
      DIMENSION X(20),Y(20),F(20,11),A(20,21),C(20)
      WRITE(6,190)
190   FORMAT(//,'ENTER ORDER OF POLYNOMIAL TO BE FIT, M = ')
C** NUMBER OF COEFFICIENTS WILL EQUAL ORDER OF FIT PLUS ONE **
      READ(5,*)MM1
      M = MM1+1
      WRITE(6,191)
191   FORMAT(//,'ENTER THE NUMBER OF DATA PAIRS TO BE ENTERED, N =')
      READ(5,*)N
      WRITE(6,200) MM1,N
200   FORMAT(' ORDER OF POLYNOMIAL = ',I2,' NO. OF DATA PAIRS= ',I3)
      DO 1 I=1,N
      WRITE(6,192)I
192   FORMAT(//,'FOR I = ',I3,'  ENTER X(I) Y(I)'
      READ(5,*)X(I),Y(I)
   1  CONTINUE
      DO 20 I = 1,N
      WRITE(6,11)
20    WRITE(6,10)X(I),Y(I)
10    FORMAT(10X,E15.6,10X,E15.6)
11    FORMAT (15X,'X',20X,'Y')
```

```
C*** USE FUNCTION SUBPROGRAMS TO DEFINE EACH POLYNOMIAL TERM ***
      DO 2 I=1,N
        F(I,1)=F1(X(I))
        F(I,2)=F2(X(I))
        F(I,3)=F3(X(I))
        F(I,4)=F4(X(I))
        F(I,5)=F5(X(I))
    2 F(I,6)=F6(X(I))
      DO 4  I=1,M
      DO 4  J=1,I
        A(I,J)=0.
      DO 3 K=1,N
    3   A(I,J)=A(I,J)+F(K,I)*F(K,J)
    4   A(J,I)=A(I,J)
      DO 5 I=1,M
        A(I,M+1)=0.
      DO 5 K=1,N
    5 A(I,M+1)=A(I,M+1)+F(K,I)*Y(K)
      MP1=M+1
      CALL MATRX(A,M,MP1,C)
      WRITE(6,6)
    6 FORMAT('1',4X,'C(1) THROUGH C(M)'/)
      WRITE(6,7)(I,C(I),I=1,M)
    7 FORMAT(' ',3X,'A(',I1,')=',E14.7)
      YSUM = 0.
      DO 85 J = 1,N
      SUM2 = 0.
      YC = 0.
      DO 80 I = 1,M
        SUM2 = C(I)*X(J)**(I-1)
   80 YC =YC + SUM2
      YDIFF2 = (Y(J) - YC)**2. + YSUM
   85 YSUM = YDIFF2
      SYX = SQRT(YSUM/(N-M))
      WRITE(6,180)SYX
  180 FORMAT(//' STANDARD ERROR OF FIT =',E12.4)
      STOP
      END
C
      FUNCTION F1(X)
      IMPLICIT REAL*8 (A-H,O-Z)
      F1=1.
      RETURN
      END
C
      FUNCTION F2(X)
      IMPLICIT REAL*8 (A-H,O-Z)
      F2=X
      RETURN
      END
      FUNCTION F3(X)
      IMPLICIT REAL*8 (A-H,O-Z)
      F3=X**2
      RETURN
      END
C
      FUNCTION F4(X)
      IMPLICIT REAL*8 (A-H,O-Z)
      F4=X**3
      RETURN
      END
C
      FUNCTION F5(X)
      IMPLICIT REAL*8 (A-H,O-Z)
      F5=X**4
      RETURN
      END
```

```
C
      FUNCTION F6(X)
      IMPLICIT REAL*8 (A-H,O-Z)
      F6=X**5
      RETURN
      END

      SUBROUTINE MATRX(A,N,M,X)
      IMPLICIT REAL*8 (A-H,O-Z)
      DIMENSION A(20,21),X(20)
      DO 20 I=1,N
C         WRITE(6,25)(I,J,A(I,J),J=1,M)
   25 FORMAT(2I3,F12.4)
   20 CONTINUE
C        *** CALCULATE FIRST ROW OF UPPER UNIT TRIANGULAR MATRIX
      DO 30 J=2,M
   30 A(1,J)=A(1,J)/A(1,1)
C        **** CALCULATE OTHER ELEMENTS OF U AND L MATRICES
      DO 80 I=2,N
       J=I
      DO 50 II=J,N
       SUM1=0.
       JM1=J-1
      DO 40 K=1,JM1
   40 SUM1=SUM1+A(II,K)*A(K,J)
   50 A(II,J)=A(II,J)-SUM1
      IP1=I+1
      DO 70 JJ=IP1,M
       SUM1=0.
       IM1=I-1
      DO 60 K=1,IM1
   60 SUM1=SUM1+A(I,K)*A(K,JJ)
   70 A(I,JJ)=(A(I,JJ)-SUM1)/A(I,I)
   80 CONTINUE
C        SOLVE FOR X(I) BY BACKSUBSTITUTION
      X(N)=A(N,N+1)
      L=N-1
      DO 100 NN=1,L
       SUM1=0.
       I=N-NN
       IP1=I+1
      DO 90 J=IP1,N
   90 SUM1=SUM1+A(I,J)*X(J)
  100 X(I)=A(I,M)-SUM1
      RETURN
      END
```

A.2 DISCRETE FOURIER TRANSFORM

The following program computes the discrete Fourier transform of the discrete variable, $x(r\, \delta t)$. In using this program, variable x can be a discrete time series called from a data file or a series can be generated within the program from an internal user defined function, $x(t)$. When a data file is used, the program will request its name and data length. The program outputs the frequency spectrum of the data set in terms of frequency versus amplitude and the power spectrum of frequency versus power. The resolution between each frequency, RESOL, is determined by the time increment between each data point 1/SR, where SR is the sample rate, and the number of data points, N, such that RESOL = SR/N. Note that the sample rate, SR, must be more than twice the highest frequency contained in $x(t)$. See Chapter 6 for proper selection of N and δt.

The number of data points, N, has been restricted to be of size 2^M, where M is an integer. It should be noted that the program computes the discrete Fourier transform X_k by direct computation using the numerical approximation

$$X_k = A_k - iB_k = \frac{1}{N} \sum_{r=1}^{N} x(r\,\delta t)e^{-i2\pi kr/N} \qquad k = 0, 1, 2, \ldots$$

which yields the power at each kth frequency

$$|X_k|^2 = C_k^2/2$$

where C_k is the amplitude of the kth frequency,

$$C_k = \sqrt{(A_k^2 + B_k^2)}$$

This program is provided for demonstration purposes only. As the value of N is increased, the DFT will take an increasingly longer time to execute. Faster algorithms are available that utilize the fast Fourier transform (FFT) method for computing the DFT.

```
      PROGRAM DFT
      DIMENSION X(1024),A(513),B(513),AMP(513),POWER(K)
      CHARACTER*12 FILDAT
C
C              M          an integer (with dimension used, M<=10)
C              N          number of data points (N = 2**M)
C              X          data set variable
C              FREQ       frequency [Hz]
C              AMP        spectral amplitudes
C              POWER      spectral power
C
      WRITE(*,100)
      WRITE(*,101)
100   FORMAT(' IF DATA ON STORAGE DEVICE, ENTER 0')
101   FORMAT(' IF DATA IS TO BE GENERATED BY FUNCTION, ENTER 1')
      READ(5,*) NTYPE
C
  2   WRITE(*,104)
      WRITE(*,106)
      WRITE(*,107)
104   FORMAT(' EXECUTION TIMES SLOWWWWS DOWN WITH LARGER M')
106   FORMAT(' MAXIMUM VALUE FOR M IS 10')
107   FORMAT(' ENTER VALUE FOR M WHERE NUMBER OF DATA = 2**M ')
      READ(5,*) M
      IF(M.GT.10) GO TO 2
      N=2**M
      AN=FLOAT(N)
      TWOPI=2.*3.14159
      IF(NTYPE.EQ.1) GO TO 5
C
C  THE FOLLOWING BLOCK WILL READ DATA FROM A STORAGE DEVICE
C
      WRITE(*,110)
110   FORMAT(' ENTER FILE NAME OF STORED DATA (e.g. B:FILE.DAT) ')
      READ(5,111) FILDAT
111   FORMAT(A12)
      OPEN(3,FILE=FILDAT)
      READ(3,*) (X(L), L=1,N)
      CLOSE(3)
      WRITE(*,115)
```

```
    115 FORMAT(' ENTER THE TIME INCREMENT BETWEEN THE DATA POINTS')
        READ(5,*)TINC
        SR=1./TINC
        GO TO 7
  C
  C   FOLLOWING BLOCK WILL GENERATE A DATA ARRAY FROM THE FUNCTION
  C       X(t) = 10. SIN (6.28*t) = 10 SIN (2*PI*1*t)
  C   THE ARRAY X(L) IS GENERATED BY MEASURING X(t) AT A RATE OF
  C   10 TIMES PER SECOND (i.e. SR = 10) OR A TIME INCREMENT OF EVERY
  C   0.1 SECONDS (i.e. TINC = 0.1). IT IS IMPORTANT THAT SR > 2F
  C
  C   SUBSTITUTE YOUR OWN FUNCTION BETWEEN STARS
  C   ****************************************************************
  C BEGIN FUNCTION INFORMATION
      5 F= 1.0
        SR = 10.
        DELT=1./SR
        TIME=0.
        DO 10 L=1,N
          X(L) = 10.*SIN(TWOPI*F*TIME)
          TIME = TIME + DELT
     10 CONTINUE
  C END FUNCTION INFORMATION
  C
  C   ****************************************************************
  C
  C INITIALIZE THE ARRAYS
      7 WRITE(*,120)
    120 FORMAT(14X,' FREQUENCY (HZ) ',10X,' AMPLITUDE ')
        DO 30 K=1,N/2
          TWOPIN=TWOPI/AN
          A(K)=0.
          B(K)=0.
          TK=TWOPIN*K
  C BEGIN THE DISCRETE FOURIER TRANSFORM
        DO 20 LR=1,N
  C COMPUTE THE FOURIER COEFFICIENTS
            A(K)=A(K)+X(LR)*COS(LR*TK)
     20     B(K)=B(K)+X(LR)*SIN(LR*TK)
  C FINISHED DISCRETE FOURIER TRANSFORM
          AMP(K)=SQRT(2.*(A(K)**2+B(K)**2))/AN
          POWER(K)= 0.5*AMP(K)**2
          RESOL = SR/AN
          FREQ=K*RESOL
          PRINT 200, K,FREQ,AMP(K),POWER(K)
     30 CONTINUE
    200 FORMAT(I4,10X,F7.3,18X,F7.3)
    STOP
    END
```

APPENDIX B
A GUIDE FOR TECHNICAL WRITING

Test results will typically be presented in written form in a *technical report* of some type. Technical reports can usually be classified into one of three formats:

1. Executive summary
2. Lab report
3. Formal report

The executive summary is usually a one- or two-page brief stating the reasons for the tests conducted and important results and conclusions. This format is used when the report is intended for upper level management who will plan corporate decisions around the input from the report. Clarity of purpose and conclusions are crucial.

Laboratory reports are internal reports accessed by members of your engineering group. The reports should document laboratory tests in detail. These reports should include the reasons for the tests, the pretest planning, the methods and test conditions used, and a discussion of the results and conclusions. Data in the form of tables and graphs would be cited in the report and attached, including an appendix containing raw data.

Formal reports are intended for a technical audience outside of the immediate group. Accordingly, these reports should include and/or cite sufficient background information so that the reader can understand the purpose of the tests, the manner in which the tests were analyzed, and the results of the tests. Details of the test procedure that might go into a laboratory report are usually omitted, but a formal report must present enough information for the reader to follow the logic of the tests and to interpret the test results. Data should be presented only in the form of well-prepared tables and graphs cited from within the report. Each company or test laboratory has its own format for these reports. However, sample guidelines for the preparation of a formal technical report are provided below.

This appendix is intended to provide some basic guidelines for technical reporting, but is not comprehensive or complete. Further information concerning technical reporting may be found in numerous guides for technical writing.[1]

A GUIDE FOR TECHNICAL WRITING[2]

The increasingly important role of engineers in a technologically advanced society necessitates the development of effective communication skills for engineers. Engineers not only should develop the skills necessary to communicate with both technical and nontechnical audiences, but also should participate more aggressively in technical public policy decisions. In the future, engineers must actively seek to communicate more effectively concerning technical issues.

This appendix provides a brief guide to constructing a technical report, primarily for a technical audience. The format of this appendix is representative of most technical reporting formats in industry and for a variety of technical publications. The major headings, such as Abstract and Introduction, normally appear in the order they are described here. The intent of this guide is to provide an understanding of the expected and generally accepted organization of a technical report.

Abstract

Effective technical communication abilities are increasingly important in a technologically advancing society. The purpose of this document is to serve as a format example for a technical report and to provide specific procedures and ideas for generating sound technical reports. Guidelines for preparing each section of a report and detailed ideas for presentation of results in plots and tables are provided.

Introduction

In a 1980 study, the U.S. Department of Education and the National Science Foundation concluded that the majority of Americans are moving toward "virtual scientific and technological illiteracy" [1]. The technical person's ability to communicate effectively with both technical and nontechnical segments of society is essential for addressing the myriads of technological problems and decisions facing our society. In addition, career advancement in the engineering profession is largely based on how well a person communicates.

---

[1]Further information on the various aspects of technical writing may be found in *Reporting Technical Information*, 6th ed., by K. W. Houp and T. E. Pearsall, Macmillan, New York, 1988, or *Effective Writing for Engineers, Managers, and Scientists*, 2d ed., by H. J. Tichy with S. Foudrinier, Wiley Interscience, New York, 1988.
[2]Adapted with permission from M. H. Henry and H. K. Lonsdale, The researcher's writing guide, *Journal of Membrane Science* 13:101–107, 1983.

It is especially important for engineers to develop effective technical writing skills that enable them to convey the results and significance of their work to a variety of audiences. The primary form of writing for practicing engineers and engineering managers (as well as researchers) is the technical report. In general, this is directed toward a technically knowledgeable audience. The purpose of this guide is to provide reasonable ideas and suggestions for producing clear and concise technical manuscripts.

Specific Writing Steps

The following steps are recommended for effective technical writing:

1. Collect all of your data, organize your thoughts, and sketch figures and tables. The scope of the material to be presented and the particular audience for whom you are writing should be kept in mind while deciding what to present.
2. Arrange your materials in the order in which you plan to present them.
3. Establish a thesis that adequately covers your material, and create a detailed outline. Expand your ideas to at least the level where topics and some sentences have been formulated. The use of this outline will help you avoid making a chronological or "stream of consciousness" presentation; the order in which you did the work and the order in which you report it will rarely match.
4. Before you begin writing, study the format or style guide for the specific publication.
5. Don't expect to create a perfect product the first time through. In writing the first draft, concentrate on organization and clearly presenting your ideas and your thesis for the particular audience for which you are writing. Write a first draft as rapidly as possible to take advantage of your continuity of thought. The draft can be revised and polished later.
6. Read the draft and, as necessary for clarity and accuracy, rewrite. Make sure that your writing clearly says what you intended; don't read your intentions into whatever is on the paper. As a last step, proofread the manuscript. Reading and proofing should seldom be attempted at the same time. Few writers can successfully perform both functions simultaneously, because the word-by-word study needed for proofing interferes with the fluidity required for reading comprehension.

Writing Technical Papers

The key to effectively communicating the results of your work is to have a clear understanding of your current knowledge; forget all previous misconceptions, mistakes, bad data, and so on. Take a fresh look at the results and organize your thoughts to present the work in the best way. There are four basic building blocks for constructing a technical report: written text, figures, tables, and appendixes. Each figure, table, and appendix must be described and explicitly referred to in the text. The following guidelines for preparing each section of a technical report provide the basics necessary for sound reporting procedures.

The Abstract—A Summary of the Entire Report

An Abstract must be a complete, concise distillation of the full report, and, as such, should always be written last. It should include a brief (one sentence) introduction to the subject, a statement of the problem, highlights of the results (quantitative, if possible), and major conclusions. It must stand alone without citing figures or tables. A concise, clear approach is essential, since most Abstracts are less than 250 words.

The Introduction—Why Did You Do What You Did?

An Introduction generally identifies the subject of the report, provides the necessary background information including appropriate literature review, and, in general, provides the reader with a clear rationale for the work described. The Introduction does not contain results, and generally does not contain equations. The use of figures, tables, and citations should be limited in the Introduction.

Analysis—What Does Theory Have to Say?

An Analysis section describes a proposed theory or a descriptive model. It does not contain results, nor should extreme mathematical details be provided. Sufficient detail (mathematical or otherwise) should be provided for the reader to clearly understand the physical assumptions associated with a theory or model.

Experimental Program—What Did You Measure and How?

The Experimental Program section is intended to describe how experimental results were obtained. As a rule of thumb, provide sufficient details to allow the experiment to be conducted by someone else. Do not give instructions or commands to the reader; rather report what was done. If a list of equipment is included in the report, it should be a table in the body of the report, or should be placed in an appendix. Uncertainty analysis information can be described either here or in the Results section, or both. In cases where both an analysis and experiment are described, these two sections of the report should complement and support each other. The relationship of the analysis to the experiment should be clearly stated.

Results and Discussion—So What Did You Find?

Results of your work must be presented, as well as discussed, in this section of the report. Data must be interpreted to be useful to most readers. When presenting your results remember that even though you are usually writing to an experienced technical audience, what may be clear to you may not be obvious to the reader. Assuming too much knowledge can be a big mistake,

FIGURE B.1 Comparison of experimentally measured and computed (PACKBED) results for axial temperature distributions in a thermal energy storage bed.

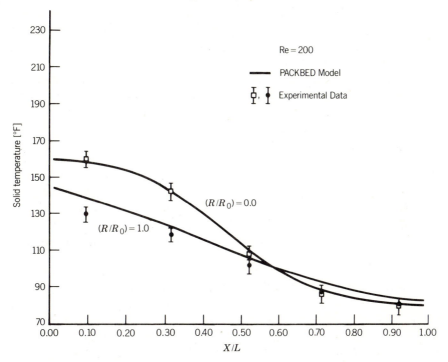

so explain your results even if it seems unnecessary. If you can't figure them out, say so: "The mechanism is unclear and we are continuing to examine this phenomenon." Often the most important vehicles for the clear presentation of results are figures and tables. Each of the figures and tables should be numbered and have a descriptive title. Column heads in tables should accurately describe the data that appear in that column. Each table and figure must be explicitly and individually referenced and described in the text of the Results section. Since you have spent significant time in preparing the plots and tables you are intimately familiar with their trends and implications; the reader needs your insight to understand the results as well as you.

Figure B.1 provides an example of an appropriate plot. Each plot must have a figure number and a caption, and should allow the reader to understand what is being presented by looking at the plot and reading the caption. Using the following suggestions will result in effective plots:

1. The independent parameter is always plotted on the x axis; the dependent parameter is always plotted on the y axis.
2. Try to arrange your figure so that the y axis is the long axis. That way, the figure will be oriented conveniently on the page and the reader need not turn the page sideways to read the graph.
3. Try to use at least four tics, but fewer than ten, for each coordinate. Multiples of 1, 2, or 5 are good increments for axis numera-

tion. Avoid 0, 3, 6, 9, 12, or 0, 4, 8, 12, 16, or similar divisions because they make interpolation unnecessarily difficult. And it is hard to envision the justification for such strange increments as 0, 7, 14.

4. In drawing smooth curves through experimental data points, try to follow these rules:
   (a) Always show the data points.
   (b) Do not extend a curve beyond the ends of the data points. If for some valid reason you find it necessary to extend the line, use a dashed line to indicate that it is some sort of extrapolation. If appropriate, indicate the curve-fit equations.
   (c) If you are certain that $y = 0$ when $x = 0$, put the curve through the origin. But if you are uncertain, don't use the origin; stop the line slightly below the lowest point. Some judgment is required here.

5. Use a minimum variety of symbols for your data. Usually filled and open circles, filled and open squares, and filled and open triangles (i.e., six different symbols) are all that one graph can handle. If you have more than six types of information, break the plot up into two or more plots.

6. To compare experimentally determined data with theoretical predictions, show the experimental points as symbols and the theory as a smooth curve. Clear labeling is essential for this kind of plot.

7. Whenever possible, label curves individually, even if it means using arrows to match the lines with the labels. Use a legend in your figure only as a last resort.

8. Remember that the figure should contain enough information to stand alone, especially if there's a chance it will later be used by itself (e.g., as a slide). However, each figure must be discussed in the text.

9. In labeling axes and providing legends, try to use the written name for the quantity plotted, and then present its units in parentheses. For example, use "Time (days)" instead of merely "Days." Abbreviations such as mg, cm, and h do not take periods.

10. Position the data on the graph (i.e., adjust the scales) so that the curves are not bunched near the top, bottom, or one of the sides. Don't run the scale up to 100% if your data only go up to 38%. The only time white space serves a valid function in a graph is when you are comparing it to another graph that makes use of the full scale.

11. In writing captions, avoid lead-ins like "Plot of hydrocortisone released . . ." or "This is a plot of . . .". Simply say "Hydrocortisone released vs . . .".

12. The limits of uncertainty for experimentally determined data are essential in constructing a plot. Uncertainty limits can be indicated for a number of measured points on a given plot using interval bands about the mean value, as shown in Figure B.1.

The visual impact of a plot conveys considerable information about the relationship between the plotted dependent and independent variables. If hand drawn, plots should be constructed by pencil or by technical pen using a straight edge, french curve, label maker, and symbol maker. However, a number of relatively low-cost data-presentation software packages are available for personal computers which are fast and easy to use for the construction and manipulation of good quality plots. Several forms of plotting formats (or graph paper) are discussed below.

Rectangular Coordinate Format: In rectangular grid format, both the ordinate and the abscissa have uniformly sized divisions providing a linear scale. This is the most common format used for constructing plots and establishing the form of the relationship between the independent and dependent variable.

Semilog Coordinate Format In a semilog format, one coordinate has a linear scale and one coordinate has a logarithmic scale. Logarithmic scales are advantageous when one of the variables spans more than one order of magnitude. In particular, the semilog format is appropriate when the data approximately follow a relationship of the forms $y = ae^x$ or $y = a \times 10^x$ since a linear curve will result. The natural logarithmic operation can also be conveyed on a logarithmic scale since the relation $\ln y = 2.3 \log y$ is simply a scaling operation.

Full-log Coordinate Format The full-log or log–log format has logarithmic scales for both axes. Such a format is preferred when both variables contain data values that span more than one order of magnitude. With data that follow a trend of the form $y = ax^n$, a linear curve can be obtained in log–log format.

Table B.1 is an example of the proper way to present results in a tabular form. Each table must be explicitly described in the text of the Results section, and each table must have a title.

Conclusions—What Do I Now Know?

The Conclusions section is where you should concisely restate your answer to the question: "What do I know now?" It is not a place to offer new facts,

TABLE B.1   Characteristics of Thermistor Anemometer 11

Velocity [ft/s]	EMF [V]
0.467	3.137
0.950	3.240
2.13	3.617
3.20	3.811
3.33	3.876
4.25	3.985
5.00	4.141
6.67	4.299
8.33	4.484
10.0	4.635
12.0	4.780

nor should it contain another rendition of experimental results or rationale. In a short summary, restate why the work was done, how it was done and provide a conclusion to the work. An appropriate conclusion might be: "The temperature measuring system calibrated in this study was found to indicate the correct temperature over the range of 30 to 250 °F with no more than a ±1 °F uncertainty at 95% confidence." It would not normally be appropriate to conclude, "The temperature measuring system was tested and worked extremely well." Conclusions should be clear and concise statements of the important findings of a particular study; most conclusions require some quantitative aspect to be useful.

References

The references cited in a formal list should be available to the reader and described in sufficient detail for the reader to obtain the source with a reasonable effort. References for this document provide a format guide.

Writing Tips

1. Accuracy is important, but so is consistency. Responsibly varied word choice and sentence construction are beneficial in avoiding monotony, but avoid variation for the sake of variation in your use of substantive terms or definitions. Define all nonstandard terms the first time they are used and stick to those terms and definitions throughout all writing on that subject. If you find that those terms or definitions need to be changed, make a point of clearly informing the reader. Don't leave the reader with unanswered questions: If a subject is worth introducing, it's worth discussing thoroughly enough for the reader to fully understand. Err on the side of clarity if you must err, so that if a particular construction is questionable, add the extra words that make it longer but guarantee its clarity.

2. Don't overdo significant figures. This is one of the surest ways of convincing the astute reader that you are an amateur. How many figures can you reproduce for a given measurement? Use that number, plus one more.

3. Avoid the use of contractions and possessives in formal scientific papers. On occasion, jargon serves a useful function—one of neatly describing or labeling an otherwise troublesome concept or process. But jargon should be used only when your audience will understand it and a simple substitution doesn't exist.

4. Style is important in effective communication of technical information. Technical writing is normally in the third person, to focus attention on the subject matter at hand. The style of technical writing should allow clear and unbiased descriptions in all aspects of the report. The active voice is preferred where possible.

Conclusions

Technical authors must be cognizant of primary goals in presenting the results of their writing. This report has outlined the essential features germane to technical report writing and is a format example. An outline for each section of the report was stated as the essential starting point for effective writing. A polished and professional product can be produced only through careful and persevering revision of the first and subsequent drafts.

References

1. Naisbitt, J., *Megatrends*, Warner Books, New York, 1984. (See also, U.S. report fears most Americans will become scientific illiterates, *New York Times*, Oct. 23, 1984.)
2. Henry, M. H., and H. K. Lonsdale, The researcher's writing guide, *Journal of Membrane Science* 13:101–107, 1983.

APPENDIX C

PROPERTY DATA AND CONVERSION FACTORS

TABLE C.1 Properties of Pure Metals and Selected Alloys

	Density [kg/m^3]	Modulus of Elasticity [GPa]	Coefficient of Thermal Expansion [10^{-6} m/m-K]	Thermal Conductivity [W/m-K]	Electrical Resistivity [10^{-6} Ω-cm]
Pure Metals					
Aluminum	2 698.9	62	23.6	247	2.655
Beryllium	1 848	275	11.6	190	4.0
Chromium	7 190	248	6.2	67	13.0
Copper	8 930	125	16.5	398	1.673
Gold	19 302	78	14.2	317.9	2.01
Iron	7 870	208.2	15.0	80	9.7
Lead	11 350	12.4	26.5	33.6	20.6
Magnesium	1 738	40	25.2	418	4.45
Molybdenum	10 220	312	5.0	142	8.0
Nickel	8 902	207	13.3	82.9	6.84
Palladium	12 020	—	11.76	70	10.8
Platinum	21 450	130.2	9.1	71.1	10.6
Rhodium	12 410	293	8.3	150.0	4.51
Silicon	2 330	112.7	5.0	83.68	1×10^5
Silver	10 490	71	19.0	428	1.47
Tin	5 765	41.6	20.0	60	11.0
Titanium	4 507	99.2	8.41	11.4	42.0
Zinc	7 133	74.4	15.0	113	5.9
Alloys					
Aluminum (2024,T6)	2 770	72.4	22.9	151	4.5
Brass (C36000)	8 500	97	20.5	115	6.6
Brass (C86500)	8 300	105	21.6	87	8.3
Bronze (C90700)	8 770	105	18	71	1.5
Constantan annealed (55% Cu 45% Ni)	8 920	—	—	19	44.1
Steel (AISI 1010)	7 832	200	12.6	60.2	20
Stainless Steel (Type 316)	8 238	190	—	14.7	—

Source: Compiled from *Metals Handbook* 9th ed., American Society for Metals, Metals Park, OH, 1978, and other sources.

TABLE C.2 Thermophysical Properties of Selected Metallic Solids

Composition	Melting point [K]	Properties at 300 K				Properties at various temperatures [K]							
		ρ [kg/m³]	c_p [J/kg·K]	k [W/m·K]	$\alpha \times 10^4$ [m²/s]	k [W/m·K]				c_p [J/kg·K]			
						100	200	400	600	100	200	400	600
Aluminum													
Pure	933	2702	903	237	97.1	302	237	240	231	482	796	949	1033
Alloy 2024-T6 (4.5% Cu, 1.5% Mg, 0.6% Mn)	775	2770	875	177	73.0	65	163	186	186	473	787	925	1042
Alloy 195, cast (4.5% Cu)	—	2790	883	168	68.2	—	—	174	185	—	—	—	—
Chromium	2118	7160	449	93.7	29.1	159	111	90.9	80.7	192	384	484	542
Copper													
Pure	1358	8933	385	401	117	482	413	393	379	252	356	397	417
Commercial bronze (90% Cu, 10% Al)	1293	8800	420	52	14	—	42	52	59	—	785	460	545
Phosphor gear bronze (89% Cu, 11% Sn)	1104	8780	355	54	17	—	41	65	74	—	—	—	—
Cartridge brass (70% Cu, 30% Zn)	1188	8530	380	110	33.9	75	95	137	149	—	360	395	425
Constantan (55% Cu, 45% Ni)	1493	8920	384	23	6.71	17	19	—	—	237	362	—	—
Iron													
Pure	1810	7870	447	80.2	23.1	134	94.0	69.5	54.7	216	384	490	574
Armco (99.75%)	—	7870	447	72.7	20.7	95.6	80.6	65.7	53.1	215	384	490	574
Carbon steels													
Plain carbon (Mn ≤ 1%, Si ≤ 0.1%)	—	7854	434	60.5	17.7	—	—	56.7	48.0	—	—	487	559
AISI 1010	—	7832	434	63.9	18.8	—	—	58.7	48.8	—	—	487	559

Carbon–silicon (Mn ≤ 1%, 0.1% < Si ≤ 0.6%)	—	7817	446	51.9	14.9	—	—	49.8	44.0	—	—	501	582
Carbon–manganese–silicon (1% < Mn ≤ 1.65%, 0.1% < Si ≤ 0.6%)	—	8131	434	41.0	11.6	—	—	42.2	39.7	—	—	487	559
Chromium (low) steels ½ Cr-¼ Mo-Si (0.18% C, 0.65% Cr, 0.23% Mo, 0.6% Si)	—	7822	444	37.7	10.9	—	—	38.2	36.7	—	—	492	575
1 Cr–½ Mo (0.16% C, 1% Cr, 0.54% Mo, 0.39% Si)	—	7858	442	42.3	12.2	—	—	42.0	39.1	—	—	492	575
1 Cr–V (0.2% C, 1.02% Cr, 0.15% V)	—	7836	443	48.9	14.1	—	—	46.8	42.1	—	—	492	575
Stainless steels													
AISI 302	—	8055	480	15.1	3.91	—	—	17.3	20.0	—	—	512	559
AISI 304	1670	7900	477	14.9	3.95	9.2	12.6	16.6	19.8	272	402	515	557
AISI 316	—	8238	468	13.4	3.48	—	—	15.2	18.3	—	—	504	550
AISI 347	—	7978	480	14.2	3.71	—	—	15.8	18.9	—	—	513	559
Lead	601	11340	129	35.3	24.1	39.7	36.7	34.0	31.4	118	125	132	142
Magnesium	923	1740	1024	156	87.6	169	159	153	149	649	934	1074	1170
Molybdenum	2894	10240	251	138	53.7	179	143	134	126	141	224	261	275
Nickel													
Pure	1728	8900	444	90.7	23.0	164	107	80.2	65.6	232	383	485	592
Nichrome (80% Ni, 20% Cr)	1672	8400	420	12	3.4	—	—	14	16	—	—	480	525
Inconel X–750 (73% Ni, 15% Cr, 6.7% Fe)	1665	8510	439	11.7	3.1	8.7	10.3	13.5	17.0	—	372	473	510

TABLE C.3   Thermophysical Properties of Saturated Water (Liquid)

T [K]	ρ [kg/m^3]	c_p [kJ/kg·K]	$\mu \times 10^6$ [N·s/m^2]	k [W/m·K]	Pr	$\beta \times 10^6$ [K^{-1}]
273.15	1000	4.217	1750	0.569	12.97	−68.05
275.0	1000	4.211	1652	0.574	12.12	−32.74
280	1000	4.198	1422	0.582	10.26	46.04
285	1000	4.189	1225	0.590	8.70	114.1
290	999	4.184	1080	0.598	7.56	174.0
295	998	4.181	959	0.606	6.62	227.5
300	997	4.179	855	0.613	5.83	276.1
305	995	4.178	769	0.620	5.18	320.6
310	993	4.178	695	0.628	4.62	361.9
315	991	4.179	631	0.634	4.16	400.4
320	989	4.180	577	0.640	3.77	436.7
325	987	4.182	528	0.645	3.42	471.2
330	984	4.184	489	0.650	3.15	504.0
335	982	4.186	453	0.656	2.89	535.5
340	979	4.188	420	0.660	2.66	566.0
345	977	4.191	389	0.664	2.46	595.4
350	974	4.195	365	0.668	2.29	624.2
355	971	4.199	343	0.671	2.15	652.3
360	967	4.203	324	0.674	2.02	679.9
365	963	4.209	306	0.677	1.90	707.1
370	961	4.214	289	0.679	1.79	728.7
373.15	958	4.217	279	0.680	1.73	750.1
400	937	4.256	217	0.688	1.34	896
450	890	4.40	152	0.678	0.99	
500	831	4.66	118	0.642	0.86	
550	756	5.24	97	0.580	0.88	
600	649	7.00	81	0.497	1.14	
647.3	315	00	45	0.238	00	

Formulas for interpolation (T = absolute temperature)
$$f(T) = A + BT + CT^2 + DT^3$$

$f(T)$	A	B	C	D	Standard deviation, σ
			$273.15 < T < 373.15$ K		
ρ	766.17	1.80396	-3.4589×10^{-3}		0.5868
c_p	5.6158	-9.0277×10^{-3}	14.177×10^{-6}		4.142×10^{-3}
k	−0.4806	5.84704×10^{-3}	-0.733188×10^{-5}		0.481×10^{-3}
			$273.15 < T < 320$ K		
$\mu \times 10^6$	0.239179×10^6	-2.23748×10^3	7.03318	-7.40993×10^{-3}	4.0534×10^{-6}
$\beta \times 10^6$	-57.2544×10^3	530.421	−1.64882	1.73329×10^{-3}	1.1498×10^{-6}
			$320 < T < 373.15$ K		
$\mu \times 10^6$	35.6602×10^3	−272.757	0.707777	-0.618833×10^{-3}	1.0194×10^{-6}
$\beta \times 10^6$	-11.1377×10^3	84.0903	−0.208544	0.183714×10^{-3}	1.2651×10^{-6}

Source: From F. P. Incropera and D. P. DeWitt, *Fundamentals of Heat and Mass Transfer*, Wiley, New York, 1985.

TABLE C.4 Thermophysical Properties of Air

T [K]	ρ [kg/m³]	c_p [kJ/kg·K]	$\mu \times 10^7$ [N·s/m²]	$\nu \times 10^4$ [m²/s]	$k \times 10^3$ [W/m·K]	$\alpha \times 10^5$ [m²/s]	Pr
200	1.7458	1.007	132.5	7.590	18.1	10.3	0.737
250	1.3947	1.006	159.6	11.44	22.3	15.9	0.720
300	1.1614	1.007	184.6	15.89	26.3	22.5	0.707
350	0.9950	1.009	208.2	20.92	30.0	29.9	0.700
400	0.8711	1.014	230.1	26.41	33.8	38.3	0.690
450	0.7740	1.021	250.7	32.39	37.3	47.2	0.686
500	0.6964	1.030	270.1	38.79	40.7	56.7	0.684
550	0.6329	1.040	288.4	45.57	43.9	66.7	0.683
600	0.5804	1.051	305.8	52.69	46.9	76.9	0.685
650	0.5356	1.063	322.5	60.21	49.7	87.3	0.690
700	0.4975	1.075	338.8	68.10	52.4	98.0	0.695
750	0.4643	1.087	354.6	76.37	54.9	109.	0.702
800	0.4354	1.099	369.8	84.93	57.3	120.	0.709
850	0.4097	1.110	384.3	93.80	59.6	131.	0.716
900	0.3868	1.121	398.1	102.9	62.0	143.	0.720
950	0.3666	1.131	411.3	112.2	64.3	155.	0.723
1000	0.3482	1.141	424.4	121.9	66.7	168.	0.726

Formulas for Interpolation ($T = absolute\ temperature$)

$$\rho = \frac{348.59}{T} \quad (\sigma = 9 \times 10^{-4})$$

$$f(T) = A + BT + CT^2 + DT^3$$

$f(T)$	A	B	C	D	Standard deviation, σ
c_p	1.0507	-3.645×10^{-4}	8.388×10^{-7}	-3.848×10^{-10}	4×10^{-4}
$\mu \times 10^7$	13.554	0.6738	-3.808×10^{-4}	1.183×10^{-7}	0.4192
$k \times 10^3$	-2.450	0.1130	-6.287×10^{-5}	1.891×10^{-8}	0.1198
$\alpha \times 10^8$	-11.064	7.04×10^{-2}	1.528×10^{-4}	-4.476×10^{-8}	0.4417
Pr	0.8650	-8.488×10^{-4}	1.234×10^{-6}	-5.232×10^{-10}	1.623×10^{-3}

Source: From F. P. Incropera and D. P. DeWitt, *Fundamentals of Heat and Mass Transfer*, Wiley, New York, 1985.

FIGURE C.1   Absolute viscosities of certain gases and
liquids. (From V. L. Streeter and E. B. Wylie, *Fluid
Mechanics*, 8th ed., McGraw-Hill, New York, 1985.)

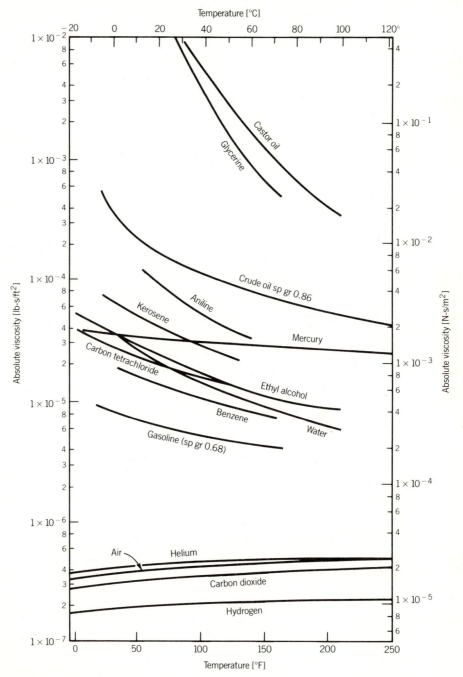

FIGURE C.2 Kinematic viscosities of certain gases and liquids. The gases are at standard pressure. (From V. L. Streeter and E. B. Wylie, *Fluid Mechanics*, 8th ed., McGraw-Hill, New York, 1985.)

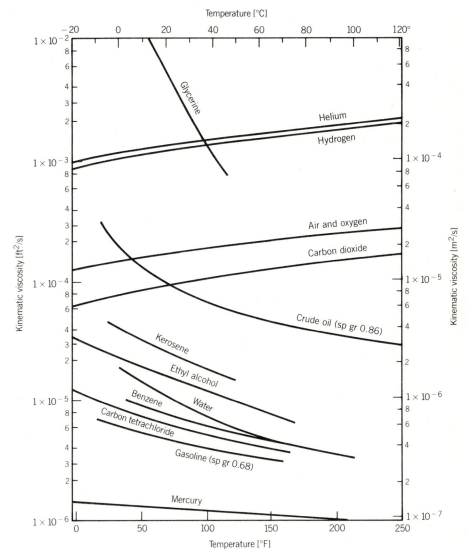

INDEX

Acceleration measurement, 435, 441
Accelerometer, 441
Accuracy, 13
Acoustic wave velocity, 352, 357
Aliasing, 227
Ammeter, 194
Amontons, Guillaume, 259
Amplifiers, 215
  differential, 217
  inverting, 216
  linear scaling, 215
  logarithmic, 216
  noninverting, 217
Amplitude, 37, 38, 45
  sampling, 231
Analog-to-digital:
  conversion, 236
  conversion error, 238
  converter, 236
    dual-slope, 242
  dynamic range, 236
  quantization error, 237
  ramp converter, 241
  saturation error, 237
  successive approximation, 240
Apparent strain, 462
Average:
  analog signal, 186
  complex waveform, 188
  digital signal, 187
  discrete time signal, 187

Balance, 432
  unequal arm, 434
Barometer, 330
Bernoulli, Daniel, 372
Bessel filter, see Filter, Bessel
Beta ratio, 379
Bias, 14
Bias error, see Error, bias
Bimetallic thermometer, see Thermometer,
  bimetallic
Binary coded decimal, 234
Binary numbers, 233
  straight code, 234
Bits, 233

Block, 8
Boulliau, Ismael, 260
Brewster, Sir David, 477
Bridge circuits, 202
  balanced, 203
  Callendar–Griffiths, 269
  deflection, 205
  four-wire, 270
  Mueller, 270
  null, 204
  three-wire, 269
  Wheatstone, 202, 269, 458
Bridge constant, 465. See also Bridge circuits,
  Wheatstone
Buffer, 247
Bus, 247
Butterworth filter, see Filter, Butterworth

Calibration, 11, 26
  curve, 11, 14
  dynamic, 12
  error, 143, 155
  random, 17
  sequence, 17
  static, 12
  temperature, see Temperature, calibration
Callendar–Griffiths bridge, see Bridge circuits,
  Callendar–Griffiths
Capillary tube viscometer, 396
Castelli, Benedetto, 372
Catch and weigh flow measurement, 410
Cathetometer, 426
Celsius, Anders, 260
Central tendency, 107, 186
Complex periodic waveform, 35, 38
Concomitant methods, 11, 25, 145
  fn, 145
Confidence interval, 106
Conservation of mass, 373
Control:
  surface, 372
  volume, 372
Coriolis, Gaspurd de, 406
Coriolis, flow meter, 409
  turndown, 409
  uncertainty, 409

Correlation, 11
Correlation coefficient, 127
Coupled systems, 96
Critically damped system, 82
Critical pressure ratio, 389
Current:
    alternating, 194
    ammeter, 194
    analog measurement, 190
    direct dc, 191

D'Arsonval movement, 193
Data acquisition, 244
    error, 143, 155
    system components, 245
Data reduction error, 143, 155
dc component, 186
Dead volume, 351
Deadweight tester, 336
    uncertainty, 337
Degree, 261
Degrees of freedom, 118, 122, 157, 163. *See
    also* Welch-Satterwaite estimate
Design-stage uncertainty, 147, 150
Deviation plot, 14
DFT, *see* Fourier transform, discrete
Dial indicator, 423
Digital:
    devices, 233
    signals, 233
    voltage measurement, 235
    voltmeter, 243
Digital signal, *see* Signal, digital
Digital-to-analog converter (D/A), 235
Dimension, 20, 24
    current, 22
    length, 21
    mass, 20
    resistance, 22
    temperature, 22
    time, 21
    voltage, 22
Discharge coefficient, 379
Discrete Fourier transform, 53, 227, 489
    amplitude ambiguity, 231
    leakage, 231
    resolution, 231
Discrete time signal, 32, 187
Displacement measurement, 426
    LVDT, 427
    potentiometer, 198, 426
Distortion, 93
Doppler, Johann, 360
Doppler:
    anemometry, 360
    effect, 360
    shift, 361
Dynamic error, 80

Electrodynamometer, 194
Electromagnetic flow meter, 397
    uncertainty, 399
Electromotive force (emf), 282
End standard, *see* Standard, end

Error:
    absolute, 13
    bias, 14, 106, 141, 144, 156
    hysteresis, 17
    linearity, 17
    loading, 210
    precision, 14
    sensitivity, 19
    temperature measurement, 304, 305
        conduction, 307
        immersion, 307
        insertion, 304
        radiation, 309
        recovery, 313
        sources, 305
    zero, 19
Error fraction, 71
Euler, Leonhard, 372
Euler, formulas, 42
Expansion factor, 379, 382

Fahrenheit, Gabriel D., 259
Feedback, control, 4
FFT, *see* Fourier transform, fast
Filter:
    analog, 219
    band, 89
    band pass, 220
    Bessel, 220
    Butterworth, 220
    design, 221, 224
    high-pass, 220
    low-pass, 219
    multiple, 222
    notch, 220
    roll-off, 220
Finite statistics, 117
First-order systems, 69
First-order uncertainty, 172
Flow coefficient, 379, 381, 386
Flow meter, *see also individual types*
    accuracy, 393
    placement, 391
    pressure loss, 383, 391
    prover, 410
Flow nozzle, 377, 383
    ASME long radius, 384
    discharge coefficient, 384
    expansion factor, 383, 384
    pressure loss, 383, 384
    pressure taps, 384
    uncertainty, 384
Forcing function, 65
Fourier:
    analysis, 36, 37
    coefficients, 40
    cosine series, 43, 45
    frequency content, 46
    harmonics, 50
    series, 39, 42
    sine series, 43, 45
Fourier transform, 50
    amplitude, 51
    discrete, 53. *See also* Discrete Fourier
        transform

fast, 53, 489
inverse, 51
Francis, James, 383
Frequency, 37
    alias, 227
    analyzer, 363
    bandwidth, 80, 88
    circular, 38
    counter, 363
    cut-off, 219
    distribution, 107
    fundamental, 44
    interval, 53
    natural, 38
    Nyquist, 228
    power spectrum, 52
    response, 78, 80, 87
        determination of, 80
    spectrum, 52
    tracker, 363
Frontinius, Sextus, 372
Function:
    aperiodic, 35
    even, 42
    odd, 43
    periodic, 40

Galvanometer, 193
Gauge block, 418, 419
    accuracy, 420
Gauge factor, 455
Gauge length, 454
Gaussian distribution, *see* Normal distribution
General model (measurement system), 65
Gosset, William, 118

Hero of Alexandria, 371
Herschel, Clemens, 383
Higher-order uncertainty, 171
Histogram, 107
Hooke's law, 448
Hot-film sensor, 358
Hot-wire sensor, 358

Impact pressure probe, 349
Impedance matching, 210
Infinite statistics, 110
Instrument uncertainty, 151, 172
Interference, 5
Interferometer, 423
Interlaboratory comparison, 145
International Practical Temperature Scale of
    1968 (IPTS-68), 263
International System of Units, 20
International Temperature Scale of 1990
    (ITS-90), 262
Interpolation, *see* Temperature, interpolation
Isoclinics, 478
ITS-90, *see* International Temperature Scale of
    1990 (ITS-90)

Kiel probe, 349
King's law, 358
Kline–McClintock, 150

Laminar flow meter, 377, 395
    flow coefficient, 396
    turndown, 397
    uncertainty, 397
Laser Doppler anemometer, 361
Leibniz, Gottfried Willhelm, 46
Length measurement, 417
    dial indicator, 423
    micrometer, 422
    optical methods, 423
    ruler, 419
    tape, 419
Linear measurement, 418. *See also* Length
    measurement
Linear variable differential transformer, 427
    excitation, 429
Line standard, *see* Standard, line
Linnaeus, Carolis, 260
Liquid-in-glass thermometer, *see*
    Thermometer, liquid-in-glass
Loading error, 210
    interstage, 210, 212
    process, 210
    voltage dividing circuit, 210
LVDT, *see* Linear variable differential
    transformer

Mach number, 315, 357
McLeod gauge, 329
    uncertainty, 330
Magnetic field, 192
Magnitude, signal, 31
Magnitude ratio, 79, 87, 98, 441
Manometer, 331
    inclined, 333
    micromanometer, 333
    uncertainty, 333, 334
    U-tube, 332
Mass flow meter, 405
    coriolis, 406
    thermal, 405
Mass flow rate, 371, 373, 389, 405
Mass measurement, 429
Mean value, 106, 110, 118
    analog signal, 186
    complex waveform, 188
    digital signal, 187
    discrete time signal, 187
Measurand, 106
Measured variable, 2, 106
Measurement, 2
    bias limit, 156
    plan, 4, 25
    precision index, 156
    system, 63
Method of least squares, 124, 487
Metrology, 417
Micrometer, 422
Modified three sigma test, 131
Modulus of elasticity, 448
Moire strain measurement, *see* Strain
    measurement, Moire
Mueller bridge, *see* Bridge circuits, Mueller
Multiple measurement uncertainty, 147, 155
Multiplexer, analog, 245

National Institute of Standards and
    Technology, *see* NIST
Newton, Sir Isaac, 259, 372
NIST, 21, 24
    RTD, 269
    temperature, 263
Noise, 5
Normal distribution, 110
Normal error function, 111, 114
Nth-order uncertainty, 172
Nyquist frequency, *see* Frequency, Nyquist

Obstruction meter, 377
    selection, 391
Ohmmeter, 202
Optical fiber thermometer, 303
Optical pyrometer, 263, 303
Optical strain measurements, 475
Orifice meter, 377, 380
    discharge coefficient, 381
    expansion factor, 382
    pressure loss, 382, 383
    tap location, 381
    uncertainty, 381
Oscilloscope, 197
Outliers, 131
Output, 3
Over damped system, 82

Parameter, 5
    control, 5, 26
Peltier, Jean Charles Athanase, 284
Peltier:
    coefficient, 284
    effect, 283
Pendulum scale, 434
Period, 38, 40
Phase:
    angle, 38, 45, 52
    linearity, 93
    shift, 78, 86, 98
Photoelastic strain measurements, *see* Strain
    measurement, photoelastic
Piezoresistance coefficient, 451
Pitot-static tube, 355
    uncertainty, 356
Pitot tube, 349
Planck, Max, 299
Platinum resistance thermometer, *see*
    Thermometer, liquid-in-glass, platinum
    resistance
Poiseulle, Jean, 396
Poisson's ratio, 449
Polariscope, 478
Pooled sample mean, 122
Pooled standard deviation, 122
    of the means, 123
Pooled statistics, 122
Positive displacement flow meter, 404
    uncertainty, 404
Potentiometer, 198. *See also* Displacement
    measurement, potentiometer
Prandtl tube, 350, 355

Precision, 14, 118
    error, 141, 146, 156
Precision estimate:
    of the slope, 127
    of the zero intercept, 128
Precision interval, 120, 133, 149
Pressure, 325
    absolute, 325
    differential, 326
    dynamic, 355
    elements, 338
        bellows, 340
        Bourdon tube, 339
        capacitance, 344
        capsule, 340
        diaphragm, 343
        strain gauge, 343
    freestream, 349
    gauge, 326
    head, 327
    reference standards, 329, 330, 331, 336
    static, 349
    total, 349
    gauges, *see* Pressure, transducer
    reference instruments, 329
    transducer, 338
        Bourdon tube, 339
        capacitance, 344
        compliance, 354
        dead volume, 351
        diaphragm, 343
        linear variable displacement, 340
        piezoelectric, 345
        piezoresistive, 344
        potentiometric, 340
        uncertainty, 340, 346
Probability, 107, 111
    density, 107
    function, 109, 111
Pyranometer, 302

Quantization, 33, 236
    error, 237

Radiation:
    blackbody, 299
    detectors, 301
    emissive power, 299
    shield, 312
    temperature measurement, 298
    wavelength distribution, 300
Radiometer, 301
Randomization methods, 6, 8
Random test, 6
Random variable, 106
Range, 13
Rayleigh relation, 358
Recovery factor, 314
Reference junction, 286
Register, 233
Regression analysis, 123
Repeatability, 19. *See also* Precision
Repetition, 10, 119, 122
Replication, 10, 119, 122

Resistance:
  analog measurement, 201
  temperature detector, 266
    construction, 267
    platinum, 268
    thin-film, 275
Resistivity, electrical, 266
  temperature coefficient of, 268, 269
Resonance, 88
  band, 88
  frequency, 88
Result:
  bias limit, 163
  precision index, 162
Reynolds number, 374
Ringing frequency, 83
Rise time, 71, 84
  determination of, 84
Root-mean-square value:
  analog signal, 185
  digital signal, 188
  discrete time signal, 188
Root-sum-square method, 147
Rotameter, 402
  turndown, 403
  uncertainty, 403
Rotating vane flow meter, 404
RTD, *see* Resistance, temperature detector
Ruge, Arthur, 451

Sample, 106
  mean value, 118
  standard deviation, 118
  variance, 118
Sampling, 187, 225
  rate, 225
  theorem, 226
  time period, 187, 188, 190
Second order systems, 81
Second power law, 150
Seebeck, Thomas Johann, 283
Seebeck coefficient, 283
Seebeck effect, 283
Seismic transducer, 436
Sensitivity index, 150
Sensor, 2
Settling time, 84
  determination of, 84
Shock measurement, 441
  piezoelectric, 441
    signal, 32
Signal:
  ac component, 186
  analog, 186
  averaging period, 190
  characteristics, 186
  conditioning, 3, 215, 245
  dc component, 186, 191
  digital, 187, 233
  discrete time, 187
  frequency, 31, 37
  input, 31, 36
  magnitude or amplitude, 31, 37
  output, 31
  sampling, 225

  waveform, 31
Signal classification:
  analog, 32
  aperiodic, 35
  deterministic, 34
  digital, 33
  discrete time, 32
  dynamic, 34
  nondeterministic, 35
  static, 34
  steady periodic, 35
Simmons, Edward, 451
Single-measurement uncertainty, 147, 171
Sonic nozzle, 389, 410
  discharge coefficient, 390
  uncertainty, 390
Source:
  bias limit, 156
  precision index, 156
Stagnation streamline, 348
Standard, 11, 20
  end, 418
  length, 418
  line, 418
Standard atmosphere, 325
Standard deviation, 110
  of the means, 120
Standard error of the fit, 125
Standardized normal variate, 111
Standards, 20
  hierarchy, 23
  primary, 20
  secondary, 20
  temperature, 262
  test, 25
Static pressure, see Pressure, static
Static sensitivity, 12, 67, 98
  determination of, 67
Steady response, 70
Steady signal, 70
Step function, 69
Strain, 448
  lateral, 449
Strain gauge, 450
  bridge constant, 465
  circuits, 458
  configuration, 469
  dynamic response, 472
  factor, 455
  foil, 453
  hysteresis, 471
  length, 454
  metallic, 451
  resistance, 450
  semiconductor, 457
  temperature compensation, 463
  transverse sensitivity, 455
  rosettes, 472
Strain measurement:
  Moire, 480
  photoelastic, 477
  resistance, 450
Strain optical coefficient, 478
Stress, 447
Student-*t* distribution, 118

Superposition, 95
System response, 63

Technical writing, 493
Temperature, 260
    absolute, 262
    calibration, 261
    dynamic, 313
    fixed points, 262, 263
    interpolation, 261, 262
    scales, 259, 261, 262
    stagnation, 313
    static, 313
    thermodynamic, 259, 262
    total, 313
Temperature measurement:
    electrical resistance, 266
    radiative, 298
    thermal expansion, 264
    thermoelectric, 282
Test plan, see Measurement, plan
Thermal anemometer, 358
Thermal expansion (ruler), 420
Thermal flow meter, 405
    turndown, 405
    uncertainty, 405
Thermistor, 266, 275
    circuit, 276
    dissipation constant, 277
    zero-power resistance, 277
Thermocouple, 282
    circuit, 282, 285, 296
    laws, 285
    reference junction, 286, 287
    standards, 287
    temperature movement, 286
    tolerance limits, 291
    type, 288
    voltage, 289
Thermoelectric temperature:
    measurement, 282
Thermometer, 262
    bimetallic, 265
    electrical resistance, 266
    liquid-in-glass, 262, 264
        complete immersion, 264
        partial immersion, 264
        total immersion, 264
        platinum resistance, 263
    radiation, 298
Thermopile, 296
Thomson, William, (Lord Kelvin), 284, 451
Thomson coefficient, 284
Thomson effect, 283
Three sigma test, 131
Time constant, 69
    determination of, 72
Time delay, 78
Time response, 70
Torricelli, Evangelista, 330
Total pressure, see Pressure, total
Transducer, 2
Transfer function, 91, 98

Transient response, 70, 82
Transmission band, 88
Transmission effects on pressure, 350
    gases, 352
    liquids, 354
Trigonometric, series, 40
True value, 106
Turbine meter, 403
    turndown, 403
    uncertainty, 403
Turndown, 393
t variable, 118

Uncertainty, 16
Uncertainty analysis, 25, 141
Under-damped system, 82
Unit, 20, 24
U.S. Engineering Unit System, 21

Variable:
    controlled, 4, 26
    dependent, 4, 26
    extraneous, 5, 8, 26
    independent, 4
Variance, 110
Velocity, freestream, 349
Velocity measurement, 354
    instrument selection, 364
    uncertainty, 364
Velocity of approach factor, 379
Vena contracta, 378
Venturi, Giovanni, 383
Venturi meter, 377, 382
    discharge coefficient, 383
    expansion factor, 382, 383
    uncertainty, 383
Vernier caliper, 422
Vernier scale, 422
Vibration measurement, 435, 438
Vibrometer, 438
Vinci, Leonardo da, 372
Voltage:
    analog measurement, 196
    digital measurement, 235, 243
    follower, 218
    oscilloscope, see Oscilloscope
    potentiometer, see Potentiometer
Voltage divider, 198
    loading error, 210
Voltmeter, digital, 243
Volume flow rate, 371, 373, 378
von Karman, Theodore, 399
Vortex flow meter, 399
    shedder shape, 400, 401
    turndown, 402
    uncertainty, 402

Waveform, 31
    analog, 32
    classification, 32
    complex periodic, 35, 36, 38, 39
    digital, 33

discrete time, 32
   Fourier analysis, 37
Weisbach, Julius, 389
Welch-Satterwaite estimate, 157, 163
Wheatstone bridge, *see* Bridge circuits,
   Wheatstone

Wobble flow meter, 404
Word (digital), 233

Zero order system, 69
Zero order uncertainty, 151, 171
z variable, 111, 118